Toyota RAV4
Automotive Repair Manual

**by Jeff Killingsworth
and John H Haynes**

Member of the Guild of Motoring Writers

Models covered:

Toyota RAV4 - 2013 through 2018

*Does not cover information specific to
RAV4 EV (Electric Vehicle) models*

(92083-2Z2)

Haynes Group Limited
Haynes North America, Inc.
www.haynes.com

Acknowledgements

Technical writers who contributed to this project include Demien Hurst and Scott 'Gonzo" Weaver. Mechanical work and photography provided by Mark Henderson.

A book in the Haynes Automotive Repair Manual Series

ISBN-13: 978-1-62092-325-2
ISBN-10: 1-62092-325-4

Library of Congress Control Number: 2018951172

18-288

Contents

Haynes mechanic and photographer with a 2015 RAV4

About this manual

Its purpose

The purpose of this manual is to help you get the best value from your vehicle. It can do so in several ways. It can help you decide what work must be done, even if you choose to have it done by a dealer service department or a repair shop; it provides information and procedures for routine maintenance and servicing; and it offers diagnostic and repair procedures to follow when trouble occurs.

We hope you use the manual to tackle the work yourself. For many simpler jobs, doing it yourself may be quicker than arranging an appointment to get the vehicle into a shop and making the trips to leave it and pick it up. More importantly, a lot of money can be saved by avoiding the expense the shop must

pass on to you to cover its labor and overhead costs. An added benefit is the sense of satisfaction and accomplishment that you feel after doing the job yourself.

Using the manual

The manual is divided into Chapters. Each Chapter is divided into numbered Sections, which are headed in bold type between horizontal lines. Each Section consists of consecutively numbered paragraphs.

The reference numbers used in illustration captions pinpoint the pertinent Section and the Step within that Section. That is, illustration 3.2 means the illustration refers to Section 3 and Step (or paragraph) 2 within that Section.

Procedures, once described in the text, are not normally repeated. When it's necessary to refer to another Chapter, the reference will be given as Chapter and Section number. Cross references given without use of the word "Chapter" apply to Sections and/or paragraphs in the same Chapter. For example, "see Section 8" means in the same Chapter.

References to the left or right side of the vehicle assume you are sitting in the driver's seat, facing forward.

Even though we have prepared this manual with extreme care, neither the publisher nor the author can accept responsibility for any errors in, or omissions from, the information given.

NOTE

A **Note** provides information necessary to properly complete a procedure or information which will make the procedure easier to understand.

CAUTION

A **Caution** provides a special procedure or special steps which must be taken while completing the procedure where the Caution is found. Not heeding a Caution can result in damage to the assembly being worked on.

WARNING

A **Warning** provides a special procedure or special steps which must be taken while completing the procedure where the Warning is found. Not heeding a Warning can result in personal injury.

Introduction

Toyota RAV4 models are available in a four-door body style with a rear liftgate.

These vehicles are equipped with a 2.5L four-cylinder engine equipped with electronic fuel injection.

The engine drives the front wheels through a six-speed automatic transaxle via independent driveaxles. On AWD models, the rear wheels are also propelled via a transfer unit, driveshaft, rear differential, and two rear driveaxles. An "active torque control 4WD"

system is used with an electric control coupling, incorporated into the front part of the rear differential. This system only transmits the amount of torque needed, based on the sensor information. The system can be set on AUTO which puts it in "all-wheel drive mode" (AWD) or LOCK mode, which is a true 4WD permitting all of the torque to be transmitted to the rear wheels.

Suspension is independent on all models, with MacPherson struts used at the

front end and trailing arms, control arms, coil springs and telescopic shock absorbers at the rear. The electrically assisted rack-and-pinion steering unit is mounted on the suspension crossmember.

The brakes are vented disc in the front and solid disc at the rear, with power assist standard. All models are equipped with an Anti-lock Brake System (ABS).

Vehicle identification numbers

1 Modifications are a continuing and unpublicized process in vehicle manufacturing. Since spare parts manuals and lists are compiled on a numerical basis, the individual vehicle numbers are essential to correctly identify the component required.

Vehicle Identification Number (VIN)

2 This very important identification number is stamped on a plate attached to the dashboard inside the windshield on the driver's side of the vehicle (see illustration). The VIN also appears on the Vehicle Certificate of Title and Registration. It contains information such as where and when the vehicle was manufactured, the model year and the body style.

Manufacturer's Certification Regulation label

3 The Manufacturer's Certification Regulation label is attached to the driver's side door jamb (see illustration). The label contains the name of the manufacturer, the month and year of production, the Gross Vehicle Weight Rating (GVWR), the Gross Axle Weight Rating (GAWR) and the certification statement.

VIN model year code

4 Counting from the left, the model year code letter designation is the 10th character. On all models covered by this manual the model year codes are:

D 2013
E 2014
F 2015
G 2016
H 2017
J 2018

Engine number

5 The engine code number is stamped into a machined pad on the driver's end of the cylinder block, on the front (radiator) side (see illustration).

Transaxle number

6 The transaxle number is located on the top of the transaxle bellhousing (see illustration).

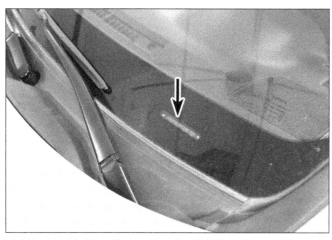

3.2 The Vehicle Identification Number (VIN) is visible through the driver's side of the windshield

3.3 Location of the Manufacturer's Certification Regulation label

3.5 Location of the engine identification number

3.6 Location of the transaxle identification number

Recall information

1   Vehicle recalls are carried out by the manufacturer in the rare event of a possible safety-related defect. The vehicle's registered owner is contacted at the address on file at the Department of Motor Vehicles and given the details of the recall. Remedial work is carried out free of charge at a dealer service department.

2   If you are the new owner of a used vehicle which was subject to a recall and you want to be sure that the work has been carried out, it's best to contact a dealer service department and ask about your individual vehicle - you'll need to furnish them your Vehicle Identification Number (VIN).

3   The table below is based on information provided by the National Highway Traffic Safety Administration (NHTSA), the body which oversees vehicle recalls in the United States. The recall database is updated constantly.

Recall date	Recall campaign number	Models affected	Concern
Dec. 30, 2014	14V828000	2014, 2015 RAV4	On some models, certain accessories such as running boards or other items were incorrectly installed. The accessory attaching fasteners were not tightened with the proper torque, possibly causing the accessory to detach from the vehicle.
January 16, 2015	15V011000	2013, 2014, 2015 RAV4	On some models equipped with an accessory trailer light module, the software within the module may incorrectly detect an electrical short and turn off the electrical current and the trailer lights, which increases risk of a collision.
March 13, 2015	15V144000	2014, 2015 RAV4	On some models, a component of the electric power steering (EPS) electronic control unit (ECU) may have been damaged during the manufacturing process. Over time, this damage may result in failure of the electric power steering system, increasing the risk of a crash.
April 7, 2016	16V198000	2016 RAV4	Some models could be equipped with an ABS actuator that may have a damaged O-ring which could result in improper brake fluid pressure control during ABS, Traction Control, or Stability Control activation, which could cause loss of vehicle control, increading the risk of a crash.
April 22, 2016	16V236000	2016 RAV4	On some models, the Load Carrying Capacity Modification Label may not reflect the correct added weight of the installed accessories. This could result in the vehicle being overloaded, increasing the risk of a crash.
May 2, 2017	17V295000	2017 RAV4	On some models, the spare tire air pressure was not adjusted to the proper pressure as stated on the Tire Pressure Label. A spare tier that is not inflated to the proper value may not perform as intended, incrasing the risk of a crash.
Dec. 20, 2017	17V831000	2017, 2018 RAV4	On some models, the Load Carrying Capacity Modification Label may not reflect the correct added weight of the installed accessories. This could result in the vehicle being overloaded, increasing the risk of a crash.

Buying parts

Replacement parts are available from many sources, which generally fall into one of two categories - authorized dealer parts departments and independent retail auto parts stores. Our advice concerning these parts is as follows:

Retail auto parts stores: Good auto parts stores will stock frequently needed components which wear out relatively fast, such as clutch components, exhaust systems, brake parts, tune-up parts, etc. These stores often supply new or reconditioned parts on an exchange basis, which can save a considerable amount of money. Discount auto parts stores are often very good places to buy materials and parts needed for general vehicle maintenance such as oil, grease, filters, spark plugs, belts, touch-up paint, bulbs, etc. They also usually sell tools and general accessories, have convenient hours, charge lower prices and can often be found not far from home.

Authorized dealer parts department: This is the best source for parts which are unique to the vehicle and not generally available elsewhere (such as major engine parts, transmission parts, trim pieces, etc.).

Warranty information: If the vehicle is still covered under warranty, be sure that any replacement parts purchased - regardless of the source - do not invalidate the warranty!

To be sure of obtaining the correct parts, have engine and chassis numbers available and, if possible, take the old parts along for positive identification.

Maintenance techniques, tools and working facilities

Maintenance techniques

There are a number of techniques involved in maintenance and repair that will be referred to throughout this manual. Application of these techniques will enable the home mechanic to be more efficient, better organized and capable of performing the various tasks properly, which will ensure that the repair job is thorough and complete.

Fasteners

Fasteners are nuts, bolts, studs and screws used to hold two or more parts together. There are a few things to keep in mind when working with fasteners. Almost all of them use a locking device of some type, either a lockwasher, locknut, locking tab or thread adhesive. All threaded fasteners should be clean and straight, with undamaged threads and undamaged corners on the hex head where the wrench fits. Develop the habit of replacing all damaged nuts and bolts with new ones. Special locknuts with nylon or fiber inserts can only be used once. If they are removed, they lose their locking ability and must be replaced with new ones.

Rusted nuts and bolts should be treated with a penetrating fluid to ease removal and prevent breakage. Some mechanics use turpentine in a spout-type oil can, which works quite well. After applying the rust penetrant, let it work for a few minutes before trying to loosen the nut or bolt. Badly rusted fasteners may have to be chiseled or sawed off or removed with a special nut breaker, available at tool stores.

If a bolt or stud breaks off in an assembly, it can be drilled and removed with a special tool commonly available for this purpose. Most automotive machine shops can perform this task, as well as other repair procedures, such as the repair of threaded holes that have been stripped out.

Flat washers and lockwashers, when removed from an assembly, should always be replaced exactly as removed. Replace any damaged washers with new ones. Never use a lockwasher on any soft metal surface (such as aluminum), thin sheet metal or plastic.

Fastener sizes

For a number of reasons, automobile manufacturers are making wider and wider use of metric fasteners. Therefore, it is important to be able to tell the difference between standard (sometimes called U.S. or SAE) and metric hardware, since they cannot be interchanged.

All bolts, whether standard or metric, are sized according to diameter, thread pitch and length. For example, a standard 1/2 - 13 x 1 bolt is 1/2 inch in diameter, has 13 threads per inch and is 1 inch long. An M12 - 1.75 x 25 metric bolt is 12 mm in diameter, has a thread pitch of 1.75 mm (the distance between threads) and is 25 mm long. The two bolts are nearly identical, and easily confused, but they are not interchangeable.

In addition to the differences in diameter, thread pitch and length, metric and standard bolts can also be distinguished by examining the bolt heads. To begin with, the distance across the flats on a standard bolt head is measured in inches, while the same dimension on a metric bolt is sized in millimeters (the same is true for nuts). As a result, a standard wrench should not be used on a metric bolt and a metric wrench should not be used on a standard bolt. Also, most standard bolts have slashes radiating out from the center of the head to denote the grade or strength of the bolt, which is an indication of the amount of torque that can be applied to it. The greater the number of slashes, the greater the strength of the bolt. Grades 0 through 5 are commonly used on automobiles. Metric bolts have a property class (grade) number, rather than a slash, molded into their heads to indicate bolt strength. In this case, the higher the number, the stronger the bolt. Property class numbers 8.8, 9.8 and 10.9 are commonly used on automobiles.

Strength markings can also be used to distinguish standard hex nuts from metric hex nuts. Many standard nuts have dots stamped into one side, while metric nuts are marked with a number. The greater the number of

dots, or the higher the number, the greater the strength of the nut.

Metric studs are also marked on their ends according to property class (grade). Larger studs are numbered (the same as metric bolts), while smaller studs carry a geometric code to denote grade.

It should be noted that many fasteners, especially Grades 0 through 2, have no distinguishing marks on them. When such is the case, the only way to determine whether it is standard or metric is to measure the thread pitch or compare it to a known fastener of the same size.

Standard fasteners are often referred to as SAE, as opposed to metric. However, it should be noted that SAE technically refers to a non-metric fine thread fastener only. Coarse thread non-metric fasteners are referred to as USS sizes.

Since fasteners of the same size (both standard and metric) may have different

strength ratings, be sure to reinstall any bolts, studs or nuts removed from your vehicle in their original locations. Also, when replacing a fastener with a new one, make sure that the new one has a strength rating equal to or greater than the original.

Tightening sequences and procedures

Most threaded fasteners should be tightened to a specific torque value (torque is the twisting force applied to a threaded component such as a nut or bolt). Overtightening the fastener can weaken it and cause it to break, while undertightening can cause it to eventually come loose. Bolts, screws and studs, depending on the material they are made of and their thread diameters, have specific torque values, many of which are noted in the Specifications at the beginning of each Chapter. Be sure to follow the torque recommen-

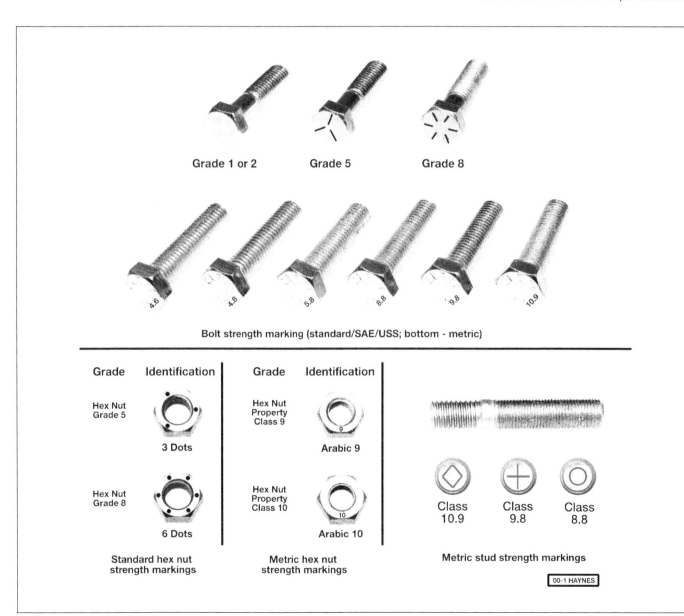

Grade 1 or 2   Grade 5   Grade 8

Bolt strength marking (standard/SAE/USS; bottom - metric)

Grade	Identification	Grade	Identification
Hex Nut Grade 5	3 Dots	Hex Nut Property Class 9	Arabic 9
Hex Nut Grade 8	6 Dots	Hex Nut Property Class 10	Arabic 10

Standard hex nut strength markings

Metric hex nut strength markings

Class 10.9   Class 9.8   Class 8.8

Metric stud strength markings

00-1 HAYNES

dations closely. For fasteners not assigned a specific torque, a general torque value chart is presented here as a guide. These torque values are for dry (unlubricated) fasteners threaded into steel or cast iron (not aluminum). As was previously mentioned, the size and grade of a fastener determine the amount of torque that can safely be applied to it. The figures listed here are approximate for Grade 2 and Grade 3 fasteners. Higher grades can tolerate higher torque values.

Fasteners laid out in a pattern, such as cylinder head bolts, oil pan bolts, differential cover bolts, etc., must be loosened or tightened in sequence to avoid warping the component. This sequence will normally be shown in the appropriate Chapter. If a specific pattern is not given, the following procedures can be used to prevent warping.

Initially, the bolts or nuts should be assembled finger-tight only. Next, they should be tightened one full turn each, in a criss-cross or diagonal pattern. After each one has been tightened one full turn, return to the first one and tighten them all one-half turn, following the same pattern. Finally, tighten each of them one-quarter turn at a time until each fastener has been tightened to the proper torque. To loosen and remove the fasteners, the procedure would be reversed.

Metric thread sizes	Ft-lbs	Nm
M-6	6 to 9	9 to 12
M-8	14 to 21	19 to 28
M-10	28 to 40	38 to 54
M-12	50 to 71	68 to 96
M-14	80 to 140	109 to 154

Pipe thread sizes		
1/8	5 to 8	7 to 10
1/4	12 to 18	17 to 24
3/8	22 to 33	30 to 44
1/2	25 to 35	34 to 47

U.S. thread sizes		
1/4 - 20	6 to 9	9 to 12
5/16 - 18	12 to 18	17 to 24
5/16 - 24	14 to 20	19 to 27
3/8 - 16	22 to 32	30 to 43
3/8 - 24	27 to 38	37 to 51
7/16 - 14	40 to 55	55 to 74
7/16 - 20	40 to 60	55 to 81
1/2 - 13	55 to 80	75 to 108

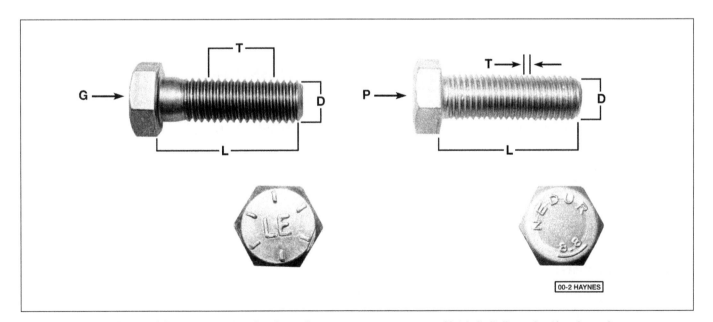

Standard (SAE and USS) bolt dimensions/grade marks

G   Grade marks (bolt strength)
L   Length (in inches)
T   Thread pitch (number of threads per inch)
D   Nominal diameter (in inches)

Metric bolt dimensions/grade marks

P   Property class (bolt strength)
L   Length (in millimeters)
T   Thread pitch (distance between threads in millimeters)
D   Diameter

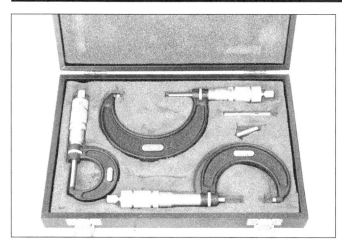

Micrometer set

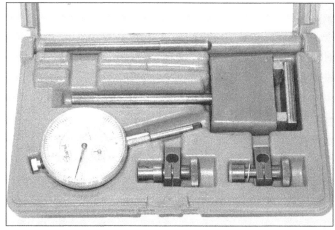

Dial indicator set

Component disassembly

Component disassembly should be done with care and purpose to help ensure that the parts go back together properly. Always keep track of the sequence in which parts are removed. Make note of special characteristics or marks on parts that can be installed more than one way, such as a grooved thrust washer on a shaft. It is a good idea to lay the disassembled parts out on a clean surface in the order that they were removed. It may also be helpful to make sketches or take instant photos of components before removal.

When removing fasteners from a component, keep track of their locations. Sometimes threading a bolt back in a part, or putting the washers and nut back on a stud, can prevent mix-ups later. If nuts and bolts cannot be returned to their original locations, they should be kept in a compartmented box or a series of small boxes. A cupcake or muffin tin is ideal for this purpose, since each cavity can hold the bolts and nuts from a particular area (i.e. oil pan bolts, valve cover bolts, engine mount bolts, etc.). A pan of this type is especially helpful when working on assemblies with very small parts, such as the carburetor, alternator, valve train or interior dash and trim pieces. The cavities can be marked with paint or tape to identify the contents.

Whenever wiring looms, harnesses or connectors are separated, it is a good idea to identify the two halves with numbered pieces of masking tape so they can be easily reconnected.

Gasket sealing surfaces

Throughout any vehicle, gaskets are used to seal the mating surfaces between two parts and keep lubricants, fluids, vacuum or pressure contained in an assembly.

Many times these gaskets are coated with a liquid or paste-type gasket sealing compound before assembly. Age, heat and pressure can sometimes cause the two parts to stick together so tightly that they are very difficult to separate. Often, the assembly can

be loosened by striking it with a soft-face hammer near the mating surfaces. A regular hammer can be used if a block of wood is placed between the hammer and the part. Do not hammer on cast parts or parts that could be easily damaged. With any particularly stubborn part, always recheck to make sure that every fastener has been removed.

Avoid using a screwdriver or bar to pry apart an assembly, as they can easily mar the gasket sealing surfaces of the parts, which must remain smooth. If prying is absolutely necessary, use an old broom handle, but keep in mind that extra clean up will be necessary if the wood splinters.

After the parts are separated, the old gasket must be carefully scraped off and the gasket surfaces cleaned. Stubborn gasket material can be soaked with rust penetrant or treated with a special chemical to soften it so it can be easily scraped off. **Caution:** *Never use gasket removal solutions or caustic chemicals on plastic or other composite components.* A scraper can be fashioned from a piece of copper tubing by flattening and sharpening one end. Copper is recommended because it is usually softer than the surfaces to be scraped, which reduces the chance of gouging the part. Some gaskets can be removed with a wire brush, but regardless of the method used, the mating surfaces must be left clean and smooth. If for some reason the gasket surface is gouged, then a gasket sealer thick enough to fill scratches will have to be used during reassembly of the components. For most applications, a non-drying (or semi-drying) gasket sealer should be used.

Hose removal tips

Warning: *If the vehicle is equipped with air conditioning, do not disconnect any of the A/C hoses without first having the system depressurized by a dealer service department or a service station.*

Hose removal precautions closely parallel gasket removal precautions. Avoid scratching or gouging the surface that the

hose mates against or the connection may leak. This is especially true for radiator hoses. Because of various chemical reactions, the rubber in hoses can bond itself to the metal spigot that the hose fits over. To remove a hose, first loosen the hose clamps that secure it to the spigot. Then, with slip-joint pliers, grab the hose at the clamp and rotate it around the spigot. Work it back and forth until it is completely free, then pull it off. Silicone or other lubricants will ease removal if they can be applied between the hose and the outside of the spigot. Apply the same lubricant to the inside of the hose and the outside of the spigot to simplify installation.

As a last resort (and if the hose is to be replaced with a new one anyway), the rubber can be slit with a knife and the hose peeled from the spigot. If this must be done, be careful that the metal connection is not damaged.

If a hose clamp is broken or damaged, do not reuse it. Wire-type clamps usually weaken with age, so it is a good idea to replace them with screw-type clamps whenever a hose is removed.

Tools

A selection of good tools is a basic requirement for anyone who plans to maintain and repair his or her own vehicle. For the owner who has few tools, the initial investment might seem high, but when compared to the spiraling costs of professional auto maintenance and repair, it is a wise one.

To help the owner decide which tools are needed to perform the tasks detailed in this manual, the following tool lists are offered: *Maintenance and minor repair, Repair/overhaul* and *Special.*

The newcomer to practical mechanics should start off with the *maintenance and minor repair* tool kit, which is adequate for the simpler jobs performed on a vehicle. Then, as confidence and experience grow, the owner can tackle more difficult tasks, buying additional tools as they are needed. Eventually the basic kit will be expanded into the *repair and overhaul* tool set. Over a period of time, the

Dial caliper

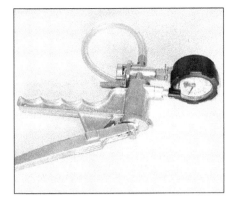

Hand-operated vacuum pump

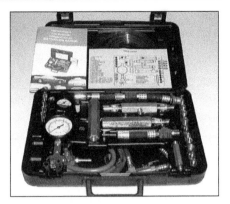

Fuel pressure gauge set

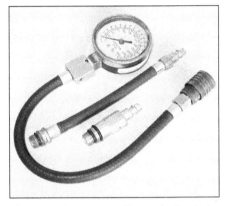

Compression gauge with spark plug hole
adapter

Damper/steering wheel puller

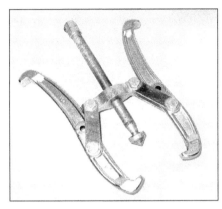

General purpose puller

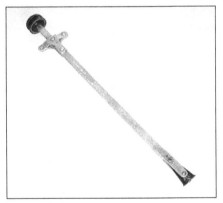

Hydraulic lifter removal tool

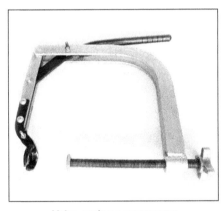

Valve spring compressor

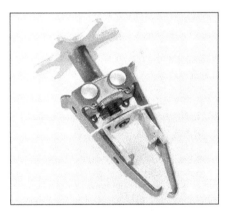

Valve spring compressor

Ridge reamer

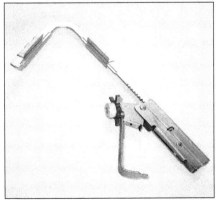

Piston ring groove cleaning tool

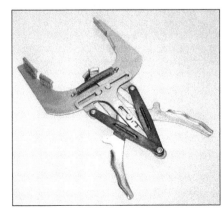

Ring removal/installation tool

Ring compressor

Cylinder hone

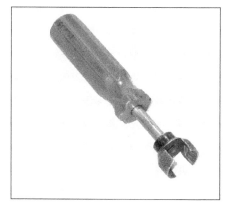

Brake hold-down spring tool

Torque angle gauge

Clutch plate alignment tool

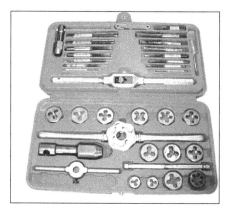

Tap and die set

experienced do-it-yourselfer will assemble a tool set complete enough for most repair and overhaul procedures and will add tools from the special category when it is felt that the expense is justified by the frequency of use.

Maintenance and minor repair tool kit

The tools in this list should be considered the minimum required for performance of routine maintenance, servicing and minor repair work. We recommend the purchase of combination wrenches (box-end and open-end combined in one wrench). While more expensive than open end wrenches, they offer the advantages of both types of wrench.

Combination wrench set (1/4-inch to 1 inch or 6 mm to 19 mm)
Adjustable wrench, 8 inch
Spark plug wrench with rubber insert
Spark plug gap adjusting tool
Feeler gauge set
Brake bleeder wrench
Standard screwdriver (5/16-inch x 6 inch)
Phillips screwdriver (No. 2 x 6 inch)
Combination pliers - 6 inch
Hacksaw and assortment of blades
Tire pressure gauge
Grease gun
Oil can
Fine emery cloth

Wire brush
Battery post and cable cleaning tool
Oil filter wrench
Funnel (medium size)
Safety goggles
Jackstands (2)
Drain pan

Note: *If basic tune-ups are going to be part of routine maintenance, it will be necessary to purchase a good quality stroboscopic timing light and combination tachometer/dwell meter. Although they are included in the list of special tools, it is mentioned here because they are absolutely necessary for tuning most vehicles properly.*

Repair and overhaul tool set

These tools are essential for anyone who plans to perform major repairs and are in addition to those in the maintenance and minor repair tool kit. Included is a comprehensive set of sockets which, though expensive, are invaluable because of their versatility, especially when various extensions and drives are available. We recommend the 1/2-inch drive over the 3/8-inch drive. Although the larger drive is bulky and more expensive, it has the capacity of accepting a very wide range of large sockets. Ideally, however, the mechanic should have a 3/8-inch drive set and a 1/2-inch drive set.

Socket set(s)
Reversible ratchet

Extension - 10 inch
Universal joint
Torque wrench (same size drive as sockets)
Ball peen hammer - 8 ounce
Soft-face hammer (plastic/rubber)
Standard screwdriver (1/4-inch x 6 inch)
Standard screwdriver (stubby - 5/16-inch)
Phillips screwdriver (No. 3 x 8 inch)
Phillips screwdriver (stubby - No. 2)
Pliers - vise grip
Pliers - lineman's
Pliers - needle nose
Pliers - snap-ring (internal and external)
Cold chisel - 1/2-inch
Scribe
Scraper (made from flattened copper tubing)
Centerpunch
Pin punches (1/16, 1/8, 3/16-inch)
Steel rule/straightedge - 12 inch
Allen wrench set (1/8 to 3/8-inch or 4 mm to 10 mm)
A selection of files
Wire brush (large)
Jackstands (second set)
Jack (scissor or hydraulic type)

Note: *Another tool which is often useful is an electric drill with a chuck capacity of 3/8-inch and a set of good quality drill bits.*

Special tools

The tools in this list include those which are not used regularly, are expensive to buy, or which need to be used in accordance with their manufacturer's instructions. Unless these tools will be used frequently, it is not very economical to purchase many of them. A consideration would be to split the cost and use between yourself and a friend or friends. In addition, most of these tools can be obtained from a tool rental shop on a temporary basis.

This list primarily contains only those tools and instruments widely available to the public, and not those special tools produced by the vehicle manufacturer for distribution to dealer service departments. Occasionally, references to the manufacturer's special tools are included in the text of this manual. Generally, an alternative method of doing the job without the special tool is offered. However, sometimes there is no alternative to their use. Where this is the case, and the tool cannot be purchased or borrowed, the work should be turned over to the dealer service department or an automotive repair shop.

Valve spring compressor
Piston ring groove cleaning tool
Piston ring compressor
Piston ring installation tool
Cylinder compression gauge
Cylinder ridge reamer
Cylinder surfacing hone
Cylinder bore gauge
Micrometers and/or dial calipers
Hydraulic lifter removal tool
Balljoint separator
Universal-type puller
Impact screwdriver
Dial indicator set
*Stroboscopic timing light (inductive
    pick-up)*
Hand operated vacuum/pressure pump
Tachometer/dwell meter
Universal electrical multimeter
Cable hoist
*Brake spring removal and installation
    tools*
Floor jack

Buying tools

For the do-it-yourselfer who is just starting to get involved in vehicle maintenance and repair, there are a number of options available when purchasing tools. If maintenance and minor repair is the extent of the work to be done, the purchase of individual tools is satisfactory. If, on the other hand, extensive work is planned, it would be a good idea to purchase a modest tool set from one of the large retail chain stores. A set can usually be bought at a substantial savings over the individual tool prices, and they often come with a tool box. As additional tools are needed, add-on sets, individual tools and a larger tool box can be purchased to expand the tool selection. Building a tool set gradually allows the cost of the

tools to be spread over a longer period of time and gives the mechanic the freedom to choose only those tools that will actually be used.

Tool stores will often be the only source of some of the special tools that are needed, but regardless of where tools are bought, try to avoid cheap ones, especially when buying screwdrivers and sockets, because they won't last very long. The expense involved in replacing cheap tools will eventually be greater than the initial cost of quality tools.

Care and maintenance of tools

Good tools are expensive, so it makes sense to treat them with respect. Keep them clean and in usable condition and store them properly when not in use. Always wipe off any dirt, grease or metal chips before putting them away. Never leave tools lying around in the work area. Upon completion of a job, always check closely under the hood for tools that may have been left there so they won't get lost during a test drive.

Some tools, such as screwdrivers, pliers, wrenches and sockets, can be hung on a panel mounted on the garage or workshop wall, while others should be kept in a tool box or tray. Measuring instruments, gauges, meters, etc. must be carefully stored where they cannot be damaged by weather or impact from other tools.

When tools are used with care and stored properly, they will last a very long time. Even with the best of care, though, tools will wear out if used frequently. When a tool is damaged or worn out, replace it. Subsequent jobs will be safer and more enjoyable if you do.

How to repair damaged threads

Sometimes, the internal threads of a nut or bolt hole can become stripped, usually from overtightening. Stripping threads is an all-too-common occurrence, especially when working with aluminum parts, because aluminum is so soft that it easily strips out.

Usually, external or internal threads are only partially stripped. After they've been cleaned up with a tap or die, they'll still work. Sometimes, however, threads are badly damaged. When this happens, you've got three choices:

1) *Drill and tap the hole to the next suitable oversize and install a larger diameter bolt, screw or stud.*
2) *Drill and tap the hole to accept a threaded plug, then drill and tap the plug to the original screw size. You can also buy a plug already threaded to the original size. Then you simply drill a hole to the specified size, then run the threaded plug into the hole with a bolt and jam nut. Once the plug is fully seated, remove the jam nut and bolt.*

3) *The third method uses a patented thread repair kit like Heli-Coil or Slimsert. These easy-to-use kits are designed to repair damaged threads in straight-through holes and blind holes. Both are available as kits which can handle a variety of sizes and thread patterns. Drill the hole, then tap it with the special included tap. Install the Heli-Coil and the hole is back to its original diameter and thread pitch.*

Regardless of which method you use, be sure to proceed calmly and carefully. A little impatience or carelessness during one of these relatively simple procedures can ruin your whole day's work and cost you a bundle if you wreck an expensive part.

Working facilities

Not to be overlooked when discussing tools is the workshop. If anything more than routine maintenance is to be carried out, some sort of suitable work area is essential.

It is understood, and appreciated, that many home mechanics do not have a good workshop or garage available, and end up removing an engine or doing major repairs outside. It is recommended, however, that the overhaul or repair be completed under the cover of a roof.

A clean, flat workbench or table of comfortable working height is an absolute necessity. The workbench should be equipped with a vise that has a jaw opening of at least four inches.

As mentioned previously, some clean, dry storage space is also required for tools, as well as the lubricants, fluids, cleaning solvents, etc. which soon become necessary.

Sometimes waste oil and fluids, drained from the engine or cooling system during normal maintenance or repairs, present a disposal problem. To avoid pouring them on the ground or into a sewage system, pour the used fluids into large containers, seal them with caps and take them to an authorized disposal site or recycling center. Plastic jugs, such as old antifreeze containers, are ideal for this purpose.

Always keep a supply of old newspapers and clean rags available. Old towels are excellent for mopping up spills. Many mechanics use rolls of paper towels for most work because they are readily available and disposable. To help keep the area under the vehicle clean, a large cardboard box can be cut open and flattened to protect the garage or shop floor.

Whenever working over a painted surface, such as when leaning over a fender to service something under the hood, always cover it with an old blanket or bedspread to protect the finish. Vinyl covered pads, made especially for this purpose, are available at auto parts stores.

Jacking and towing

Jacking

Warning: *The jack supplied with the vehicle should only be used for changing a tire or placing jackstands under the frame. Never work under the vehicle or start the engine while this jack is being used as the only means of support.*

1   The vehicle should be on level ground. Place the shift lever in Park, if you have an automatic, or Reverse if you have a manual transaxle. Block the wheel diagonally opposite the wheel being changed. Set the parking brake.

2   Remove the spare tire and jack from stowage. Remove the wheel cover and trim ring (if so equipped) with the tapered end of the lug nut wrench by inserting and twisting the handle and then prying against the back of the wheel cover. Loosen, but do not remove, the lug nuts (one-half turn is sufficient).

3   Place the scissors-type jack under the vehicle and adjust the jack height until it engages with the reinforced area of the rocker panel seam. There is a front and rear jacking point on each side of the vehicle (see illustrations).

4   Turn the jack handle clockwise until the tire clears the ground. Remove the lug nuts and pull the wheel off, then install the spare.

5   Install the lug nuts with the beveled edges facing in. Tighten them snugly. Don't attempt to tighten them completely until the vehicle is lowered or it could slip off the jack. Turn the jack handle counterclockwise to lower the vehicle. Remove the jack and tighten the lug nuts in a diagonal pattern.

6   Install the cover (and trim ring, if used) and be sure it's snapped into place all the way around.

7   Stow the tire, jack and wrench. Unblock the wheels.

Towing

8   Two-wheel drive models can be towed from the front with the front wheels off the ground, using a wheel lift type tow truck or with the front wheels on a dolly. Four wheel drive models must be towed with all four wheels off the ground. A sling-type tow truck cannot be used, as body damage will result. The best way to tow the vehicle is with a flat-bed car carrier.

9   In an emergency the vehicle can be towed a short distance with a tow strap, cable or chain attached to the towing eyelet that can be threaded into the one of the holes behind the small round covers in the front bumper (see illustrations). The driver must remain in the vehicle to operate the steering and brakes (remember that power steering and power brakes will not work with the engine off).

Note: *The towing eyelet is stored in the tool pouch (located under the rear compartment floor panel), along with the jack handle and lug wrench.*

7.3a The cutout in the jack head must engage with the rocker panel seam, between the two notches

7.3b Front and rear jacking points

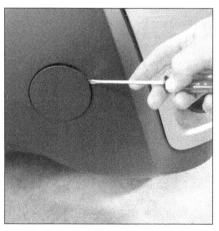

7.9a Pry out the plastic cover, being careful not to scratch the bumper . . .

7.9b . . . thread the towing eyelet into the hole behind the bumper cover . . .

7.9c . . . then tighten the eyelet securely using the handle of the lug wrench

Booster battery (jump) starting

1 Observe these precautions when using a booster battery to start a vehicle:

a) *Before connecting the booster battery, make sure the ignition switch is in the Off position.*
b) *Turn off the lights, heater and other electrical loads.*
c) *Your eyes should be shielded. Safety goggles are a good idea.*
d) *Make sure the booster battery is the same voltage as the dead one in the vehicle.*
e) *The two vehicles MUST NOT TOUCH each other!*
f) *Make sure the transaxle is in Neutral (manual) or Park (automatic).*
g) *If the booster battery is not a maintenance-free type, remove the vent caps and lay a cloth over the vent holes.*

2 Connect the red jumper cable to the positive (+) terminals of each battery (see illustration).
3 Connect one end of the black jumper cable to the negative (-) terminal of the booster battery. The other end of this cable should be connected to a good ground on the vehicle to be started, such as a bolt or bracket on the body.
4 Start the engine using the booster battery, then, with the engine running at idle speed, disconnect the jumper cables in the reverse order of connection.

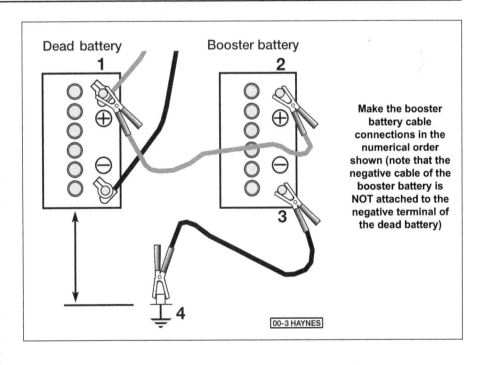

Make the booster battery cable connections in the numerical order shown (note that the negative cable of the booster battery is NOT attached to the negative terminal of the dead battery)

Automotive chemicals and lubricants

A number of automotive chemicals and lubricants are available for use during vehicle maintenance and repair. They include a wide variety of products ranging from cleaning solvents and degreasers to lubricants and protective sprays for rubber, plastic and vinyl.

Cleaners

Carburetor cleaner and choke cleaner is a strong solvent for gum, varnish and carbon. Most carburetor cleaners leave a dry-type lubricant film which will not harden or gum up. Because of this film it is not recommended for use on electrical components.

Brake system cleaner is used to remove brake dust, grease and brake fluid from the brake system, where clean surfaces are absolutely necessary. It leaves no residue and often eliminates brake squeal caused by contaminants.

Electrical cleaner removes oxidation, corrosion and carbon deposits from electrical contacts, restoring full current flow. It can also be used to clean spark plugs, carburetor jets, voltage regulators and other parts where an oil-free surface is desired.

Demoisturants remove water and moisture from electrical components such as alternators, voltage regulators, electrical connectors and fuse blocks. They are non-conductive and non-corrosive.

Degreasers are heavy-duty solvents used to remove grease from the outside of the engine and from chassis components. They can be sprayed or brushed on and, depending on the type, are rinsed off either with water or solvent.

Lubricants

Motor oil is the lubricant formulated for use in engines. It normally contains a wide variety of additives to prevent corrosion and reduce foaming and wear. Motor oil comes in various weights (viscosity ratings) from 0 to 50. The recommended weight of the oil depends on the season, temperature and the demands on the engine. Light oil is used in cold climates and under light load conditions. Heavy oil is used in hot climates and where high loads are encountered. Multi-viscosity oils are designed to have characteristics of both light and heavy oils and are available in a number of weights from 0W-20 to 20W-50.

Gear oil is designed to be used in differentials, manual transmissions and other areas where high-temperature lubrication is required.

Chassis and wheel bearing grease is a heavy grease used where increased loads and friction are encountered, such as for wheel bearings, balljoints, tie-rod ends and universal joints.

High-temperature wheel bearing grease is designed to withstand the extreme temperatures encountered by wheel bearings in disc brake equipped vehicles. It usually contains molybdenum disulfide (moly), which is a dry-type lubricant.

White grease is a heavy grease for metal-to-metal applications where water is a problem. White grease stays soft under both low and high temperatures (usually from -100 to +190-degrees F), and will not wash off or dilute in the presence of water.

Assembly lube is a special extreme pressure lubricant, usually containing moly, used to lubricate high-load parts (such as main and rod bearings and cam lobes) for initial start-up of a new engine. The assembly lube lubricates the parts without being squeezed out or washed away until the engine oiling system begins to function.

Silicone lubricants are used to protect rubber, plastic, vinyl and nylon parts.

Graphite lubricants are used where oils cannot be used due to contamination problems, such as in locks. The dry graphite will lubricate metal parts while remaining uncontaminated by dirt, water, oil or acids. It is electrically conductive and will not foul electrical contacts in locks such as the ignition switch.

Moly penetrants loosen and lubricate frozen, rusted and corroded fasteners and prevent future rusting or freezing.

Heat-sink grease is a special electrically non-conductive grease that is used for mounting electronic ignition modules where it is essential that heat is transferred away from the module.

Sealants

RTV sealant is one of the most widely used gasket compounds. Made from silicone, RTV is air curing, it seals, bonds, waterproofs, fills surface irregularities, remains flexible, doesn't shrink, is relatively easy to remove, and is used as a supplementary sealer with almost all low and medium temperature gaskets.

Anaerobic sealant is much like RTV in that it can be used either to seal gaskets or to form gaskets by itself. It remains flexible, is solvent resistant and fills surface imperfections. The difference between an anaerobic sealant and an RTV-type sealant is in the curing. RTV cures when exposed to air, while an anaerobic sealant cures only in the absence of air. This means that an anaerobic sealant cures only after the assembly of parts, sealing them together.

Thread and pipe sealant is used for sealing hydraulic and pneumatic fittings and vacuum lines. It is usually made from a Teflon compound, and comes in a spray, a paint-on liquid and as a wrap-around tape.

Chemicals

Anti-seize compound prevents seizing, galling, cold welding, rust and corrosion in fasteners. High-temperature ant-seize, usually made with copper and graphite lubricants, is used for exhaust system and exhaust manifold bolts.

Anaerobic locking compounds are used to keep fasteners from vibrating or working loose and cure only after installation, in the absence of air. Medium strength locking compound is used for small nuts, bolts and screws that may be removed later. High-strength locking compound is for large nuts, bolts and studs which aren't removed on a regular basis.

Oil additives range from viscosity index improvers to chemical treatments that claim to reduce internal engine friction. It should be noted that most oil manufacturers caution against using additives with their oils.

Gas additives perform several functions, depending on their chemical makeup. They usually contain solvents that help dissolve gum and varnish that build up on carburetor, fuel injection and intake parts. They also serve to break down carbon deposits that form on the inside surfaces of the combustion chambers. Some additives contain upper cylinder lubricants for valves and piston rings, and others contain chemicals to remove condensation from the gas tank.

Miscellaneous

Brake fluid is specially formulated hydraulic fluid that can withstand the heat and pressure encountered in brake systems. Care must be taken so this fluid does not come in contact with painted surfaces or plastics. An opened container should always be resealed to prevent contamination by water or dirt.

Weatherstrip adhesive is used to bond weatherstripping around doors, windows and trunk lids. It is sometimes used to attach trim pieces.

Undercoating is a petroleum-based, tar-like substance that is designed to protect metal surfaces on the underside of the vehicle from corrosion. It also acts as a sound-deadening agent by insulating the bottom of the vehicle.

Waxes and polishes are used to help protect painted and plated surfaces from the weather. Different types of paint may require the use of different types of wax and polish. Some polishes utilize a chemical or abrasive cleaner to help remove the top layer of oxidized (dull) paint on older vehicles. In recent years many non-wax polishes that contain a wide variety of chemicals such as polymers and silicones have been introduced. These non-wax polishes are usually easier to apply and last longer than conventional waxes and polishes.

Conversion factors

Length (distance)

Inches (in)	X	25.4	= Millimeters (mm)	X	0.0394	= Inches (in)
Feet (ft)	X	0.305	= Meters (m)	X	3.281	= Feet (ft)
Miles	X	1.609	= Kilometers (km)	X	0.621	= Miles

Volume (capacity)

Cubic inches (cu in; in^3)	X	16.387	= Cubic centimeters (cc; cm^3)	X	0.061	= Cubic inches (cu in; in^3)
Imperial pints (Imp pt)	X	0.568	= Liters (l)	X	1.76	= Imperial pints (Imp pt)
Imperial quarts (Imp qt)	X	1.137	= Liters (l)	X	0.88	= Imperial quarts (Imp qt)
Imperial quarts (Imp qt)	X	1.201	= US quarts (US qt)	X	0.833	= Imperial quarts (Imp qt)
US quarts (US qt)	X	0.946	= Liters (l)	X	1.057	= US quarts (US qt)
Imperial gallons (Imp gal)	X	4.546	= Liters (l)	X	0.22	= Imperial gallons (Imp gal)
Imperial gallons (Imp gal)	X	1.201	= US gallons (US gal)	X	0.833	= Imperial gallons (Imp gal)
US gallons (US gal)	X	3.785	= Liters (l)	X	0.264	= US gallons (US gal)

Mass (weight)

Ounces (oz)	X	28.35	= Grams (g)	X	0.035	= Ounces (oz)
Pounds (lb)	X	0.454	= Kilograms (kg)	X	2.205	= Pounds (lb)

Force

Ounces-force (ozf; oz)	X	0.278	= Newtons (N)	X	3.6	= Ounces-force (ozf; oz)
Pounds-force (lbf; lb)	X	4.448	= Newtons (N)	X	0.225	= Pounds-force (lbf; lb)
Newtons (N)	X	0.1	= Kilograms-force (kgf; kg)	X	9.81	= Newtons (N)

Pressure

Pounds-force per square inch (psi; lbf/in^2; lb/in^2)	X	0.070	= Kilograms-force per square centimeter (kgf/cm^2; kg/cm^2)	X	14.223	= Pounds-force per square inch (psi; lbf/in^2; lb/in^2)
Pounds-force per square inch (psi; lbf/in^2; lb/in^2)	X	0.068	= Atmospheres (atm)	X	14.696	= Pounds-force per square inch (psi; lbf/in^2; lb/in^2)
Pounds-force per square inch (psi; lbf/in^2; lb/in^2)	X	0.069	= Bars	X	14.5	= Pounds-force per square inch (psi; lbf/in^2; lb/in^2)
Pounds-force per square inch (psi; lbf/in^2; lb/in^2)	X	6.895	= Kilopascals (kPa)	X	0.145	= Pounds-force per square inch (psi; lbf/in^2; lb/in^2)
Kilopascals (kPa)	X	0.01	= Kilograms-force per square centimeter (kgf/cm^2; kg/cm^2)	X	98.1	= Kilopascals (kPa)

Torque (moment of force)

Pounds-force inches (lbf in; lb in)	X	1.152	= Kilograms-force centimeter (kgf cm; kg cm)	X	0.868	= Pounds-force inches (lbf in; lb in)
Pounds-force inches (lbf in; lb in)	X	0.113	= Newton meters (Nm)	X	8.85	= Pounds-force inches (lbf in; lb in)
Pounds-force inches (lbf in; lb in)	X	0.083	= Pounds-force feet (lbf ft; lb ft)	X	12	= Pounds-force inches (lbf in; lb in)
Pounds-force feet (lbf ft; lb ft)	X	0.138	= Kilograms-force meters (kgf m; kg m)	X	7.233	= Pounds-force feet (lbf ft; lb ft)
Pounds-force feet (lbf ft; lb ft)	X	1.356	= Newton meters (Nm)	X	0.738	= Pounds-force feet (lbf ft; lb ft)
Newton meters (Nm)	X	0.102	= Kilograms-force meters (kgf m; kg m)	X	9.804	= Newton meters (Nm)

Vacuum

Inches mercury (in. Hg)	X	3.377	= Kilopascals (kPa)	X	0.2961	= Inches mercury
Inches mercury (in. Hg)	X	25.4	= Millimeters mercury (mm Hg)	X	0.0394	= Inches mercury

Power

Horsepower (hp)	X	745.7	= Watts (W)	X	0.0013	= Horsepower (hp)

Velocity (speed)

Miles per hour (miles/hr; mph)	X	1.609	= Kilometers per hour (km/hr; kph)	X	0.621	= Miles per hour (miles/hr; mph)

Fuel consumption*

Miles per gallon, Imperial (mpg)	X	0.354	= Kilometers per liter (km/l)	X	2.825	= Miles per gallon, Imperial (mpg)
Miles per gallon, US (mpg)	X	0.425	= Kilometers per liter (km/l)	X	2.352	= Miles per gallon, US (mpg)

Temperature

Degrees Fahrenheit = (°C x 1.8) + 32   Degrees Celsius (Degrees Centigrade; °C) = (°F - 32) x 0.56

*It is common practice to convert from miles per gallon (mpg) to liters/100 kilometers (l/100km), where mpg (Imperial) x l/100 km = 282 and mpg (US) x l/100 km = 235

DECIMALS to MILLIMETERS

Decimal	mm	Decimal	mm
0.001	0.0254	0.500	12.7000
0.002	0.0508	0.510	12.9540
0.003	0.0762	0.520	13.2080
0.004	0.1016	0.530	13.4620
0.005	0.1270	0.540	13.7160
0.006	0.1524	0.550	13.9700
0.007	0.1778	0.560	14.2240
0.008	0.2032	0.570	14.4780
0.009	0.2286	0.580	14.7320
		0.590	14.9860
0.010	0.2540		
0.020	0.5080		
0.030	0.7620		
0.040	1.0160	0.600	15.2400
0.050	1.2700	0.610	15.4940
0.060	1.5240	0.620	15.7480
0.070	1.7780	0.630	16.0020
0.080	2.0320	0.640	16.2560
0.090	2.2860	0.650	16.5100
		0.660	16.7640
0.100	2.5400	0.670	17.0180
0.110	2.7940	0.680	17.2720
0.120	3.0480	0.690	17.5260
0.130	3.3020		
0.140	3.5560		
0.150	3.8100		
0.160	4.0640	0.700	17.7800
0.170	4.3180	0.710	18.0340
0.180	4.5720	0.720	18.2880
0.190	4.8260	0.730	18.5420
		0.740	18.7960
0.200	5.0800	0.750	19.0500
0.210	5.3340	0.760	19.3040
0.220	5.5880	0.770	19.5580
0.230	5.8420	0.780	19.8120
0.240	6.0960	0.790	20.0660
0.250	6.3500		
0.260	6.6040		
0.270	6.8580	0.800	20.3200
0.280	7.1120	0.810	20.5740
0.290	7.3660	0.820	21.8280
		0.830	21.0820
0.300	7.6200	0.840	21.3360
0.310	7.8740	0.850	21.5900
0.320	8.1280	0.860	21.8440
0.330	8.3820	0.870	22.0980
0.340	8.6360	0.880	22.3520
0.350	8.8900	0.890	22.6060
0.360	9.1440		
0.370	9.3980		
0.380	9.6520		
0.390	9.9060	0.900	22.8600
0.400	10.1600	0.910	23.1140
0.410	10.4140	0.920	23.3680
0.420	10.6680	0.930	23.6220
0.430	10.9220	0.940	23.8760
0.440	11.1760	0.950	24.1300
0.450	11.4300	0.960	24.3840
0.460	11.6840	0.970	24.6380
0.470	11.9380	0.980	24.8920
0.480	12.1920	0.990	25.1460
0.490	12.4460	1.000	25.4000

FRACTIONS to DECIMALS to MILLIMETERS

Fraction	Decimal	mm	Fraction	Decimal	mm
1/64	0.0156	0.3969	33/64	0.5156	13.0969
1/32	0.0312	0.7938	17/32	0.5312	13.4938
3/64	0.0469	1.1906	35/64	0.5469	13.8906
1/16	0.0625	1.5875	9/16	0.5625	14.2875
5/64	0.0781	1.9844	37/64	0.5781	14.6844
3/32	0.0938	2.3812	19/32	0.5938	15.0812
7/64	0.1094	2.7781	39/64	0.6094	15.4781
1/8	0.1250	3.1750	5/8	0.6250	15.8750
9/64	0.1406	3.5719	41/64	0.6406	16.2719
5/32	0.1562	3.9688	21/32	0.6562	16.6688
11/64	0.1719	4.3656	43/64	0.6719	17.0656
3/16	0.1875	4.7625	11/16	0.6875	17.4625
13/64	0.2031	5.1594	45/64	0.7031	17.8594
7/32	0.2188	5.5562	23/32	0.7188	18.2562
15/64	0.2344	5.9531	47/64	0.7344	18.6531
1/4	0.2500	6.3500	3/4	0.7500	19.0500
17/64	0.2656	6.7469	49/64	0.7656	19.4469
9/32	0.2812	7.1438	25/32	0.7812	19.8438
19/64	0.2969	7.5406	51/64	0.7969	20.2406
5/16	0.3125	7.9375	13/16	0.8125	20.6375
21/64	0.3281	8.3344	53/64	0.8281	21.0344
11/32	0.3438	8.7312	27/32	0.8438	21.4312
23/64	0.3594	9.1281	55/64	0.8594	21.8281
3/8	0.3750	9.5250	7/8	0.8750	22.2250
25/64	0.3906	9.9219	57/64	0.8906	22.6219
13/32	0.4062	10.3188	29/32	0.9062	23.0188
27/64	0.4219	10.7156	59/64	0.9219	23.4156
7/16	0.4375	11.1125	15/16	0.9375	23.8125
29/64	0.4531	11.5094	61/64	0.9531	24.2094
15/32	0.4688	11.9062	31/32	0.9688	24.6062
31/64	0.4844	12.3031	63/64	0.9844	25.0031
1/2	0.5000	12.7000	1	1.0000	25.4000

Safety first!

Regardless of how enthusiastic you may be about getting on with the job at hand, take the time to ensure that your safety is not jeopardized. A moment's lack of attention can result in an accident, as can failure to observe certain simple safety precautions. The possibility of an accident will always exist, and the following points should not be considered a comprehensive list of all dangers. Rather, they are intended to make you aware of the risks and to encourage a safety conscious approach to all work you carry out on your vehicle.

Essential DOs and DON'Ts

DON'T rely on a jack when working under the vehicle. Always use approved jackstands to support the weight of the vehicle and place them under the recommended lift or support points.

DON'T attempt to loosen extremely tight fasteners (i.e. wheel lug nuts) while the vehicle is on a jack - it may fall.

DON'T start the engine without first making sure that the transmission is in Neutral (or Park where applicable) and the parking brake is set.

DON'T remove the radiator cap from a hot cooling system - let it cool or cover it with a cloth and release the pressure gradually.

DON'T attempt to drain the engine oil until you are sure it has cooled to the point that it will not burn you.

DON'T touch any part of the engine or exhaust system until it has cooled sufficiently to avoid burns.

DON'T siphon toxic liquids such as gasoline, antifreeze and brake fluid by mouth, or allow them to remain on your skin.

DON'T inhale brake lining dust - it is potentially hazardous (see *Asbestos* below).

DON'T allow spilled oil or grease to remain on the floor - wipe it up before someone slips on it.

DON'T use loose fitting wrenches or other tools which may slip and cause injury.

DON'T push on wrenches when loosening or tightening nuts or bolts. Always try to pull the wrench toward you. If the situation calls for pushing the wrench away, push with an open hand to avoid scraped knuckles if the wrench should slip.

DON'T attempt to lift a heavy component alone - get someone to help you.

DON'T rush or take unsafe shortcuts to finish a job.

DON'T allow children or animals in or around the vehicle while you are working on it.

DO wear eye protection when using power tools such as a drill, sander, bench grinder, etc. and when working under a vehicle.

DO keep loose clothing and long hair well out of the way of moving parts.

DO make sure that any hoist used has a safe working load rating adequate for the job.

DO get someone to check on you periodically when working alone on a vehicle.

DO carry out work in a logical sequence and make sure that everything is correctly assembled and tightened.

DO keep chemicals and fluids tightly capped and out of the reach of children and pets.

DO remember that your vehicle's safety affects that of yourself and others. If in doubt on any point, get professional advice.

Steering, suspension and brakes

These systems are essential to driving safety, so make sure you have a qualified shop or individual check your work. Also, compressed suspension springs can cause injury if released suddenly - be sure to use a spring compressor.

Airbags

Airbags are explosive devices that can **CAUSE** injury if they deploy while you're working on the vehicle. Follow the manufacturer's instructions to disable the airbag whenever you're working in the vicinity of airbag components.

Asbestos

Certain friction, insulating, sealing, and other products - such as brake linings, brake bands, clutch linings, torque converters, gaskets, etc. - may contain asbestos or other hazardous friction material. Extreme care must be taken to avoid inhalation of dust from such products, since it is hazardous to health. If in doubt, assume that they do contain asbestos.

Fire

Remember at all times that gasoline is highly flammable. Never smoke or have any kind of open flame around when working on a vehicle. But the risk does not end there. A spark caused by an electrical short circuit, by two metal surfaces contacting each other, or even by static electricity built up in your body under certain conditions, can ignite gasoline vapors, which in a confined space are highly explosive. Do not, under any circumstances, use gasoline for cleaning parts. Use an approved safety solvent.

Always disconnect the battery ground (-) cable at the battery before working on any part of the fuel system or electrical system. Never risk spilling fuel on a hot engine or exhaust component. It is strongly recommended that a fire extinguisher suitable for use on fuel and electrical fires be kept handy in the garage or workshop at all times. Never try to extinguish a fuel or electrical fire with water.

Fumes

Certain fumes are highly toxic and can quickly cause unconsciousness and even death if inhaled to any extent. Gasoline vapor falls into this category, as do the vapors from some cleaning solvents. Any draining or pouring of such volatile fluids should be done in a well ventilated area.

When using cleaning fluids and solvents, read the instructions on the container carefully. Never use materials from unmarked containers.

Never run the engine in an enclosed space, such as a garage. Exhaust fumes contain carbon monoxide, which is extremely poisonous. If you need to run the engine, always do so in the open air, or at least have the rear of the vehicle outside the work area.

The battery

Never create a spark or allow a bare light bulb near a battery. They normally give off a certain amount of hydrogen gas, which is highly explosive.

Always disconnect the battery ground (-) cable at the battery before working on the fuel or electrical systems.

If possible, loosen the filler caps or cover when charging the battery from an external source (this does not apply to sealed or maintenance-free batteries). Do not charge at an excessive rate or the battery may burst.

Take care when adding water to a non maintenance-free battery and when carrying a battery. The electrolyte, even when diluted, is very corrosive and should not be allowed to contact clothing or skin.

Always wear eye protection when cleaning the battery to prevent the caustic deposits from entering your eyes.

Household current

When using an electric power tool, inspection light, etc., which operates on household current, always make sure that the tool is correctly connected to its plug and that, where necessary, it is properly grounded. Do not use such items in damp conditions and, again, do not create a spark or apply excessive heat in the vicinity of fuel or fuel vapor.

Secondary ignition system voltage

A severe electric shock can result from touching certain parts of the ignition system (such as the spark plug wires) when the engine is running or being cranked, particularly if components are damp or the insulation is defective. In the case of an electronic ignition system, the secondary system voltage is much higher and could prove fatal.

Hydrofluoric acid

This extremely corrosive acid is formed when certain types of synthetic rubber, found in some O-rings, oil seals, fuel hoses, etc. are exposed to temperatures above 750-degrees F (400-degrees C). The rubber changes into a charred or sticky substance containing the acid. *Once formed, the acid remains dangerous for years. If it gets onto the skin, it may be necessary to amputate the limb concerned.*

When dealing with a vehicle which has suffered a fire, or with components salvaged from such a vehicle, wear protective gloves and discard them after use.

Troubleshooting

Contents

This section provides an easy reference guide to the more common problems which may occur during the operation of your vehicle. These problems and their possible causes are grouped under headings denoting various components or systems, such as Engine, Cooling system, etc. They also refer you to the chapter and/or section which deals with the problem.

Remember that successful troubleshooting is not a mysterious black art practiced only by professional mechanics. It is simply the result of the right knowledge combined with an intelligent, systematic approach to the problem. Always work by a process of elimination, starting with the simplest solution and working through to the most complex - and never overlook the obvious. Anyone can run the gas tank dry or leave the lights on overnight, so don't assume that you are exempt from such oversights.

Finally, always establish a clear idea of why a problem has occurred and take steps to ensure that it doesn't happen again. If the electrical system fails because of a poor connection, check the other connections in the system to make sure that they don't fail as well. If a particular fuse continues to blow, find out why - don't just replace one fuse after another. Remember, failure of a small component can often be indicative of potential failure or incorrect functioning of a more important component or system.

Engine

1 Engine will not rotate when attempting to start

1 Battery terminal connections loose or corroded (Chapter 1).
2 Battery discharged or faulty (Chapters 1 and 5).
3 Automatic transaxle not completely engaged in Park (Chapter 7A)
4 Broken, loose or disconnected wiring in the starting circuit (Chapters 5 and 12).
5 Starter motor pinion jammed in flywheel ring gear (Chapter 5).
6 Starter solenoid faulty (Chapter 5).
7 Starter motor faulty (Chapter 5).
8 Ignition switch faulty (Chapter 12).
9 Starter pinion or flywheel teeth worn or broken (Chapter 5).

2 Engine rotates but will not start

1 Fuel tank empty.
2 Battery discharged (engine rotates slowly) (Chapter 5).
3 Battery terminal connections loose or corroded (Chapter 1).
4 Leaking fuel injector(s), faulty fuel pump, pressure regulator, etc. (Chapter 4).
5 Broken timing chain (Chapter 2A).
6 Ignition system problem (Chapter 5).
7 Worn, faulty or incorrectly gapped spark plugs (Chapter 1).
8 Loose distributor is changing ignition timing (Chapter 5).
9 Defective crankshaft or camshaft sensor (Chapter 6).

3 Engine hard to start when cold

1 Battery discharged or low (Chapter 1).
2 Malfunctioning fuel system (Chapter 4).
3 Faulty coolant temperature sensor or intake air temperature sensor (Chapter 6).
4 Faulty ignition system (Chapter 5).

4 Engine hard to start when hot

1 Air filter clogged (Chapter 1).
2 Fuel not reaching the fuel injection system (Chapter 4).
3 Corroded battery connections, especially ground (Chapter 1).
4 Faulty coolant temperature sensor or intake air temperature sensor (Chapter 6).

5 Starter motor noisy or excessively rough in engagement

1 Pinion or flywheel gear teeth worn or broken (Chapter 5).
2 Starter motor mounting bolts loose or missing (Chapter 5).

6 Engine starts but stops immediately

1 Insufficient fuel reaching the fuel injector(s) (Chapters 1 and 4).
2 Problem in the engine management system (Chapter 6).
3 Vacuum leak at the gasket between the intake manifold/plenum and throttle body (Chapter 4).

7 Oil puddle under engine

1 Oil pan gasket and/or oil pan drain bolt washer leaking (Chapter 2A).
2 Oil pressure sending unit leaking (Chapter 2B).
3 Valve cover leaking (Chapter 2A).
4 Engine oil seals leaking (Chapter 2A).
5 Oil pump housing leaking (Chapter 2A).

8 Engine lopes while idling or idles erratically

1 Vacuum leakage (Chapter 2B).
2 Air filter clogged (Chapter 1).
3 Malfunction in the fuel injection or engine control system (Chapters 4 and 6).

4 Leaking head gasket (Chapter 2A).
5 Timing chain and/or sprockets worn (Chapter 2A).
6 Camshaft lobes worn (Chapter 2A).

9 Engine misses at idle speed

1 Spark plugs worn or not gapped properly (Chapter 1).
2 Faulty spark plug wires (Chapter 1).
3 Vacuum leaks (Chapter 1).
4 Uneven or low compression (Chapter 2B).
5 Problem with the fuel injection system (Chapter 4).

10 Engine misses throughout driving speed range

1 Fuel filter clogged and/or impurities in the fuel system (Chapters 1 and 4).
2 Low fuel pressure (Chapter 4).
3 Faulty or incorrectly gapped spark plugs (Chapter 1).
4 Faulty emission system components (Chapter 6).
5 Low or uneven cylinder compression pressures (Chapter 2B).
6 Weak or faulty ignition system (Chapter 5).
7 Vacuum leak in fuel injection system, intake manifold, air control valve or vacuum hoses (Chapters 4 and 6).

11 Engine stumbles on acceleration

1 Spark plugs fouled (Chapter 1).
2 Problem with fuel injection or engine control system (Chapters 4 and 6).
3 Fuel filter clogged (Chapter 4).
4 Intake manifold air leak (Chapter 2A).
5 Problem with the emissions control system (Chapter 6).

12 Engine surges while holding accelerator steady

1 Intake air leak (Chapter 4).
2 Fuel pump or fuel pressure regulator

faulty (Chapter 4).
3   Problem with fuel injection system (Chapter 4).
4   Problem with the emissions control system (Chapter 6).

13 Engine stalls

1   Fuel filter clogged and/or water and impurities in the fuel system (Chapter 4).
2   Faulty emissions system components (Chapter 6).
3   Faulty or incorrectly gapped spark plugs (Chapter 1).
4   Vacuum leak in the fuel injection system, intake manifold or vacuum hoses (Chapter 2A and Chapter 4).

14 Engine lacks power

1   Obstructed exhaust system (Chapter 4).
2   Faulty or incorrectly gapped spark plugs (Chapter 1).
3   Problem with the fuel injection system (Chapter 4).
4   Restricted air filter (Chapter 1).
5   Brakes binding (Chapter 9).
6   Automatic transaxle fluid level incorrect (Chapter 1).
7   Clutch slipping (Chapter 8).
8   Fuel filter clogged and/or impurities in the fuel system (Chapters 1 and 4).
9   Emission control system not functioning properly (Chapter 6).
10   Low or uneven cylinder compression pressures (Chapter 2B).

15 Engine backfires

1   Emission control system not functioning properly (Chapter 6).
2   Problem with the fuel injection system (Chapter 4).
3   Vacuum leak at fuel injector(s), intake manifold, air control valve or vacuum hoses (Chapters 2A and 4).

16 Pinging or knocking engine sounds during acceleration or uphill

1   Incorrect grade of fuel.
2   Fuel injection system faulty (Chapter 4).
3   Improper or damaged spark plugs (Chapter 1).
4   Knock sensor defective (Chapter 6).
5   EGR valve not functioning (Chapter 6).
6   Vacuum leak (Chapter 4).

17 Engine runs with oil pressure light on

1   Low oil level (Chapter 1).
2   Idle rpm below specification (Chapter 1).
3   Short in wiring circuit (Chapter 12).
4   Faulty oil pressure sender (Chapter 2B).
5   Worn engine bearings and/or oil pump (Chapter 2B).

18 Engine diesels (continues to run) after switching off

1   Leaking fuel injector (Chapter 4).
2   Defective ignition system (Chapter 5).

Engine electrical system

19 Battery will not hold a charge

1   Drivebelt defective or tensioner defective (Chapter 1).
2   Battery electrolyte level low (Chapter 1).
3   Battery terminals loose or corroded (Chapter 1).
4   Alternator not charging properly (Chapter 5).
5   Loose, broken or faulty wiring in the charging circuit (Chapter 5).
6   Short in vehicle wiring (Chapter 12).
7   Internally defective battery (Chapters 1 and 5).

20 Alternator light fails to go out

1   Faulty alternator or charging circuit (Chapter 5).
2   Drivebelt defective or out of adjustment (Chapter 1).
3   Alternator voltage regulator inoperative (Chapter 5).

21 Alternator light fails to come on when key is turned on

1   Warning light bulb defective (Chapter 12).
2   Fault in the wiring or instrument cluster (Chapter 12).

Fuel system

22 Excessive fuel consumption

1   Dirty or clogged air filter element (Chapter 1).
2   Emissions system not functioning properly (Chapter 6).

3   Fuel injection system not functioning properly (Chapter 4).
4   Low tire pressure or incorrect tire size (Chapter 1).

23 Fuel leakage and/or fuel odor

1   Leaking fuel feed line (Chapters 1 and 4).
2   Tank overfilled.
3   Problem with fuel injection system (Chapter 4).

Cooling system

24 Overheating

1   Insufficient coolant in system (Chapter 1).
2   Drivebelt defective or out of adjustment (Chapter 1).
3   Radiator core blocked or grille restricted (Chapter 3).
4   Thermostat faulty (Chapter 3).
5   Electric coolant fan inoperative or blades broken (Chapter 3).
6   Radiator cap not maintaining proper pressure (Chapter 3).

25 Overcooling

1   Faulty thermostat (Chapter 3).
2   Inaccurate temperature gauge sending unit (Chapter 3).

26 External coolant leakage

1   Deteriorated/damaged hoses; loose clamps (Chapters 1 and 3).
2   Water pump defective (Chapter 3).
3   Leakage from radiator core or coolant reservoir (Chapter 3).
4   Engine drain or water jacket core plugs leaking (Chapter 1).

27 Internal coolant leakage

1   Leaking cylinder head gasket (Chapter 2A).
2   Cracked cylinder bore or cylinder head (Chapter 2B).

28 Coolant loss

1   Too much coolant in expansion tank (Chapter 1).
2   Coolant boiling away because of overheating (Chapter 3).
3   Internal or external leakage (Chapter 3).
4   Faulty expansion tank cap (Chapter 3).

29 Poor coolant circulation

1   Inoperative water pump (Chapter 3).
2   Restriction in cooling system (Chapters 1 and 3).
3   Drivebelt or tensioner defective (Chapter 1).
4   Thermostat sticking (Chapter 3).

Automatic transaxle

30 Fluid leakage

1   Automatic transaxle fluid is a deep red color. Fluid leaks should not be confused with engine oil, which can easily be blown onto the transaxle by air flow.
2   To pinpoint a leak, first remove all built-up dirt and grime from the transaxle housing with degreasing agents and/or steam cleaning. Then drive the vehicle at low speeds so air flow will not blow the leak far from its source. Raise the vehicle and determine where the leak is coming from. Common areas of leakage are:
  a) *Pan (Chapter 2A)*
  b) *Transaxle oil lines (Chapter 7A)*
  c) *Speed sensor (Chapter 6)*
  d) *Driveaxle oil seals (Chapter 8).*

31 Transaxle fluid brown or has a burned smell

Transaxle fluid overheated (Chapter 1).

32 General shift mechanism problems

1   Chapter 7A deals with checking and adjusting the shift cable on automatic transaxles. Common problems which may be attributed to poorly adjusted linkage are:
  a) *Engine starting in gears other than Park or Neutral.*
  b) *Indicator on shifter pointing to a gear other than the one actually being used.*
  c) *Vehicle moves when in Park.*
2   Refer to Chapter 7A for the shift cable adjustment procedure.

33 Engine will start in gears other than Park or Neutral

Park/Neutral Position (PNP) switch malfunctioning (Chapter 7A).

34 Transaxle slips, shifts roughly, is noisy or has no drive in forward or reverse gears

There are many probable causes for the above problems, but the home mechanic should be concerned with only one possibility - fluid level. Before taking the vehicle to a repair shop, check the level and condition of the fluid as described in Chapter 1. Correct the fluid level as necessary or change the fluid and filter if needed. If the problem persists, have a professional diagnose the cause.

Driveaxles

35 Clicking noise in turns

Worn or damaged outboard CV joint (Chapter 8).

36 Shudder or vibration during acceleration

1   Excessive toe-in (Chapter 10).
2   Incorrect spring heights (Chapter 10).
3   Worn or damaged inboard or outboard CV joints (Chapter 8).
4   Sticking inboard CV joint assembly (Chapter 8).

37 Vibration at highway speeds

1   Out-of-balance front wheels and/or tires.
2   Out-of-round front tires.
3   Worn CV joint(s) (Chapter 8).

Brakes

38 Vehicle pulls to one side during braking

1   Incorrect tire pressures (Chapter 1).
2   Front end out of alignment (have the front end aligned).
3   Front, or rear, tire sizes not matched to one another.
4   Restricted brake lines or hoses (Chapter 9).
5   Malfunctioning brake caliper (Chapter 9).
6   Loose suspension parts (Chapter 10).
7   Loose calipers (Chapter 9).
8   Excessive wear of brake shoe or pad material or disc/drum on one side (Chapter 9).
9   Contamination (grease or brake fluid) of brake shoe or pad material or disc/drum on one side (Chapter 9).

39 Noise (grinding or high-pitched squeal when the brakes are applied)

Front and/or rear disc brake pads worn out. Replace pads with new ones immediately (Chapter 9).

40 Brake roughness or chatter (pedal pulsates)

1   Excessive lateral runout (Chapter 9).
2   Uneven pad wear (Chapter 9).
3   Defective disc (Chapter 9).

41 Excessive brake pedal effort required to stop vehicle

1   Malfunctioning power brake booster (Chapter 9).
2   Partial system failure (Chapter 9).
3   Excessively worn pads or shoes (Chapter 9).
4   Piston in caliper stuck or sluggish (Chapter 9).
5   Brake pads or shoes contaminated with oil or grease (Chapter 9).
6   Brake disc grooved and/or glazed (Chapter 9).
7   New pads installed and not yet seated. It will take a while for the new material to seat against the disc.

42 Excessive brake pedal travel

1   Partial brake system failure (Chapter 9).
2   Insufficient fluid in master cylinder (Chapters 1 and 9).
3   Air trapped in system (Chapter 9).

43 Dragging brakes

1   Master cylinder pistons not returning correctly (Chapter 9).
2   Restricted brakes lines or hoses (Chapter 9).
3   Incorrect parking brake adjustment (Chapter 9).

44 Grabbing or uneven braking action

1   Malfunction of power brake booster unit (Chapter 9).
2   Binding brake pedal mechanism (Chapter 9).

45 Brake pedal feels spongy when depressed

1   Air in hydraulic lines (Chapter 9).
2   Master cylinder mounting bolts loose (Chapter 9).
3   Master cylinder defective (Chapter 9).

46 Brake pedal travels to the floor with little resistance

1   Little or no fluid in the master cylinder reservoir caused by leaking caliper or wheel cylinder (Chapter 9).
2   Loose, damaged or disconnected brake lines (Chapter 9).

47 Parking brake does not hold

Parking brake linkage improperly adjusted (Chapter 9).

Suspension and steering systems

48 Vehicle pulls to one side

1   Mismatched or uneven tires.
2   Broken or sagging springs (Chapter 10).
3   Wheel alignment incorrect. Have the wheels professionally aligned.
4   Front brake dragging (Chapter 9).

49 Abnormal or excessive tire wear

1   Wheel alignment out-of-specification. Have the wheels aligned.
2   Sagging or broken springs (Chapter 10).
3   Tire out-of-balance.
4   Worn strut damper (Chapter 10).
5   Overloaded vehicle.
6   Tires not rotated regularly.

50 Wheel makes a thumping noise

1   Blister or bump on tire.
2   Improper strut damper action (Chapter 10).

51 Shimmy, shake or vibration

1   Tire or wheel out-of-balance or out-of-round.
2   Worn wheel bearings (Chapter 10).
3   Worn tie-rod ends (Chapter 10).
4   Worn balljoints (Chapter 10).
5   Excessive wheel runout (Chapter 10).
6   Blister or bump on tire.

52 Hard steering

1   Worn balljoints, tie-rod ends or steering gear (Chapter 10).
2   Wheel alignment out-of-specifications. Have the wheels professionally aligned.
3   Low tire pressure(s) (Chapter 1).

53 Poor returnability of steering to center

1   Worn balljoints or tie-rod ends (Chapter 10).
2   Binding in steering column (Chapter 10).
3   Worn steering gear assembly (Chapter 10).
4   Wheel alignment out-of-specifications. Have the wheels professionally aligned.

54 Abnormal noise at the front end

1   Worn balljoints or tie-rod ends (Chapter 10).
2   Damaged strut mount (Chapter 10).
3   Worn control arm bushings or tie-rod ends (Chapter 10).
4   Loose stabilizer bar (Chapter 10).
5   Loose wheel nuts (Chapter 1).
6   Loose suspension bolts (Chapter 10).

55 Wander or poor steering stability

1   Mismatched or uneven tires.
2   Worn balljoints or tie-rod ends (Chapter 10).
3   Worn strut assemblies (Chapter 10).
4   Loose stabilizer bar (Chapter 10).
5   Broken or sagging springs (Chapter 10).
6   Wheels out of alignment. Have the wheels professionally aligned.

56 Erratic steering when braking

1   Wheel bearings worn (Chapter 10).
2   Broken or sagging springs (Chapter 10).
3   Leaking brake caliper (Chapter 10).
4   Warped brake discs (Chapter 9).

57 Excessive pitching and/or rolling around corners or during braking

1   Loose stabilizer bar (Chapter 10).
2   Worn strut dampers or mounts (Chapter 10).
3   Broken or sagging springs (Chapter 10).
4   Overloaded vehicle.

58 Suspension bottoms

1   Overloaded vehicle.
2   Sagging springs (Chapter 10).

59 Cupped tires

1   Front wheel or rear wheel alignment out-of-specifications. Have the wheels professionally aligned.
2   Worn strut dampers or shock absorbers (Chapter 10).
3   Wheel bearings worn (Chapter 10).
4   Excessive tire or wheel runout.
5   Worn balljoints (Chapter 10).

60 Excessive tire wear on outside edge

1   Inflation pressures incorrect (Chapter 1).
2   Excessive speed in turns.
3   Wheel alignment incorrect (excessive toe-in). Have professionally aligned.
4   Suspension arm bent (Chapter 10).

61 Excessive tire wear on inside edge

1   Inflation pressures incorrect (Chapter 1).
2   Wheel alignment incorrect (toe-out). Have professionally aligned.
3   Loose or damaged steering components (Chapter 10).

62 Tire tread worn in one place

1   Tires out-of-balance.
2   Damaged wheel.
3   Defective tire.

63 Excessive play or looseness in steering system

1   Wheel bearing(s) worn (Chapter 10).
2   Tie-rod end loose (Chapter 10).
3   Steering gear loose (Chapter 10).
4   Worn or loose steering intermediate shaft U-joint (Chapter 10).

64 Rattling or clicking noise in steering gear

1   Steering gear loose or defective (Chapter 10).
2   Steering gear defective.

Chapter 1
Tune-up and routine maintenance

Contents

Specifications

Recommended lubricants and fluids

Note: *Listed here are manufacturer recommendations at the time this manual was written. Manufacturers occasionally upgrade their fluid and lubricant specifications, so check with your local auto parts store for current recommendations.*

Engine oil	
Type	API "certified for gasoline engines" - ILSAC GF-5
Viscosity	0W-20
Fuel	Unleaded gasoline, 87 octane or higher
Automatic transaxle fluid	Toyota automatic transaxle fluid ATF-WS or equivalent
Transfer case lubricant (AWD models)	API GL-5, SAE 75W-85 gear oil
Rear differential lubricant (AWD models)	API GL-5, SAE 75W-85 gear oil
Brake fluid	DOT 3 brake fluid
Engine coolant	
U.S.A. models	50/50 mixture of Toyota Super Long Life Coolant or equivalent ethylene glycol based non-silicate, non-amine, non-nitrite and non-borate coolant with long-life hybrid organic acid technology (H.O.A.T.) and de-ionized water
Canada models	55-percent Toyota Super Long Life Coolant or equivalent ethylene glycol based non-silicate, non-amine, non-nitrite and non-borate coolant with long-life hybrid organic acid technology (H.O.A.T.) and 45-percent de-ionized water

Capacities*

Engine oil (including filter)	4.6 qts (4.4 liters)
Coolant	7.2 qts (6.8 liters)
Automatic transaxle (drain and refill)	Up to 3.4 qts (3.2 liters)
Transfer case (AWD models)	0.5 qt (0.5 liter)
Rear differential (AWD models)	0.58 qt (0.55 liter)

All capacities approximate. Add as necessary to bring up to appropriate level.

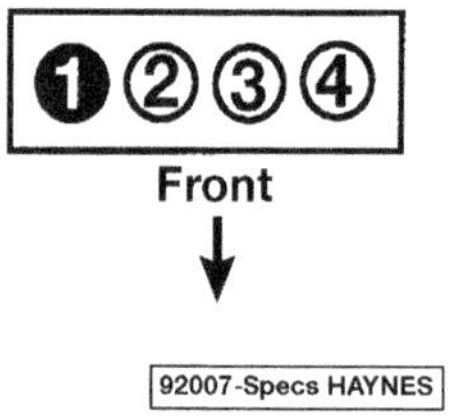

❶ ② ③ ④

Front

↓

92007-Specs HAYNES

Cylinder locations

Ignition system

Spark plug type and gap	
Type	Denso SK16HR11 or equivalent
Gap	0.043 inch (1.1 mm)
Engine firing order	1-3-4-2

Cooling system

Thermostat rating	
Starts to open	176 to 183-degrees F (80 to 84-degrees C)
Fully open	203-degrees F (95-degrees C)

Brakes

Disc brake pad lining thickness (minimum)	3/64 inch (1.0 mm)
Parking brake adjustment	6 to 8 clicks

Suspension and steering

Steering wheel freeplay limit	1.2 inches (30.0 mm)
Balljoint allowable movement	0 inch (0 mm)

Torque specifications

Note: *One foot-pound (ft-lb) of torque is equivalent to 12 inch-pounds (in-lbs) of torque. Torque values below approximately 15 ft-lbs are expressed in inch-pounds, because most foot-pound torque wrenches are not accurate at these smaller values.*

	Ft-lbs (unless otherwise indicated)	Nm
Engine oil drain plug	30	40
Engine oil filter drain plug	110 in-lbs	12.5
Engine oil filter housing	18	25
Automatic transaxle		
Pan bolts	67 in-lbs	7.5
Overflow plug	35	47
Overflow tube	7 in-lbs	0.8
Transfer unit drain and check/fill plugs	29	39
Rear differential drain and check/fill plugs	29	39
Spark plugs	18	25
Seat bolts/nuts	27	37
Drivebelt tensioner bolt	15	21
Wheel lug nuts	76	103

2.2a Engine compartment component locations

1   Brake fluid reservoir
2   No. 1 fuse/relay block
3   Air filter housing
4   Battery

5   Engine oil dipstick
6   Engine oil filler cap
7   Coolant reservoir

8   Windshield washer fluid reservoir
9   No. 2 fuse/relay block
10 Spark plug and ignition coil

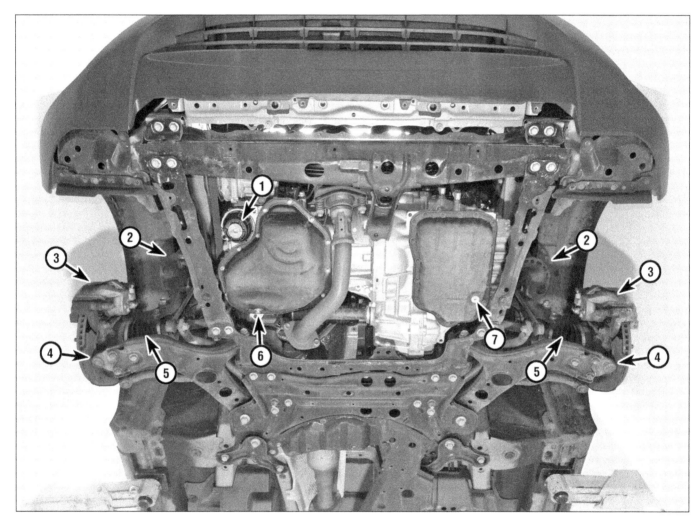

2.2b Engine compartment underside component locations

1	Engine oil filter	4	Balljoint	6	Engine oil drain plug
2	Front suspension strut unit	5	Driveaxle boot	7	Automatic transaxle fluid drain plug
3	Front disc brake caliper				

2.2c Typical rear underside component locations - 2WD models

1	Muffler	3	Shock absorber	5	Fuel tank
2	Exhaust system hanger	4	Rear disc brake		

1 Maintenance schedule

1   The maintenance intervals in this manual are provided with the assumption that you, not the dealer, will be doing the work. These are the minimum maintenance intervals recommended by the factory for vehicles that are driven daily. If you wish to keep your vehicle in peak condition at all times, you may wish to perform some of these procedures even more often. Because frequent maintenance enhances the efficiency, performance and resale value of your car, we encourage you to do so. If you drive in dusty areas, tow a trailer, idle or drive at low speeds for extended periods or drive for short distances (less than four miles) in below freezing temperatures, shorter intervals are also recommended.

2   When your vehicle is new, it should be serviced by a factory authorized dealer service department to protect the factory warranty. In many cases, the initial maintenance check is done at no cost to the owner.

Every 250 miles (400 km) or weekly, whichever comes first

Check the engine oil level (Section 4)
Check the engine coolant level (Section 4)
Check the windshield washer fluid level (Section 4)
Check the brake fluid level (Section 4)
Check the tires and tire pressures (Section 5)

Every 3000 miles (4800 km) or 3 months, whichever comes first

All items listed above plus:
Change the engine oil and oil filter (Section 6)

Every 5000 miles (8000 km)

Rotate the tires (Section 7)

Every 7500 miles (12,000 km) or 6 months, whichever comes first

Inspect (and replace, if necessary) the windshield wiper blades (Section 8)
Check and service the battery (Section 9)
Check the engine drivebelts (Section 10)
Inspect (and replace, if necessary) all underhood hoses (Section 11)
Check the cooling system (Section 12)
Check the seat belts (Section 13)

Every 15,000 miles (24,000 km) or 12 months, whichever comes first

All items listed above plus:
Inspect the brake system (Section 14)*
Check and replace, if necessary, the air filter (Section 15)*
Inspect the fuel system (Section 16)
Check the transfer case lubricant level (AWD models) (Section 17)
Inspect the suspension and steering components (Section 18)*
Check the torque of the driveshaft fasteners (AWD models) (see Chapter 8)
Check the driveaxle boots (Section 19)
Check the rear differential lubricant level (AWD models) (Section 20)
Replace the interior ventilation filter (Section 21)

Every 30,000 miles (48,000 km) or 24 months, whichever comes first

All items listed above plus:
Change the brake fluid (Section 22)
Check the automatic transaxle fluid level (Section 26)
Service the cooling system (drain, flush and refill) (Section 23)
Inspect the evaporative emissions control system (Section 24)
Inspect the exhaust system (Section 25)
Check and replace if necessary the PCV valve (Section 29)

Every 60,000 miles (96,000 km) or 48 months, whichever comes first

Change the automatic transaxle fluid and filter (Section 30)**
Change the transfer unit lubricant (AWD models) (Section 27)*
Change the rear differential lubricant (4WD models) (Section 28)*

Every 100,000 miles (161,000 km)

Replace the spark plugs (Section 31)

This item is affected by "severe" operating conditions as described below. If your vehicle is operated under "severe" conditions, perform all maintenance indicated with an asterisk (*) at 3000 mile/3 month intervals. Severe conditions are indicated if you mainly operate your vehicle under one or more of the following conditions:

Operating in dusty areas
Operating in mud/water
Towing a trailer
Idling for extended periods and/or low speed operation
Operating when outside temperatures remain below freezing and when most trips are less than 4 miles

** If operated under one or more of the following conditions, change the manual or automatic transaxle fluid and differential lubricant every 15,000 miles:

In heavy city traffic where the outside temperature regularly reaches 90-degrees F (32-degrees C) or higher
In hilly or mountainous terrain
Frequent trailer pulling

2   Introduction

1   This Chapter is designed to help the home mechanic maintain the Toyota RAV4 for peak performance, economy, safety and long life.

2   Included is a master maintenance schedule, followed by sections dealing specifically with each item on the schedule. Visual checks, adjustments, component replacement and other helpful items are included. Refer to the accompanying illustrations of the engine compartment and the underside of the vehicle for the location of various components.

3   Servicing your vehicle in accordance with the mileage/time maintenance schedule and the following Sections will provide it with a planned maintenance program that should result in a long and reliable service life. This is a comprehensive plan, so maintaining some items but not others at the specified service intervals will not produce the same results.

4   As you service your RAV4, you will discover that many of the procedures can - and should - be grouped together because of the nature of the particular procedure you're performing or because of the close proximity of two otherwise unrelated components to one another.

5   For example, if the vehicle is raised for any reason, you should inspect the exhaust, suspension, steering and fuel systems while you're under the vehicle. When you're rotating the tires, it makes good sense to check the brakes and wheel bearings since the wheels are already removed.

6   Finally, let's suppose you have to borrow or rent a torque wrench. Even if you only need to tighten the spark plugs, you might as well check the torque of as many critical fasteners as time allows.

7   The first step of this maintenance program is to prepare yourself before the actual work begins. Read through all sections pertinent to the procedures you're planning to do, then make a list of and gather together all the parts and tools you will need to do the job. If it looks as if you might run into problems during a particular segment of some procedure, seek advice from your local parts counterperson or dealer service department.

Owner's manual and VECI label information

8   Your vehicle owner's manual was written for your year and model and contains very specific information on component locations, specifications, fuse ratings, part numbers, etc. The owner's manual is an important resource for the do-it-yourselfer to have; if one was not supplied with your vehicle, it can generally be ordered from a dealer parts department.

9   Among other important information, the Vehicle Emissions Control Information (VECI) label contains specifications and procedures for tune-up adjustments (if applicable) and, in some instances, spark plugs (see Chapter 6 for more information on the VECI label). The

information on this label is the exact maintenance data recommended by the manufacturer. This data often varies by intended operating altitude, local emissions regulations, month of manufacture, etc.

10   This Chapter contains procedural details, safety information and more ambitious maintenance intervals than you might find in the manufacturer's literature. However, you may also find procedures or specifications in your owner's manual or VECI label that differ with what's printed here. In these cases, the owner's manual or VECI label can be considered correct, since it is specific to your particular vehicle.

3   Tune-up general information

1   The term tune-up is used in this manual to represent a combination of individual operations rather than one specific procedure.

2   If, from the time the vehicle is new, the routine maintenance schedule is followed closely and frequent checks are made of fluid levels and high wear items, as suggested throughout this manual, the engine will be kept in relatively good running condition and the need for additional work will be minimized.

3   More likely than not, however, there will be times when the engine is running poorly due to lack of regular maintenance. This is even more likely if a used vehicle, which has not received regular and frequent maintenance checks, is purchased. In such cases, an engine tune-up will be needed outside of the regular routine maintenance intervals.

4   The first step in any tune-up or engine diagnosis to help correct a poor running engine would be a cylinder compression check. A check of the engine compression (see Chapter 2B) will give valuable information regarding the overall performance of many internal components and should be used as a basis for tune-up and repair procedures. If, for instance, a compression check indicates serious internal engine wear, a conventional tune-up will not help the running condition of

the engine and would be a waste of time and money.

5   The following series of operations are those most often needed to bring a generally poor running engine back into a proper state of tune.

Minor tune-up

Check all engine related fluids (Section 4)
Clean, inspect and test the battery (Section 9)
Check the drivebelts (Section 10)
Check all underhood hoses (Section 11)
Check the cooling system (Section 12)
Check the air filter (Section 15)
Inspect the spark plugs

Major tune-up

All items listed under Minor tune-up, plus . . .
Replace the air filter (Section 15)
Check the fuel system (Section 16)
Replace the spark plugs (Section 31)
Check the charging system (Chapter 5)

4   Fluid level checks (every 250 miles [400 km] or weekly)

1   Fluids are an essential part of the lubrication, cooling, brake, clutch and other systems. Because these fluids gradually become depleted and/or contaminated during normal operation of the vehicle, they must be periodically replenished. See *Recommended lubricants and fluids* and *Capacities* in this Chapter's Specifications before adding fluid to any of the following components.

Note: *The vehicle must be on level ground before fluid levels can be checked.*

Engine oil

2   The engine oil level is checked with a dipstick located at the front side of the engine (see illustration). The dipstick extends through a metal tube from which it protrudes down into the engine oil pan.

3   The oil level should be checked before the vehicle has been driven, or about five minutes after the engine has been shut off.

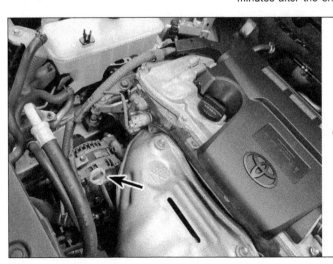

4.2 The engine oil dipstick is located at the front of the engine

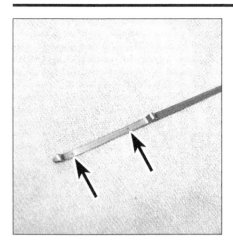

4.4 The engine oil dipstick has two dimples - keep the oil level at or near the upper dimple

4.6 The threaded oil filler cap is located on the valve cover - always make sure the area around the opening is clean before unscrewing the cap to prevent dirt from contaminating the engine

4.8 The coolant expansion tank is on the right side of the engine compartment

If the oil is checked immediately after driving the vehicle, some of the oil will remain in the upper engine components, producing an inaccurate reading on the dipstick.

4   Pull the dipstick from the tube and wipe all the oil from the end with a clean rag or paper towel. Insert the clean dipstick all the way back into its metal tube and pull it out again. Observe the oil at the end of the dipstick. At its highest point, the level should be between the two dimples (see illustration).

5   It takes one quart of oil to raise the level from the L mark to the F mark (or the lower dimple and the upper dimple) on the dipstick. Do not allow the level to drop below the L mark (or the lower dimple) or oil starvation may cause engine damage. Conversely, overfilling the engine (adding oil above the F mark or upper dimple) may cause oil fouled spark plugs, oil leaks or oil seal failures.

6   Remove the threaded cap from the valve cover to add oil (see illustration). Use a funnel to prevent spills. After adding the oil, install the filler cap hand tight. Start the engine and look carefully for any small leaks around the oil filter or drain plug. Stop the engine and check the oil level again after it has had suf-

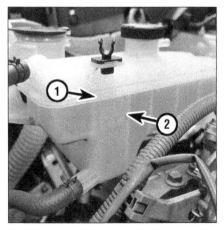

4.9 The coolant level must be between the Full (1) and Low (2) lines on the tank

ficient time to drain from the upper block and cylinder head galleys.

7   Checking the oil level is an important preventive maintenance step. A continually dropping oil level indicates oil leakage through damaged seals, from loose connections, or past worn rings or valve guides. If the oil looks milky in color or has water droplets in it, a cylinder head gasket may be blown. The engine should be checked immediately. The condition of the oil should also be checked. Each time you check the oil level, slide your thumb and index finger up the dipstick before wiping off the oil. If you see small dirt or metal particles clinging to the dipstick, the oil should be changed (see Section 6).

Engine coolant

Warning: *Do not allow antifreeze to come in contact with your skin or painted surfaces of the vehicle. Flush contaminated areas immediately with plenty of water. Don't store new coolant or leave old coolant lying around where it's accessible to children or pets - they're attracted by its sweet smell and may drink it. Ingestion of even a small amount of coolant can be fatal! Wipe up garage floor and drip pan spills immediately. Keep antifreeze containers covered and repair cooling system leaks as they're noticed.*

8   All vehicles covered by this manual are equipped with a coolant recovery system. The expansion tank is located on the right side of the engine compartment and is connected by a hose to the radiator (see illustration). The expansion tank is under the same pressure as the rest of the cooling system and is equipped with a pressure cap (there is no radiator cap on these models).

9   The coolant level should be checked regularly. It must be between the Full and Low lines on the tank (see illustration). The level will vary with the temperature of the engine. When the engine is cold, the coolant level should be at or slightly above the Low mark

on the tank. Once the engine has warmed up, the level should be at or near the Full mark. If it isn't, allow the engine to cool completely, then remove the cap from the expansion tank and add coolant to bring the level up to the Full line. Use only ethylene/glycol type coolant and water in the mixture ratio recommended by your owner's manual. Do not use supplemental inhibitor additives. If only a small amount of coolant is required to bring the system up to the proper level, water can be used. However, repeated additions of water will dilute the recommended antifreeze and water solution. In order to maintain the proper ratio of antifreeze and water, it is advisable to top up the coolant level with the correct mixture. Refer to your owner's manual for the recommended ratio.

10   If the coolant level drops within a short time after replenishment, there may be a leak in the system. Inspect the radiator, hoses, pressure cap, drain plugs and water pump. If no leak is evident, have the radiator cap or pressure cap pressure tested by a service station.

Warning: *Never remove the cap on the expansion tank (also called the expansion tank pressure cap) when the engine is running or has just been shut down, because the cooling system is hot. Escaping steam and scalding liquid could cause serious injury.*

11   If it is necessary to open the expansion tank cap, wait until the system has cooled completely, then wrap a thick cloth around the cap and slowly unscrew it. If any steam escapes, wait until the system has cooled further, then remove the cap.

12   When checking the coolant level, always note its condition. It should be relatively clear. If it is brown or rust colored, the system should be drained, flushed and refilled. Even if the coolant appears to be normal, the corrosion inhibitors wear out with use, so it must be replaced at the specified intervals.

13   Do not allow antifreeze to come in contact with your skin or painted surfaces of the vehicle. Flush contacted areas immediately with plenty of water.

4.14 The windshield/rear window washer fluid reservoir is located in the right front corner of the engine compartment

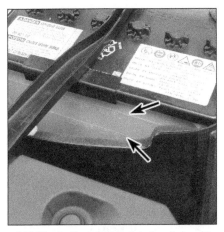

4.15 The battery electrolyte level should be between the upper and lower level in all cells

4.17 The brake fluid level should be kept between the MIN (1) and MAX (2) marks on the translucent plastic reservoir

Windshield washer fluid

14   Fluid for the windshield (and rear window) washer system is stored in a plastic reservoir which is located on the right side of the engine compartment (see illustration). In milder climates, plain water can be used to top up the reservoir, but the reservoir should be kept no more than two-thirds full to allow for expansion should the water freeze. In colder climates, the use of a specially designed windshield washer fluid, available at your dealer and any auto parts store, will help lower the freezing point of the fluid. Mix the solution with water in accordance with the manufacturer's directions on the container. Do not use regular antifreeze. It will damage the vehicle's paint.

Battery electrolyte

15   On models not equipped with a sealed battery, unscrew the filler/vent caps and check the electrolyte level. It must be between the upper and lower levels. If the level is low, add distilled water. Install and securely retighten the caps (see illustration).
Caution: *Overfilling the cells may cause electrolyte to spill over during periods of heavy charging, causing corrosion or damage.*

Brake fluid

16   The brake master cylinder is mounted on the front of the power booster unit in the engine compartment.
17   To check the fluid level, simply look at the MAX and MIN marks on the brake fluid reservoir (see illustration). The level should be at or near the maximum fill line.
18   If the level is low, wipe the top of the reservoir cover with a clean rag to prevent contamination of the brake or clutch system before lifting the cover.
19   Add only the specified brake fluid to the reservoir (refer to *Recommended lubricants and fluids* in this Chapter's Specifications or to your owner's manual). Mixing different types of brake fluid can damage the system.

Fill the brake master cylinder reservoir only to the dotted line - this brings the fluid to the correct level when you put the cover back on.
Warning: *Use caution when filling the reservoir - brake fluid can harm your eyes and damage painted surfaces. Do not use brake fluid that has been opened for more than one year (even if the cap has been on) or has been left open. Brake fluid absorbs moisture from the air. Excess moisture can cause a dangerous loss of braking.*
20   While the reservoir cap is removed, inspect the master cylinder reservoir for contamination. If deposits, dirt particles or water droplets are present, the fluid in the brake system should be changed (see Section 22 for the brake fluid replacement procedure or Chapter 8 for the clutch hydraulic system bleeding procedure).
21   After filling the reservoir to the proper level, make sure the lid is properly seated to prevent fluid leakage and/or system pressure loss.
22   The brake fluid in the master cylinder will drop slightly as the brake pads at each wheel wear down during normal operation. If the master cylinder requires repeated replen-

ishing to keep it at the proper level, this is an indication of leakage in the brake or clutch release system, which should be corrected immediately. Check all brake and clutch release lines and connections, along with the brake and clutch master cylinders, brake calipers, wheel cylinders and clutch release cylinder (see Section 14 for more information).
23   If, upon checking the master cylinder fluid level, you discover the reservoir empty or nearly empty, the brake system must be diagnosed immediately (see Chapter 9).

5   Tire and tire pressure checks (every 250 miles [400 km] or weekly)

1   Periodic inspection of the tires may spare you from the inconvenience of being stranded with a flat tire. It can also provide you with vital information regarding possible problems in the steering and suspension systems before major damage occurs.
2   Normal tread wear can be monitored with a simple, inexpensive device known as a tread depth indicator (see illustration). When

5.2 A tire tread depth indicator should be used to monitor tire wear - they are available at auto parts stores and service stations and cost very little

UNDERINFLATION

CUPPING

Cupping may be caused by:
- Underinflation and/or mechanical irregularities such as out-of-balance condition of wheel and/or tire, and bent or damaged wheel.
- Loose or worn steering tie-rod or steering idler arm.
- Loose, damaged or worn front suspension parts.

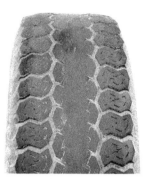

OVERINFLATION

INCORRECT TOE-IN OR EXTREME CAMBER

FEATHERING DUE TO MISALIGNMENT

5.3 This chart will help you determine the condition of your tires, the probable cause(s) of abnormal wear and the corrective action necessary

the tread depth reaches the specified minimum, replace the tire(s).

3   Note any abnormal tread wear (see illustration). Tread pattern irregularities such as cupping, flat spots and more wear on one side than the other are indications of front end alignment and/or balance problems. If any of these conditions are noted, take the vehicle to a tire shop or service station to correct the problem.

4   Look closely for cuts, punctures and embedded nails or tacks. Sometimes a tire will hold its air pressure for a short time or leak down very slowly even after a nail has embedded itself into the tread. If a slow leak persists, check the valve stem core to make sure it is tight (see illustration). Examine the tread for an object that may have embedded itself into the tire or for a plug that may have begun to leak (radial tire punctures are repaired with a plug that is installed in a puncture). If a puncture is suspected, it can be easily verified by spraying a solution of soapy water onto the puncture area (see illustration). The soapy solution will bubble if there is a leak. Unless the puncture is inordinately large, a tire shop or gas station can usually repair the punctured tire.

5   Carefully inspect the inner sidewall of each tire for evidence of brake fluid leakage. If you see any, inspect the brakes immediately.

6   Correct tire air pressure adds miles to the lifespan of the tires, improves mileage and enhances overall ride quality. Tire pressure cannot be accurately estimated by looking at a tire. A tire pressure gauge is therefore essential. Keep an accurate gauge in the glove box. The pressure gauges fitted to the nozzles of air hoses at gas stations are often inaccurate.

7   Always check tire pressure when the tires are cold. "Cold," in this case, means the vehicle has not been driven over a mile in the three hours preceding a tire pressure check. A pressure rise of four to eight pounds is not uncommon once the tires are warm.

5.4a If a tire loses air on a steady basis, check the valve core first to make sure it's snug (special inexpensive wrenches are commonly available at auto parts stores)

5.4b If the valve core is tight, raise the corner of the vehicle with the low tire and spray a soapy water solution onto the tread as the tire is turned slowly - slow leaks will cause small bubbles to appear

5.8 To extend the life of your tires, check the air pressure at least once a week with an accurate gauge (don't forget the spare!)

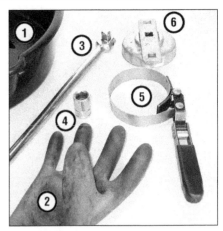

6.2 These tools are required when changing the engine oil and filter

8   Unscrew the valve cap protruding from the wheel or hubcap and push the gauge firmly onto the valve (see illustration). Note the reading on the gauge and compare this figure to the recommended tire pressure shown on the tire placard on the left door. Be sure to reinstall the valve cap to keep dirt and moisture out of the valve stem mechanism. Check all four tires and, if necessary, add enough air to bring them up to the recommended pressure levels.

9   Don't forget to keep the spare tire inflated to the specified pressure (consult your owner's manual). Note that the air pressure specified for the compact spare is significantly higher than the pressure of the regular tires.

6   Engine oil and oil filter change (every 3000 miles [4800 km] or 3 months)

1   Frequent oil changes are the best preventive maintenance the home mechanic can give the engine, because aging oil becomes diluted and contaminated, which leads to premature engine wear.

2   Make sure that you have all the necessary tools before you begin this procedure (see illustration). You should also have plenty of rags or newspapers handy for mopping up any spills.

3   Access to the underside of the vehicle is greatly improved if the vehicle can be lifted on a hoist, driven onto ramps or supported by jackstands.

Warning: *Do not work under a vehicle which is supported only by a bumper, hydraulic or scissors-type jack.*

4   If this is your first oil change, get under the vehicle and familiarize yourself with the location of the oil drain plug. The engine and exhaust components will be warm during the actual work, so try to anticipate any potential problems before the engine and accessories are hot.

5   Park the vehicle on a level spot. Start the engine and allow it to reach its normal oper-

ating temperature (the needle on the temperature gauge should be at least above the bottom mark). Warm oil and sludge will flow out more easily. Turn off the engine when it's warmed up. Remove the filler cap in the rear cam cover.

6   Raise the vehicle and support it on jackstands.

Warning: *To avoid personal injury, never get beneath the vehicle when it is supported only by a jack. The jack provided with your vehicle is designed solely for raising the vehicle to remove and replace the wheels. Always use jackstands to support the vehicle when it becomes necessary to place your body underneath the vehicle.*

7   Being careful not to touch the hot exhaust components, place the drain pan under the drain plug in the bottom of the pan and remove the plug (see illustration). You may want to wear gloves while unscrewing the plug the final few turns if the engine is really hot.

8   Allow the old oil to drain into the pan. It may be necessary to move the pan farther under the engine as the oil flow slows to a trickle. Inspect the old oil for the presence of metal shavings and chips.

9   After all the oil has drained, wipe off the drain plug with a clean rag. Even minute metal

*1   **Drain pan** - It should be fairly shallow in depth, but wide to prevent spills*

*2   **Rubber gloves** - When removing the drain plug and filter, you will get oil on your hands (the gloves will prevent burns)*

*3   **Breaker bar** - Sometimes the oil drain plug is tight, and a long breaker bar is needed to loosen it*

*4   **Socket** – To be used with the breaker bar or a ratchet (must be the correct size to fit the drain plug - six-point preferred)*

*5   **Filter wrench** - This is a metal band-type wrench, which requires clearance around the filter to be effective*

*6   **Filter wrench** - This type fits on the bottom of the filter and can be turned with a ratchet or breaker bar (different-size wrenches are available for different types of filters)*

particles clinging to the plug would immediately contaminate the new oil.

10   Clean the area around the drain plug opening, reinstall the plug and tighten it securely, but do not strip the threads.

11   Move the drain pan into position under the oil filter.

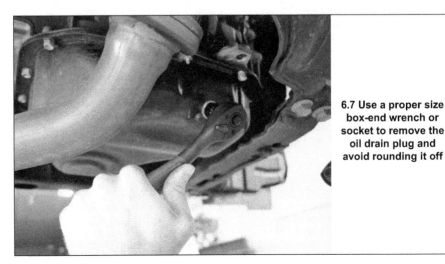

6.7 Use a proper size box-end wrench or socket to remove the oil drain plug and avoid rounding it off

6.12a Use a 3/8-inch extension to remove the oil filter housing drain plug - the entire filter housing can be removed instead, but it creates more spillage

6.12b Insert the filter drain hose and connector supplied with the new filter element and allow the oil to drain through the hose - make sure to have the drain pan centered under the filter

6.12c The main filter housing can usually be unscrewed by hand - if it's stuck you'll have to use an oil filter wrench, or special housing removal tool (but be careful as the housing can be easily damaged)

6.14a Install a new large O-ring to the element housing, then . . .

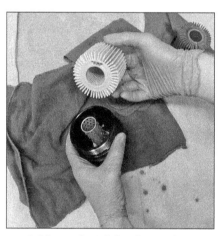

6.14b . . . install the element into the housing

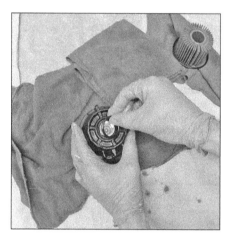

6.14c If the drain plug was removed, replace the O-ring

12   Use a ratchet with an extension to remove the plug from the bottom of the oil filter housing, being careful to minimize the inevitable spillage (see illustrations). You can then use a blunt tool to push up the inner valve and allow the canister to drain, or, if the filter came with a drain fitting attachment, use it (as it will be far less messy). Unscrew the main oil filter housing. If it's too tight to be removed by hand, you can use an oil filter wrench.
Note: *A special filter housing removal tool is available from most auto parts stores to make the job easier and as clean as possible.*
13   Remove the filter element and the large O-ring from the housing.
Note: *Don't use a metal tool to remove the O-ring, as this may scratch the soft housing.*
14   Carefully clean all components and the engine block sealing area. Install a new O-ring and filter (see illustrations), then screw the assembly back onto the engine. Tighten it to the torque listed in this Chapter's Specifi-

cations. If you removed the drain plug, clean it thoroughly, install a new O-ring and tighten the plug to the torque listed in this Chapter's Specifications. Lower the vehicle.
15   Add new oil to the engine through the oil filler cap in the valve cover. Use a funnel to prevent oil from spilling onto the top of the engine. Pour four quarts of fresh oil into the engine. Wait a few minutes to allow the oil to drain into the pan, then check the level on the oil dipstick (see Section 4 if necessary). If the oil level is at or near the F mark, install the filler cap hand tight, start the engine and allow the new oil to circulate.
16   Allow the engine to run for about a minute. While the engine is running, look under the vehicle and check for leaks at the oil pan drain plug and around the oil filter. If either is leaking, stop the engine and tighten the plug or filter slightly.
17   Wait a few minutes to allow the oil to trickle down into the pan, then recheck the

level on the dipstick and, if necessary, add enough oil to bring the level to the F mark.
18   During the first few trips after an oil change, make it a point to check frequently for leaks and proper oil level.
19   The old oil drained from the engine cannot be reused in its present state and should be discarded. Oil reclamation centers, auto garages and gas stations will normally accept the oil, which can be refined and used again. After the oil has cooled, it can be drained into a suitable container (capped plastic jugs, topped bottles, milk cartons, etc.) for transport to one of these disposal sites.

Service reminder light resetting

Monochrome display models

20   Switch the display to the trip meter and then turn the engine switch to the "LOCK" position.

21   While pressing the "DISP" button, turn the engine switch to the "ON" position.
Caution: *Do not start the engine because re-set mode will be canceled.*
22   Continue to press and hold the button until the trip meter displays "000000" and the MAINT REQD light should turn off.

Color display models
23   Press the left or right arrow of the meter control switches, and select the "Gear" symbol on the multi-information display.
24   Press the up or down arrow of the meter control switches and select "Meter Settings," then press the enter symbol.
25   Press the up or down arrow of the meter control switches and select "Scheduled Maintenance," then press the enter symbol.
26   Once in the Scheduled Maintenance screen, select "Yes" then press the enter symbol and the MAINT REQD light should turn off.

7   Tire rotation (every 5000 miles [8000 km])

1   The tires should be rotated at the specified intervals and whenever uneven wear is noticed. Since the vehicle will be raised and the tires removed anyway, check the brakes (see Section 14) at this time.
2   Radial tires must be rotated in a specific pattern (see illustrations).
3   Refer to the information in *Jacking and towing* at the front of this manual for the proper procedures to follow when raising the vehicle and changing a tire. If the brakes are to be checked, do not apply the parking brake as stated. Make sure the tires are blocked to prevent the vehicle from rolling.
4   Preferably, the entire vehicle should be raised at the same time. This can be done on a hoist or by jacking up each corner and then lowering the vehicle onto jackstands placed under the frame rails. Always use four jackstands and make sure the vehicle is firmly supported.
5   After rotation, check and adjust the tire pressures as necessary and be sure to check the lug nut tightness.

8   Windshield wiper blade inspection and replacement (every 7500 miles [12,000 km] or 6 months)

1   The windshield wiper and blade assembly should be inspected periodically for damage, loose components and cracked or worn blade elements.
2   Road film can build up on the wiper blades and affect their efficiency, so they should be washed regularly with a mild detergent solution.
3   The action of the wiping mechanism can loosen bolts, nuts and fasteners, so they

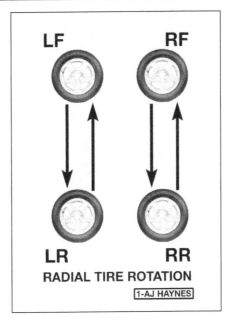

7.2a Four-tire rotation pattern

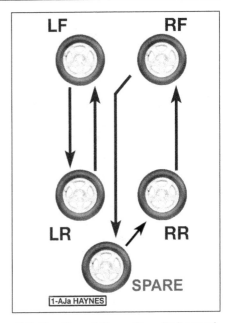

7.2b Five-tire rotation pattern (to be used only if the spare tire is the same as the other four)

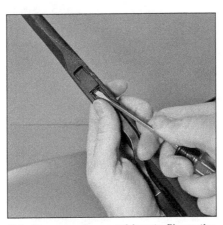

8.5a Use a small screwdriver to flip up the release lever . . .

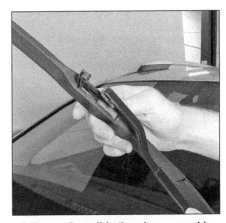

8.5b . . . then slide the wiper assembly out of the hook in the end of the wiper arm

should be checked and tightened, as necessary, at the same time the wiper blades are checked.
4   If the wiper blade elements are cracked, worn or warped, or no longer clean adequately, they should be replaced with new ones.

Windshield wiper blades
5   Remove the wiper blade assembly from the arm by raising the release lever, then sliding the assembly out of the hook in the end of the arm (see illustrations).
6   Detach the blade insert element and pull it out of the right end of the wiper frame.
7   Insert the new element end with the small protrusions into the right side of the wiper frame. Slide the element fully into place, then seat the protrusions in the end of the frames to secure it.

Rear wiper blade
8   Lift the wiper arm from the rear window.
9   Rotating the wiper blade away from the arm until "click" sound is heard.
10   Pull the wiper blade out toward the left side of the vehicle to remove it from the wiper arm
11   Insert the new wiper blade into the claw on the arm and snap the wiper blade into place on the wiper arm.

9   Battery check, maintenance and charging (every 7500 miles [12,000 km] or 6 months)

Warning: *Certain precautions must be followed when checking and servicing the battery. Hydrogen gas, which is highly flammable,*

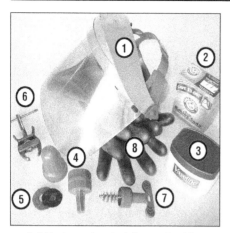

9.1 Tools and materials required for battery maintenance

1   *Face shield/safety goggles - When removing corrosion with a brush, the acidic particles can easily fly up into your eyes*
2   *Baking soda - A solution of baking soda and water can be used to neutralize corrosion*
3   *Petroleum jelly - A layer of this on the battery posts will help prevent corrosion*
4   *Battery post/cable cleaner - This wire brush cleaning tool will remove all traces of corrosion from the battery posts and cable clamps*
5   *Treated felt washers - Placing one of these on each post, directly under the cable clamps, will help prevent corrosion*
6   *Puller - Sometimes the cable clamps are very difficult to pull off the posts, even after the nut/bolt has been completely loosened. This tool pulls the clamp straight up and off the post without damage*
7   *Battery post/cable cleaner - Here is another cleaning tool which is a slightly different version of Number 4 above, but it does the same thing*
8   *Rubber gloves - Another safety item to consider when servicing the battery; remember that's acid inside the battery!*

is always present in the battery cells, so keep lighted tobacco and all other open flames and sparks away from the battery. The electrolyte inside the battery is actually diluted sulfuric acid, which will cause injury if splashed on your skin or in your eyes. It will also ruin clothes and painted surfaces. When removing the battery cables, always detach the negative cable first and hook it up last!

1   A routine preventive maintenance program for the battery in your vehicle is the only way to ensure quick and reliable starts. But before performing any battery maintenance, make sure that you have the proper equipment necessary to work safely around the battery (see illustration).

9.6a Battery terminal corrosion usually appears as light, fluffy powder

9.7a When cleaning the cable clamps, all corrosion must be removed

2   There are also several precautions that should be taken whenever battery maintenance is performed. Before servicing the battery, always turn the engine and all accessories off and disconnect the cable from the negative terminal of the battery.
3   The battery produces hydrogen gas, which is both flammable and explosive. Never create a spark, smoke or light a match around the battery. Always charge the battery in a ventilated area.
4   Electrolyte contains poisonous and corrosive sulfuric acid. Do not allow it to get in your eyes, on your skin or your clothes. Never ingest it. Wear protective safety glasses when working near the battery. Keep children away from the battery.
5   Note the external condition of the battery. If the positive terminal and cable clamp on your vehicle's battery is equipped with a rubber protector, make sure that it's not torn or damaged. It should completely cover the terminal. Look for any corroded or loose connections, cracks in the case or cover or loose hold-down clamps. Also check the entire length of each cable for cracks and frayed conductors.
6   If corrosion, which looks like white, fluffy

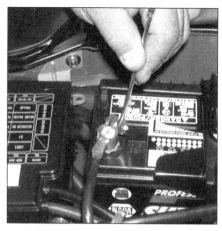

9.6b Removing a cable from the battery post with a wrench - sometimes a pair of special battery pliers are required for this procedure if corrosion has caused deterioration of the nut hex (always remove the ground (-) cable first and hook it up last!)

9.7b Regardless of the type of tool used to clean the battery posts, a clean, shiny surface should be the result

deposits (see illustration) is evident, particularly around the terminals, the battery should be removed for cleaning. Loosen the cable clamp bolts with a wrench, being careful to remove the ground cable first, and slide them off the terminals (see illustration). Then disconnect the hold-down clamp bolt and nut, remove the clamp and lift the battery from the engine compartment.
7   Clean the cable clamps thoroughly with a battery brush or a terminal cleaner and a solution of warm water and baking soda. Wash the terminals and the top of the battery case with the same solution but make sure that the solution doesn't get into the battery. When cleaning the cables, terminals and battery top, wear safety goggles and rubber gloves to prevent any solution from coming in contact with your eyes or hands. Wear old clothes too - even diluted, sulfuric acid splashed onto clothes will burn holes in them. If the terminals have been extensively corroded, clean them up with a terminal cleaner (see illustrations). Thoroughly

9.8 Make sure the battery hold-down fasteners are tight

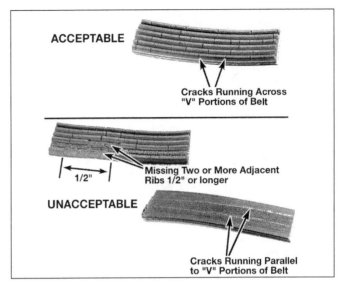

ACCEPTABLE

Cracks Running Across "V" Portions of Belt

1/2" Missing Two or More Adjacent Ribs 1/2" or longer

UNACCEPTABLE

Cracks Running Parallel to "V" Portions of Belt

10.4 Small cracks in the underside of a V-ribbed belt are acceptable - lengthwise cracks, or missing pieces that cause the belt to make noise, are cause for replacement

wash all cleaned areas with plain water.

8   Make sure that the battery tray is in good condition and the hold-down nut and bolt are tight (see illustration). If the battery is removed from the tray, make sure no parts remain in the bottom of the tray when the battery is reinstalled. When reinstalling the hold-down clamp bolt or nut, do not overtighten it.

9   Information on removing and installing the battery can be found in Chapter 5. Information on jump starting can be found at the front of this manual. For more detailed battery checking procedures, refer to the *Haynes Automotive Electrical Manual.*

10   Any metal parts of the vehicle damaged by corrosion should be covered with a zinc-based primer, then painted.

Charging

Warning: *When batteries are being charged, hydrogen gas, which is very explosive and flammable, is produced. Do not smoke or allow open flames near a charging or a recently charged battery. Wear eye protection when near the battery during charging. Also, make sure the charger is unplugged before connecting or disconnecting the battery from the charger.*

11   Slow-rate charging is the best way to restore a battery that's discharged to the point where it will not start the engine. It's also a good way to maintain the battery charge in a vehicle that's only driven a few miles between starts. Maintaining the battery charge is particularly important in the winter when the battery must work harder to start the engine and electrical accessories that drain the battery are in greater use.

12   It's best to use a one or two-amp battery charger (sometimes called a "trickle" charger). They are the safest and put the least strain on the battery. They are also the least expensive. For a faster charge, you can use

a higher amperage charger, but don't use one rated more than 1/10th the amp/hour rating of the battery. Rapid boost charges that claim to restore the power of the battery in one to two hours are hardest on the battery and can damage batteries not in good condition. This type of charging should only be used in emergency situations.

13   The average time necessary to charge a battery should be listed in the instructions that come with the charger. As a general rule, a trickle charger will charge a battery in 12 to 16 hours.

10   Drivebelt check, adjustment and replacement (every 7500 miles [12,000 km] or 6 months)

Check

1   The drivebelt is located at the right end of the engine. The good condition of the belt is critical to the operation of the engine. Because of its composition and the high stresses to which it is subjected, the drivebelt stretches and deteriorates as it gets older. It must therefore be periodically inspected.

2   A single serpentine drivebelt is used to transmit power from the crankshaft to the alternator, water pump and the air conditioning compressor.

3   With the engine stopped, inspect the full length of the drivebelt(s) for cracks and separation of the belt plies. It will be necessary to turn the engine (using a wrench or socket and breaker bar on the crankshaft pulley bolt, working clockwise only) in order to move the belt from the pulleys so that the belt can be inspected thoroughly. Check for fraying, and glazing which gives the belt a shiny appearance. Check the pulleys for nicks, cracks, distortion and corrosion.

4   Note that it is not unusual for a ribbed belt to exhibit small cracks in the edges of the belt ribs, and unless these are extensive or very deep, belt replacement is not essential (see illustration).

Note: *There is no drivebelt tension adjustment, it is adjusted automatically by a spring-loaded tensioner.*

Replacement

Warning: *Disconnect the cable from the negative terminal of the battery before performing this procedure (see Chapter 5).*

5   Loosen the lug nuts of the right front wheel. Raise the vehicle, support it securely on jackstands and remove the right front wheel.

6   Open the hood, remove the coolant expansion tank mounting bolts and set the tank out of the way (see Chapter 3, if necessary).

7   Remove the inner fender splash shield for access to the tensioner (see illustration).

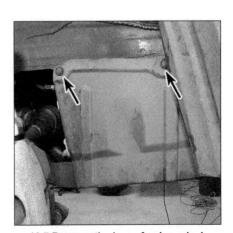

10.7 Remove the inner fender splash shield fasteners and shield

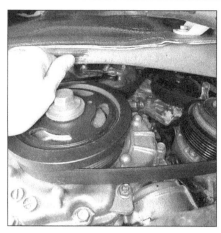

10.8 To release the tension on the drivebelt, place a long wrench or socket and ratchet on the hex-shaped boss, then turn the tensioner clockwise

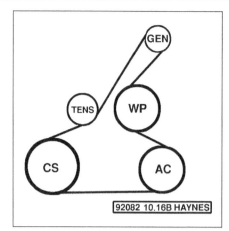

10.9 Drivebelt routing diagram

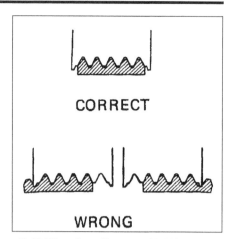

10.10 When installing a multi-ribbed belt, make sure it is centered - it must not overlap either edge of the pulley

8   Place a wrench on the hexagon portion of the tensioner and rotate the belt tensioner clockwise to release the tension. Slip the belt off the pulleys (see illustration). Slowly release the tensioner.
Note: *Working room is very limited. It may be necessary to use a box-end wrench.*
9   Route the new belt over the pulleys (see illustration), again rotating the tensioner to allow the belt to be installed, then release the belt tensioner.
10   Make sure the belt is properly centered in the pulleys (see illustration).

Tensioner replacement

Warning: *Disconnect the cable from the negative terminal of the battery before performing this procedure (see Chapter 5).*
Note: *It may be necessary to remove the alternator to access the tensioner (see Chapter 5).*
11   To replace a tensioner that can't properly tension the belt, or one that exhibits binding or a worn-out bearing/pulley, remove the drivebelt, then unscrew the mounting bolt (see illustration).
Note: *On 2017 and later models, the tensioner*

bolt is covered by a dust cap; use a screwdriver to pop the cap off of the tensioner to access the mounting bolt.
12   Installation is the reverse of the removal procedure. Tighten the bolt to the torque value listed in this Chapter's Specifications.
13   Install the drivebelt.

---

11   Underhood hose check and replacement (every 7500 miles [12,000 km] or 6 months)

Caution: *Never remove air conditioning components or hoses until the system has been depressurized.*

General

1   High temperatures in the engine compartment can cause the deterioration of the rubber and plastic hoses used for engine, accessory and emission systems operation. Periodic inspection should be made for cracks, loose clamps, material hardening and leaks.
2   Information specific to the cooling system hoses can be found in Section 12.

10.11 Drivebelt tensioner mounting bolt

3   Some, but not all, hoses are secured to the fittings with clamps. Where clamps are used, check to be sure they haven't lost their tension, allowing the hose to leak. If clamps aren't used, make sure the hose has not expanded and/or hardened where it slips over the fitting, allowing it to leak.

Vacuum hoses

4   It's quite common for vacuum hoses, especially those in the emissions system, to be color coded or identified by colored stripes molded into them. Various systems require hoses with different wall thickness, collapse resistance and temperature resistance. When replacing hoses, be sure the new ones are made of the same material.
5   Often the only effective way to check a hose is to remove it completely from the vehicle. If more than one hose is removed, be sure to label the hoses and fittings to ensure correct installation.
6   When checking vacuum hoses, be sure to include any plastic T-fittings in the check. Inspect the fittings for cracks and the hose where it fits over the fitting for distortion, which could cause leakage.
7   A small piece of vacuum hose (1/4-inch inside diameter) can be used as a stethoscope to detect vacuum leaks. Hold one end of the hose to your ear and probe around vacuum hoses and fittings, listening for the hissing sound characteristic of a vacuum leak.
Warning: *When probing with the vacuum hose stethoscope, be very careful not to come into contact with moving engine components such as the drivebelts, cooling fan, etc.*

Fuel hose

Warning: *Gasoline is extremely flammable, so take extra precautions when you work on any part of the fuel system. Don't smoke or allow open flames or bare light bulbs near the work area, and don't work in a garage where a gas-type appliance (such as a water heater or a clothes dryer) is present. Since gasoline is carcinogenic, wear fuel-resistant gloves when* there's a possibility of being exposed to fuel, and, if you spill any fuel on your skin, rinse it off immediately with soap and water. Mop up any spills immediately and do not store fuel-soaked rags where they could ignite. The fuel system is under constant pressure, so, if any fuel lines are to be disconnected, the fuel pressure in the system must be relieved first (see Chapter 4, Section 3). When you perform any kind of work on the fuel system, wear safety glasses and have a Class B type fire extinguisher on hand.

8   Check all rubber fuel lines for deterioration and chafing. Check especially for cracks in areas where the hose bends and just before fittings, such as where a hose attaches to the fuel filter.

9   High quality fuel line, specifically designed for fuel injection systems, must be used for fuel line replacement.
Warning: *Never use anything other than the proper fuel line for fuel line replacement.*

10   Spring-type clamps are commonly used on fuel lines. These clamps often lose their tension over a period of time, and can be sprung during removal. Replace all spring-type clamps with screw clamps whenever a hose is replaced.

Metal lines

11   Sections of metal line are often used for fuel line between the fuel pump and fuel injection unit. Check carefully to be sure the line has not been bent or crimped and that cracks have not started in the line.

12   If a section of metal fuel line must be replaced, only seamless steel tubing should be used, since copper and aluminum tubing don't have the strength necessary to withstand normal engine vibration.

13   Check the metal brake lines where they enter the master cylinder and brake proportioning unit (if used) for cracks in the lines or loose fittings. Any sign of brake fluid leakage calls for an immediate thorough inspection of the brake system.

---

12   Cooling system check (every 7500 miles [12,000 km] or 6 months)

---

1   Many major engine failures can be attributed to a faulty cooling system. If the vehicle is equipped with an automatic transaxle, the cooling system also cools the transaxle fluid and thus plays an important role in prolonging transaxle life.

2   The cooling system should be checked with the engine cold. Do this before the vehicle is driven for the day or after the engine has been shut off for at least three hours.
Warning: *Never remove the cooling system pressure cap when the engine is running or has just been shut down, because the cooling system is hot. Escaping steam and scalding liquid could cause serious injury. Wait until the engine has cooled completely.*

3   Slowly unscrew the pressure cap from the expansion tank. If you hear a hissing sound (indicating there is still pressure in the system), wait until it stops. Thoroughly clean the cap, inside and out, with clean water. Also clean the opening on the expansion tank. All traces of corrosion or residue should be removed. The coolant inside the expansion tank should be relatively transparent. If not, the system should be drained and refilled (see Section 23). If the coolant level isn't up to the top, add additional antifreeze/coolant mixture (see Section 4).

4   Carefully check the large upper and lower radiator hoses along with the smaller diameter heater hoses which run from the engine to the firewall. Inspect each hose along its entire length, replacing any hose which is cracked, swollen or shows signs of deterioration. Cracks may become more apparent if the hose is squeezed (see illustration). Regardless of condition, it's a good idea to replace hoses with new ones every two years.

5   Make sure that all hose connections are tight. A leak in the cooling system will usually show up as white or rust colored deposits on the areas adjoining the leak. If wire-type clamps are used at the ends of the hoses, it may be a good idea to replace them with more secure screw-type clamps.

6   Use compressed air or a soft brush to remove bugs, leaves, etc., from the front of the radiator or air conditioning condenser. Be careful not to damage the delicate cooling fins or cut yourself on them.

7   Every other inspection, or at the first indication of cooling system problems, have the cap and system pressure tested. If you don't have a pressure tester, most gas stations and repair shops will do this for a minimal charge.

---

13   Seat belt check (every 7500 miles [12,000 km] or 6 months)

---

1   Check the seat belts, buckles, latch plates and guide loops for obvious damage and signs of wear. Seat belts that exhibit fraying along the edges should be replaced.

2   Where the seat belt receptacle bolts to the floor of the vehicle, check that the bolts are secure.

3   See if the seat belt reminder light comes on when the key is turned to the Run or Start position. A chime should also sound.

---

14   Brake check (every 15,000 miles [24,000 km] or 12 months)

Warning: *Dust created by the brake system is harmful to your health. Never blow it out with compressed air and don't inhale any of it. An approved filtering mask should be worn when working on the brakes. Do not, under any circumstances, use petroleum-based solvents to clean brake parts. Use brake system cleaner only!*

Check for a chafed area that could fail prematurely.

Check for a soft area indicating the hose has deteriorated inside.

Overtightening the clamp on a hardened hose will damage the hose and cause a leak.

Check each hose for swelling and oil-soaked ends. Cracks and breaks can be located by squeezing the hose.

12.4 Hoses, like drivebelts, have a habit of failing at the worst possible time - to prevent the inconvenience of a blown radiator or heater hose, inspect them carefully as shown here

Note: *For detailed photographs of the brake system, refer to Chapter 9.*

1   In addition to the specified intervals, the brakes should be inspected every time the wheels are removed or whenever a defect is suspected. Any of the following symptoms could indicate a potential brake system defect: The vehicle pulls to one side when the brake pedal is depressed; the brakes make squealing or dragging noises when applied; brake travel is excessive; the pedal pulsates; brake fluid leaks, usually onto the inside of the tire or wheel.

2   The disc brake pads have built-in wear indicators which should make a high-pitched squealing or scraping noise when they are worn to the replacement point. When you hear this noise, replace the pads immediately or expensive damage to the discs can result.

3   Loosen the wheel lug nuts.

14.6 You will find an inspection hole like this in each caliper through which you can view the thickness of remaining friction material for the inner pad

15.1a Unlatch these two clips . . .

15.1b . . . pull the cover out of the way and lift the element out

4   Raise the vehicle and place it securely on jackstands.
5   Remove the wheels (see *Jacking and towing* at the front of this manual, or your owner's manual, if necessary).

Disc brakes

6   There are two pads - an outer and an inner - in each caliper. The inner pads are visible through small inspection holes in each caliper (see illustration). The outer pads are more easily viewed at the edge of the caliper.
7   Check the pad thickness by looking at each end of the caliper and through the inspection hole in the caliper body. If the lining material is less than the thickness listed in this Chapter's Specifications, replace the pads.
Note: *Keep in mind that the lining material is riveted or bonded to a metal backing plate and the metal portion is not included in this measurement.*
8   If it is difficult to determine the exact thickness of the remaining pad material by the above method, or if you are at all concerned about the condition of the pads, remove the caliper(s), then remove the pads for further inspection (see Chapter 9).
9   Once the pads are removed from the calipers, clean them with brake cleaner and re-measure them.
10   Measure the disc thickness with a micrometer to make sure that it still has service life remaining. If any disc is thinner than the specified minimum thickness, replace it (see Chapter 9). Even if the disc has service life remaining, check its condition. Look for scoring, gouging and burned spots. If these conditions exist, remove the disc and have it resurfaced (see Chapter 9).
11   Before installing the wheels, check all brake lines and hoses for damage, wear, deformation, cracks, corrosion, leakage, bends and twists, particularly in the vicinity of the rubber hoses at the calipers.
12   Check the clamps for tightness and the

connections for leakage. Make sure that all hoses and lines are clear of sharp edges, moving parts and the exhaust system. If any of the above conditions are noted, repair, reroute or replace the lines and/or fittings as necessary (see Chapter 9).

Brake booster check

13   Sit in the driver's seat and perform the following sequence of tests.
14   With the brake fully depressed, start the engine - the pedal should move down a little when the engine starts.
15   With the engine running, depress the brake pedal several times - the travel distance should not change.
16   Depress the brake, stop the engine and hold the pedal in for about 30 seconds - the pedal should neither sink nor rise.
17   Restart the engine, run it for about a minute and turn it off. Then firmly depress the brake several times - the pedal travel should decrease with each application.
18   If your brakes do not operate as described above when the preceding tests are performed, the brake booster has failed. Refer to Chapter 9 for the replacement procedure.

Parking brake

19   Pull up on the parking brake lever with a force of approximately 44 lbs. and count the number of clicks. The adjustment is correct if you hear the specified number of clicks as listed in this Chapter's Specifications. If you hear more or fewer clicks, it's time to adjust the parking brake (see Chapter 9).
20   An alternative method of checking the parking brake is to park the vehicle on a steep hill with the parking brake set and the transaxle in Neutral (be sure to stay in the vehicle for this procedure). If the parking brake cannot prevent the vehicle from rolling, it is in need of adjustment (see Chapter 9).

15   Air filter replacement (every 15,000 miles [24,000 km] or 12 months)

1   To remove the air filter, release the spring clips that keep the two halves of the air filter housing together, then pull the cover away. Noting how the filter is installed, remove the air filter element (see illustrations).
2   Inspect the outer surface of the filter element. If it is dirty, replace it. If it is only moderately dusty, it can be reused by blowing it clean from the back to the front surface with compressed air. Because it is a pleated paper type filter, it cannot be washed or oiled. If it cannot be cleaned satisfactorily with compressed air, discard and replace it. While the cover is off, be careful not to drop anything down into the housing.
Caution: *Never drive the vehicle with the air cleaner removed. Excessive engine wear could result and backfiring could even cause a fire under the hood.*
3   Wipe out the inside of the filter housing.
4   Place the new filter into the air cleaner housing, making sure it seats properly.
5 Installation of the cover is the reverse of removal.

16   Fuel system check (every 15,000 miles [24,000 km] or 12 months)

Warning: *Gasoline is extremely flammable, so take extra precautions when you work on any part of the fuel system. Don't smoke or allow open flames or bare light bulbs near the work area, and don't work in a garage where a gas-type appliance (such as a water heater or a clothes dryer) is present. Since gasoline is carcinogenic, wear fuel-resistant gloves when there's a possibility of being exposed to fuel, and, if you spill any fuel on your skin, rinse*

16.2 Use a small screwdriver to carefully pry out the old gasket - take care not to damage the cap

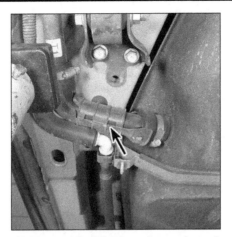

16.5 Inspect the filler hose for cracks and make sure the clamps are tight

17.2 Remove the check/fill plug and use your finger as a dipstick to check the transfer unit lubricant level

A   *Check/fill plug*
B   *Drain plug*

it off immediately with soap and water. Mop up any spills immediately and do not store fuel-soaked rags where they could ignite. The fuel system is under constant pressure, so, if any fuel lines are to be disconnected, the fuel pressure in the system must be relieved first. When you perform any kind of work on the fuel system, wear safety glasses and have a Class B type fire extinguisher on hand.

1   If you smell gasoline while driving or after the vehicle has been sitting in the sun, inspect the fuel system immediately.

2   Remove the fuel filler cap and inspect it for damage and corrosion. The gasket should have an unbroken sealing imprint. If the gasket is damaged or corroded, remove it and install a new one (see illustration).

3   Inspect the fuel feed lines for cracks. Make sure all fuel line fittings are dry.

4   Since some components of the fuel system - the fuel tank and part of the fuel feed and return lines, for example - are underneath the vehicle, they can be inspected more easily with the vehicle raised on a hoist. If that's not possible, raise the vehicle and support it securely on jackstands.

5   With the vehicle raised and safely supported, inspect the fuel tank and filler neck for punctures, cracks and other damage. The connection between the filler neck and the tank is particularly critical. Sometimes a rubber filler neck will leak because of loose clamps or deteriorated rubber (see illustration). These are problems a home mechanic can usually rectify.

Warning: *Do not, under any circumstances, try to repair a fuel tank (except rubber components).*

6   Carefully check all rubber hoses and metal lines leading away from the fuel tank. Check for loose connections, deteriorated hoses, crimped lines and other damage. Carefully inspect the lines from the tank to the fuel injection system. Repair or replace damaged sections as necessary (see Chapter 4).

17   Transfer unit lubricant level check (AWD models) (every 15,000 miles [24,000 km] or 12 months)

1   Raise the vehicle and support it securely on jackstands.

2   Using the appropriate wrench, unscrew the check/fill plug from the transfer unit (see illustration).

3   Use your little finger to reach inside the housing to feel the lubricant level. The level should be at or near the bottom of the plug hole. If it isn't, add the recommended lubricant through the plug hole with a syringe or squeeze bottle.

4   Install the plug and tighten it to the torque listed in this Chapter's Specifications. Check for leaks after the first few miles of driving.

18   Steering and suspension check (every 15,000 miles [24,000 km] or 12 months)

Note: *For detailed illustrations of the steering and suspension components, refer to Chapter 10.*

With the wheels on the ground

1   With the vehicle stopped and the front wheels pointed straight ahead, rock the steering wheel gently back and forth. If freeplay (see illustration) is excessive, a front wheel bearing, steering shaft universal joint or lower arm balljoint is worn or the steering gear is out of adjustment or broken. Refer to Chapter 10 for the appropriate repair procedure.

2   Other symptoms, such as excessive vehicle body movement over rough roads, swaying (leaning) around corners and binding as the steering wheel is turned, may indicate faulty steering and/or suspension components.

3   Check the shock absorbers by pushing down and releasing the vehicle several times at each corner. If the vehicle does not come back to a level position within one or two bounces, the shocks/struts are worn and must be replaced. When bouncing the vehicle up and down, listen for squeaks and noises from the suspension components. Additional information on suspension components can be found in Chapter 10.

Under the vehicle

4   Raise the vehicle with a floor jack and support it securely on jackstands. See *Jacking and towing* at the front of this manual for the proper jacking points.

5   Check the tires for irregular wear patterns and proper inflation. See Section 5 in this Chapter for information regarding tire wear and Chapter 10 for the wheel bearing replacement procedures.

6   Inspect the universal joint between the steering shaft and the steering gear housing. Check the steering gear housing for grease

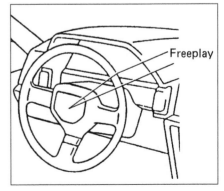

18.1 Steering wheel freeplay is the amount of travel between an initial steering input and the point at which the front wheels begin to turn (indicated by a slight resistance)

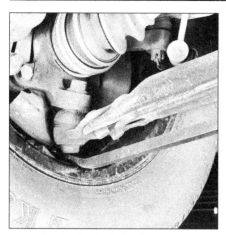

18.7 To check a balljoint for wear, try to pry the control arm up and down to make sure there is no play in the balljoint (if there is, replace it)

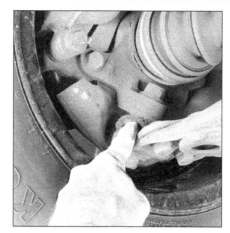

18.8 Push on the balljoint boot to check for damage

19.2 Flex the driveaxle boots by hand to check for cracks and/or leaking grease

leakage or oozing. Make sure that the dust seals and boots are not damaged and that the boot clamps are not loose. Check the steering linkage for looseness or damage. Check the tie-rod ends for excessive play. Look for loose bolts, broken or disconnected parts and deteriorated rubber bushings on all suspension and steering components. While an assistant turns the steering wheel from side to side, check the steering components for free movement, chafing and binding. If the steering components do not seem to be reacting with the movement of the steering wheel, try to determine where the slack is located.

7   Check the balljoints for wear by trying to move each control arm up and down with a pry bar (see illustration) to ensure that its balljoint has no play. If any balljoint does have play, replace it. See Chapter 10 for the balljoint replacement procedure.

8   Inspect the balljoint boots for damage and leaking grease (see illustration). Replace the balljoints with new ones if they are damaged (see Chapter 10).

9   At the rear of the vehicle, inspect the suspension arm bushings for deterioration.

19   Driveaxle boot check (every 15,000 miles [24,000 km] or 12 months)

1   The driveaxle boots are very important because they prevent dirt, water and foreign material from entering and damaging the constant velocity (CV) joints. Because it constantly pivots back and forth following the steering action of the front hub, the outer CV joint boot wears out sooner and should be inspected regularly.

2   Inspect the boots for tears and cracks as well as loose clamps (see illustration). If there is any evidence of cracks or leaking lubricant, the best solution is to replace the driveaxle with a rebuilt unit (see Chapter 8).

20   Rear differential lubricant level check (AWD models) (every 15,000 miles [24,000 km] or 12 months)

1   Raise the vehicle and support it securely on jackstands.
2   Using the appropriate wrench, unscrew the check/fill plug from the rear differential.
3   Use your little finger to reach inside the housing to feel the lubricant level. The level should be at or near the bottom of the plug hole. If it isn't, add the recommended lubricant through the plug hole with a syringe or squeeze bottle.
4   Install the plug and tighten it to the torque listed in this Chapter's Specifications. Check for leaks after the first few miles of driving.

21   Interior ventilation filter replacement (every 15,000 miles [24,000 km] or 12 months)

1   Open the glove box door, push in on the sides and allow the door to hang down (see Chapter 11, if necessary).
2   Unclip the cover from the end of the housing (see illustration).
3   Pull the filter element straight out of the case and install the new one (see illustration).
4   Installation is the reverse of removal.

22   Brake fluid change (every 30,000 miles [48,000 km] or 24 months)

Warning: *Brake fluid can harm your eyes and damage painted surfaces, so use extreme caution when handling or pouring it. Do not use brake fluid that has been standing open or is more than one year old. Brake fluid absorbs moisture from the air. Excess moisture can cause a dangerous loss of braking effectiveness.*

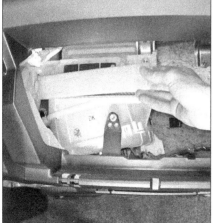

21.2 Unclip the filter cover from the housing

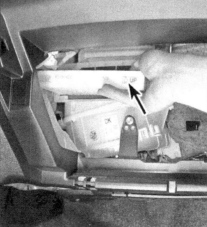

21.3 When installing the filter, make sure the directional arrows are pointing UP

23.4 You will have to remove the forward under-vehicle splash shield for access to the radiator drain fitting located at the bottom left corner of the radiator - before opening the valve, push a length of hose onto the plastic fitting to prevent the coolant from splashing

1   At the specified intervals, the brake fluid should be drained and replaced. Since the brake fluid may drip or splash when pouring it, place plenty of rags around the master cylinder to protect any surrounding painted surfaces.

2   Before beginning work, purchase the specified brake fluid (see this Chapter's Specifications).

3   Remove the cap from the master cylinder reservoir.

4   Using a hand-held suction pump or similar device, withdraw the fluid from the master cylinder reservoir.

5   Add new fluid to the master cylinder until it rises to the line indicated on the reservoir.

6   Bleed the brake system as described in Chapter 9 at all four brakes until new and uncontaminated fluid is expelled from the bleeder screw. Be sure to maintain the fluid level in the master cylinder as you perform the bleeding process. If you allow the master cylinder to run dry, air will enter the system.

7   Refill the master cylinder with fluid and check the operation of the brakes. The pedal should feel solid when depressed, with no sponginess.

Warning: *Do not operate the vehicle if you are in doubt about the effectiveness of the brake system.*

23   Cooling system servicing (draining, flushing and refilling) (every 30,000 miles [48,000 km] or 24 months)

Warning: *The engine must be completely cool before beginning this procedure.*

Warning: *Do not allow engine coolant (antifreeze) to come in contact with your skin or painted surfaces of the vehicle. Rinse off spills immediately with plenty of water. Antifreeze is highly toxic if ingested. Never leave anti-*

24.2 The EVAP canister is located under the vehicle on the left side, forward of the fuel tank (cover removed)

freeze lying around in an open container or in puddles on the floor; children and pets are attracted by it's sweet smell and may drink it. Check with local authorities on disposing of used antifreeze. Many communities have collection centers which will see that antifreeze is disposed of safely. Antifreeze is flammable under certain conditions - be sure to read the precautions on the container.

Note: *Non-toxic antifreeze is available at most auto parts stores. Although the antifreeze is non-toxic when fresh, proper disposal is still required.*

1   Periodically, the cooling system should be drained, flushed and refilled to replenish the antifreeze mixture and prevent formation of rust and corrosion, which can impair the performance of the cooling system and cause engine damage. When the cooling system is serviced, all hoses and the radiator cap should be checked and replaced if necessary.

Draining

2   Apply the parking brake and block the wheels. If the vehicle has just been driven, wait several hours to allow the engine to cool down before beginning this procedure.

3   Once the engine is completely cool, remove the expansion tank cap.

4   Move a large container under the radiator drain to catch the coolant. Attach a hose to the drain fitting to direct the coolant into the container (some models are already equipped with a hose), then open the drain fitting (a pair of pliers may be required to turn it) (see illustration).

5   While the coolant is draining, check the condition of the radiator hoses, heater hoses and clamps (refer to Section 12 if necessary). Replace any damaged clamps or hoses.

Flushing

6   Close the radiator drain valve.

7   Fill the cooling system with clean water, following the *Refilling* procedure (see Step 13).

8   Start the engine and allow it to reach normal operating temperature, then rev up the engine a few times.

9   Turn the engine off and allow it to cool completely, then drain the system as described earlier.

10   Repeat Steps 7 through 9 until the water being drained is free of contaminants.

11   In severe cases of contamination or clogging of the radiator, remove the radiator (see Chapter 3) and have a radiator repair facility clean and repair it if necessary.

12   Many deposits can be removed by the chemical action of a cleaner available at auto parts stores. Follow the procedure outlined in the manufacturer's instructions.

Note: *When the coolant is regularly drained and the system refilled with the correct antifreeze/water mixture, there should be no need to use chemical cleaners or descalers.*

Refilling

13   Close and tighten the radiator drain. Remove the drain hose.

14   Place the heater temperature control in the maximum heat position.

15   Slowly add new coolant (see this Chapter's Specifications for the proper mixture of water and antifreeze) to the expansion tank until it's full. Add coolant to the reservoir up to the lower mark.

16   Squeeze the upper and lower radiator hoses to expel air, then recheck the coolant level, adding as necessary.

17   Install the radiator cap/expansion tank cap and run the engine in a well-ventilated area at approximately 2000 rpm until the engine cooling fan comes on.

Caution: *If the coolant level drops below the Low line, allow the engine to cool completely, then add coolant.*

18   Turn the engine off and let it cool completely. Add more coolant mixture to bring the level back up between the Low and Full lines on the expansion tank.

19   Squeeze the upper radiator hose to expel air, then add more coolant mixture if necessary. Reinstall the radiator cap/expansion tank cap.

20   Start the engine, allow it to reach normal operating temperature and check for leaks.

24   Evaporative emissions control system check (every 30,000 miles [48,000 km] or 24 months)

1   The function of the evaporative emissions control system is to draw fuel vapors from the gas tank and fuel system, store them in a charcoal canister and then burn them during normal engine operation.

2   The most common symptom of a fault in the evaporative emissions control system is a strong fuel odor from under the vehicle. If a fuel odor is detected, inspect the charcoal canister. The canister is located under the vehicle on the left side, forward of the fuel tank (see illustration). Check the canister and all hoses for damage and deterioration.

3   The evaporative emissions control system is explained in more detail in Chapter 6.

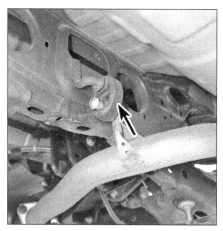

25.4 Be sure to check each exhaust system rubber hanger for damage

26.1 Remove the transaxle fluid overflow plug

26.2a If fluid flows from the hole, let it drain until it stops

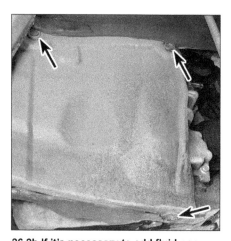

26.2b If it's necessary to add fluid, remove the inner fender splash shield fasteners and shield . . .

26.2c . . . then remove the transaxle fill plug

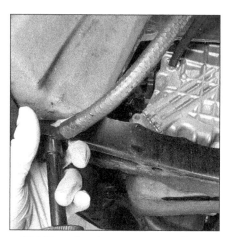

26.3 Use a hand pump connected to a hose to add fluid to the transaxle

25   Exhaust system check (every 30,000 miles [48,000 km] or 24 months)

1   With the engine cold (at least three hours after the vehicle has been driven), check the complete exhaust system from its starting point at the engine to the end of the tailpipe. This should be done on a hoist where unrestricted access is available.

2   Check the pipes and connections for evidence of leaks, severe corrosion or damage. Make sure that all brackets and hangers are in good condition and tight.

3   At the same time, inspect the underside of the body for holes, corrosion, open seams, etc., which may allow exhaust gases to enter the passenger compartment. Seal all body openings with silicone or body putty.

4   Rattles and other noises can often be traced to the exhaust system, especially the mounts and hangers. Try to move the pipes, muffler and catalytic converter. If the compo-

nents can come in contact with the body or suspension parts, secure the exhaust system with new mounts (see illustration).

26   Automatic transaxle fluid level check (every 30,000 miles [48,000 km] or 24 months)

Note: *These models are not equipped with an automatic transaxle fluid dipstick.*
Note: *The vehicle must be level for this check. If there is not enough room to crawl under the vehicle, raise both ends and support it securely on jackstands.*
Note: *The fluid temperature must be between 104 and 113-degrees F (40 to 45-degrees C) to perform this check.*

1   Set the parking brake and block the rear wheels. With the engine idling, remove the overflow plug from the bottom of the transaxle fluid pan (see illustration).

2   If the fluid runs out of the hole, allow it to

drip until it stops. If no fluid comes out of the hole, remove the fill plug (see illustrations), located on the side of the transaxle housing, near the rear.

3   Add the proper type of transmission fluid (see this Chapter's Specifications) through the fill plug hole (see illustration), until fluid flows from the overflow hole in the bottom of the pan. When the flow of fluid slows to a trickle, install the overflow plug and tighten it to the torque listed in this Chapter's Specifications.

4   Tighten the fill plug to the torque listed in this Chapter's Specifications.

27   Transfer unit lubricant change (AWD models) (every 60,000 miles [96,000 km] or 48 months)

1   Raise the front of the vehicle and support it securely on jackstands. Remove the check plug, followed by the drain plug, and allow the lubricant to drain (see illustration 17.2).

2   Reinstall the drain plug and tighten it to the torque listed in this Chapter's Specifications.
3   Add the specified type of lubricant until is even with the lower edge of the hole. Reinstall the check/fill plug and tighten it to the torque listed in this Chapter's Specifications.

28   Rear differential lubricant change (AWD models) (every 60,000 miles [96,000 km] or 72 months)

1   Raise the rear of the vehicle and support it securely on jackstands. Remove the check/fill plug from the front face of the differential.
2   Remove the drain plug and drain the lubricant (see illustration).
3   Reinstall the drain plug and tighten it to the torque listed in this Chapter's Specifications.
4   Add new lubricant until it is even with the lower edge of the filler hole. See *Recommended lubricants and fluids* for the specified lubricant type. Install the plug and tighten it to the torque listed in this Chapter's Specifications.

29   Positive Crankcase Ventilation (PCV) valve and hose check and replacement (every 60,000 miles [96,000 km] or 72 months)

1   The PCV valve is located below the intake manifold (see illustration).
2   Remove the intake manifold (see Chapter 2A) detach the hose, unscrew the valve from the separator case and replace it (see illustration).
3   When purchasing a replacement PCV valve, make sure it's for your particular vehicle and engine size. Compare the old valve with the new one to make sure they're the same.

30   Automatic transaxle fluid change (every 60,000 miles [96,000 km] or 72 months)

1   At the specified time intervals, the automatic transaxle fluid should be drained and replaced.
Note: *Although the manufacturer doesn't specify it, it is a good idea to clean the transaxle fluid strainer periodically to remove accumulated dirt and metal particles.*
2   Before beginning work, purchase the specified transaxle fluid (see *Recommended lubricants and fluids* in this Chapter's Specifications).
3   The fluid should be drained immediately after the vehicle has been driven. Hot fluid is more effective than cold fluid at removing built up sediment.
Warning: *Fluid temperature can exceed 350-degrees F in a hot transaxle. Wear protective gloves.*

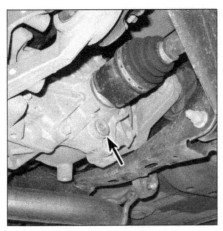

28.2 Location of the rear differential drain plug on 4WD models

29.1 The PCV valve (1) is threaded into the oil separator case (2) - (intake manifold removed)

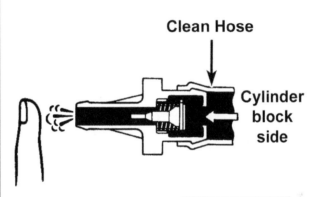

29.2 To check the PVC valve, first attach a clean length of hose to the cylinder block side of the valve and blow through it - air should pass through easily - then blow through the intake manifold side of the valve and verify that air passes through with difficulty

Clean Hose

Cylinder block side

92083-30.02 HAYNES

30.6a Remove the overflow plug . . .

4   Raise the vehicle and support it securely on jackstands. Remove the engine splash shields.
5   Move the drain pan under the vehicle, being careful not to touch any of the hot exhaust components.

6   Remove the transaxle fluid overflow plug from the fluid pan, then unscrew the overflow tube and allow the fluid to drain (see illustrations).
7   After the fluid has finished draining, reinstall the overflow tube and tighten it to the

30.6b . . . then unscrew the overflow tube . . .

30.6c . . . to drain the fluid

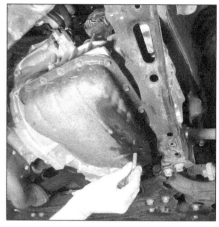

30.7a Install the overflow tube . . .

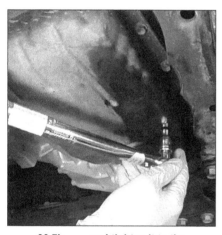

30.7b . . . and tighten it to the specified torque

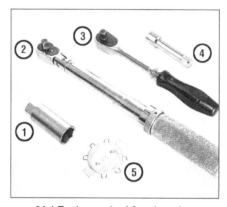

31.1 Tools required for changing spark plugs

1 **Spark plug socket** - *This will have special padding inside to protect the spark plug's porcelain insulator*
2 **Torque wrench** - *Although not mandatory, using this tool is the best way to ensure the plugs are tightened properly*
3 **Ratchet** - *Standard hand tool to fit the spark plug socket*
4 **Extension** - *Depending on model and accessories, you may need special extensions and universal joints to reach one or more of the plugs*
5 **Spark plug gap gauge** - *This gauge for checking the gap comes in a variety of styles. Make sure the gap for your engine is included*

31.7 Use a spark plug socket with a long extension to unscrew the spark plugs

torque listed in this Chapter's Specifications (see illustrations).
8 Install the overflow plug and tighten it temporarily at this time.
9 Refer to Section 26 for the refilling and fluid level checking procedure.

---

31 Spark plug check and replacement (every 100,000 miles [161,000 km])

1 Spark plug replacement requires a spark plug socket which fits onto a ratchet wrench. This socket is lined with a rubber grommet to protect the porcelain insulator of the spark plug and to hold the plug while you remove it from the spark plug hole. You will also need a wire-type feeler gauge to check and adjust the spark plug gap and a torque wrench to tighten the new plugs to the specified torque (see illustration).
2 If you are replacing the plugs, purchase the new plugs, adjust them to the proper gap and replace each plug one at a time.

Note: *When buying new spark plugs, it's essential that you obtain the correct plugs for your specific vehicle. This information can be found in this Chapter's Specifications.*

Check

3 Inspect each of the new plugs for defects. If there are any signs of cracks in the porcelain insulator of a plug, don't use it.
4 The gap between the electrodes should

be identical to that listed in this Chapter's Specifications or in your owner's manual. If the gap is incorrect, the spark plug must be replaced.
Caution: *These spark plugs have iridium-coated electrodes. Don't attempt to adjust the gap on a used iridium-coated plug.*

Removal

5 Remove the ignition coils (see Chapter 5).
6 If compressed air is available, blow any dirt or foreign material away from the spark plug area before proceeding.
7 Remove the spark plugs (see illustration).
8 Whether you are replacing the plugs at this time or intend to reuse the old plugs, compare each old spark plug with the chart on the inside back cover of this manual to determine the overall running condition of the engine.

Installation

9 Prior to installation, it's a good idea to coat the spark plug threads with anti-seize

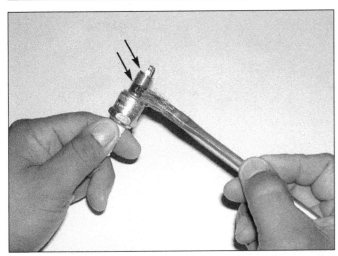

31.9a Apply a thin coat of anti-seize compound to the spark plug threads, but not in the area near the electrodes

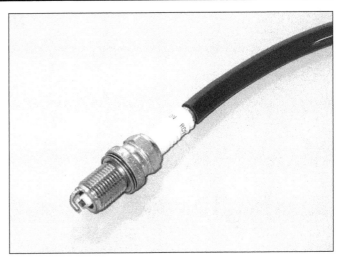

31.9b A length of snug-fitting rubber hose will save time and prevent damaged threads when installing the spark plugs

compound (see illustration). Also, it's often difficult to insert spark plugs into their holes without cross-threading them. To avoid this possibility, fit a length of snug-fitting rubber hose over the end of the spark plug (see illustration). The flexible hose acts as a universal joint to help align the plug with the plug hole. Should the plug begin to cross-thread, the hose will slip on the spark plug, preventing thread damage. Tighten the plug to the torque listed in this Chapter's Specifications.

10   Install the ignition coil (see Chapter 5), if necessary.

11   Follow the above procedure for the remaining spark plugs, replacing them one at a time to prevent mixing up the ignition coils.

Notes

Chapter 2 Part A
Engine

Contents

Specifications

General

Engine designation	2AR-FE
Displacement	152.2 cubic inches (2.5 liters)
Cylinder numbers (drivebelt end-to-transaxle end)	1-2-3-4
Firing order	1-3-4-2

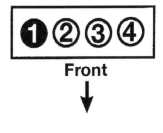

Cylinder locations

Cylinder head

Warpage limit	0.00197 inch (0.05 mm)

Timing chain

Timing chain sprocket wear limit (see illustration 6.13b)	
Crankshaft sprocket (w/chain)	2.36 inches (59.94 mm)
Camshaft sprocket (w/chain)	Not available
Timing chain stretch limit (see illustration 6.13a)	
8 links (15 pins)	5.42 inches (137.7 mm)
Timing chain guide wear limit	0.039 inch (1.0 mm)
No. 1 chain vibration damper wear limit	0.039 inch (1.0 mm)
Tensioner slipper or guide wear limit	0.039 inch (1.0 mm)

Camshaft

Journal diameter	
No. 1 journal	1.356 to 1.357 inches (34.442 to 34.468 mm)
All others	0.904 to 0.905 inch (22.961 to 22.987 mm)
Bearing oil clearance	
No.1 journal	
Intake	
Standard	0.00137 to 0.00283 inch (0.035 to 0.072 mm)
Maximum	0.00335 inch (0.085 mm)
Exhaust	
Standard	0.00193 to 0.00339 inch (0.049 to 0.086 mm)
Maximum	0.00335 inch (0.085 mm)
All others	
Standard	0.00098 to 0.00244 inch (0.025 to 0.062 mm)
Maximum	0.00335 inch (0.085 mm)
Runout limit	0.0012 inch (0.03 mm)
Thrust clearance (endplay)	0.00236 to 0.00610 inch (0.060 to 0.155 mm)

Camshaft (continued)

Lobe height
    Intake camshaft
        Standard .... 1.739 to 1.744 inches (44.163 to 44.298 mm)
        Service limit (minimum) .... 1.733 inches (44.013 mm)
    Exhaust camshaft
        Standard .... 1.738 to 1.744 inches (44.144 to 44.286 mm)
        Service limit (minimum) .... 1.732 inches (43.996 mm)

Torque specifications

Ft-lbs (unless otherwise indicated)   **Nm**

Note: *One foot-pound (ft-lb) of torque is equivalent to 12 inch-pounds (in-lbs) of torque. Torque values below approximately 15 ft-lbs are expressed in inch-pounds, because most foot-pound torque wrenches are not accurate at these smaller values.*

	Ft-lbs	Nm
Camshaft bearing cap bolts (x11) (see illustration 7.26)	12	16
Camshaft housing and bearing cap bolts (x20), (see illustration 7.30)....	20	27
Camshaft sprocket bolts..	63	85
Crankshaft pulley/vibration damper bolt	192	260
Cylinder head bolts (in sequence - see illustration 10.24)		
Step 1	27	37
Step 2	27	37
Step 3	Tighten an additional 90-degrees	
Step 4	Tighten an additional 90-degrees	
Drivebelt tensioner..	15	21
Engine balancer assembly	18	24
Engine mounts		
Passenger's side mount		
Mount-to-frame and mount-to-engine bracket	70	95
Mount spacer-to-mount bracket bolt	70	95
Mount bracket-to-timing cover (see illustration 6.30)		
Bolts 1, 2 and 3	41	55
Bolt 4	15	21
Driver's side mount		
Mount bolts	70	95
Mount through-bolt/nut..	41	56
Front mount		
Mount through-bolt/nut..	107	145
Mount-to-frame and mount-to-engine bolts	70	95
Rear mount		
Mount-to-frame and mount-to-engine bracket bolts/nut..............	70	95
Exhaust manifold nuts	26	35
Exhaust manifold brace bolt/nut	32	43
Exhaust manifold heat shield bolts	108 in-lbs	12
Exhaust pipe-to-exhaust manifold bolts...	32	43
Flywheel/driveplate bolts		
Manual transaxle..	96	130
Automatic transaxle..	72	98
Front suspension reinforcement bolts (left and right)		
Front bolts	71	96
Rear bolts	73	99
Intake manifold bolts...	21	28
Lower crankcase-to-engine block bolts	See Chapter 2B	
Oil pan bolts (tighten bolt A twice)	84 in-lbs	10
Timing chain guide bolts (stationary)...	15	21
Timing chain tensioner pivot arm bolt	15	21
Timing chain cover bolts (see illustration 6.29b)		
Bolts 1, 2, 3 and 4	41	56
All others	15	21
Timing chain cover nuts...	15	21
Timing chain tensioner	90 in-lbs	10
Upper timing chain guide bolt...	15	21
Valve cover	108 in-lbs	12

1 General information

1   This Part of Chapter 2 is devoted to engine component repair procedures, except for engine overhaul. Information concerning engine removal, installation and overhaul can be found in Chapter 2B.

2   Some of the following repair procedures are based on the assumption that the engine is installed in the vehicle. However, due to the design of the vehicle, many procedures that are normally accomplished in-vehicle require engine removal.

3   The Specifications included in this Part of Chapter 2 apply only to the procedures contained in this Part. Chapter 2B contains the general Specifications necessary for certain procedures considering engine rebuilding.

4   These engines incorporate an aluminum cylinder block with a lower crankcase to strengthen the lower half of the block. Although the cylinder head utilizes dual overhead camshafts (DOHC) with four valves per cylinder as in previous model years, it is of a new design. The camshafts are driven from a single timing chain off the crankshaft, and a Variable Valve Timing (VVT-i) system is incorporated on the intake and exhaust camshafts to increase horsepower and decrease emissions.

2 Repair operations possible with the engine in the vehicle

1   Many major repair operations can be accomplished without removing the engine from the vehicle.

2   Clean the engine compartment and the exterior of the engine with some type of degreaser before any work is done. It will make the job easier and help keep dirt out of the internal areas of the engine.

3   Depending on the components involved, it may be helpful to remove the hood to improve access to the engine as repairs are performed (refer to Chapter 11 if necessary). Cover the fenders to prevent damage to the paint. Special pads are available, but an old bedspread or blanket will also work.

4   If vacuum, exhaust, oil or coolant leaks develop, indicating a need for gasket or seal replacement, the repairs can generally be made with the engine in the vehicle. The intake and exhaust manifold gaskets, lower oil pan gasket and crankshaft oil seals are all accessible with the engine in place.

5   Exterior engine components, such as the intake and exhaust manifolds, the water pump, the starter motor, the alternator and the fuel system components can be removed for repair with the engine in place.

6   Camshaft, camshaft sprockets, rocker arm and lash adjuster servicing can also be accomplished with the engine in the vehicle.

3 Top Dead Center (TDC) for number one piston - locating

1   Top Dead Center (TDC) is the highest point in the cylinder that each piston reaches as it travels up the cylinder bore. Each piston reaches TDC on the compression stroke and again on the exhaust stroke, but TDC generally refers to piston position on the compression stroke.

2   Positioning the piston(s) at TDC is an essential part of certain procedures such as camshaft and sprocket removal.

3   Before beginning this procedure, make sure the transmission is in Park and apply the parking brake or block the rear wheels. Dis-connect the cable from the negative terminal of the battery (see Chapter 5).

4   In order to bring any piston to TDC, the crankshaft must be turned using a socket, extension and ratchet or breaker bar. When looking at the front of the engine, normal crankshaft rotation is clockwise.

5   Remove the spark plugs (see Chapter 1) and install a compression gauge in the number one spark plug hole (see illustration). It should be a gauge with a screw-in fitting and a hose at least six inches long.

6   Loosen the right front wheel lug nuts, then raise the front of the vehicle and support it securely on jackstands. Remove the wheel and, if necessary, the inner fender splash shield.

7   Rotate the crankshaft using a socket and breaker bar or ratchet while observing for pressure on the compression gauge. The moment the gauge shows pressure indicates that the number one cylinder has begun the compression stroke.

8   Once the compression stroke has begun, TDC for the compression stroke is reached by bringing the piston to the top of the cylinder.

9   Continue turning the crankshaft until the notch in the crankshaft damper is aligned with the "TDC" or the "0" mark on the timing chain cover (see illustration). At this point, the number one cylinder is at TDC on the compression stroke. If the marks are aligned but there was no compression, the piston was on the exhaust stroke; continue rotating the crankshaft 360-degrees (one turn).

10   After the number one piston has been positioned at TDC on the compression stroke, TDC for any of the remaining cylinders can be located by turning the crankshaft 180 degrees and following the firing order (refer to this Chapter's Specifications). Rotating the engine 180 degrees past TDC #1 will put the engine at TDC compression for cylinder #3.

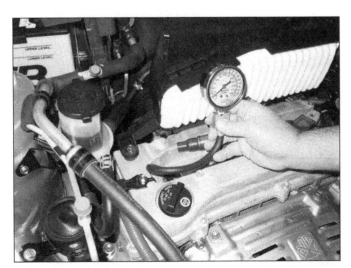

3.5 A compression gauge can be used in the number one plug hole to assist in finding TDC

3.9 Align the groove in the damper (A) with the "0" mark on the timing chain cover (B)

4.4 Valve cover fastener locations

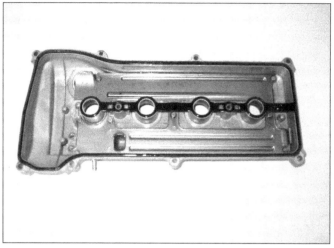

4.5 The valve cover gasket and the spark plug tube seals are incorporated into a single rubber O-ring-like seal. Press the gasket evenly into the grooves around the underside of the valve cover and the spark plug openings

4   Valve cover - removal and installation

Removal

1   Disconnect the cable from the negative battery terminal (see Chapter 5). Remove the engine cover by pulling it straight up.
2   Disconnect the electrical connectors from the ignition coils, remove the nuts securing the wiring harness to the valve cover and position the ignition coil wiring harness aside. Remove the ignition coil pack from each of the spark plugs (see Chapter 5).
3   Detach the PCV hoses from the valve cover.
4   Remove the valve cover mounting nuts, then detach the valve cover and gasket from the cylinder head (see illustration). If the valve cover is stuck to the cylinder head, bump the end with a wood block and a hammer to jar it loose. If that doesn't work, try to slip a flexible putty knife between the cylinder head and

valve cover to break the seal.
Caution: *Don't pry at the valve cover-to-cylinder head joint or damage to the sealing surfaces may occur, leading to oil leaks after the valve cover is reinstalled.*

Installation

5   Remove the valve cover gasket from the valve cover and clean the mating surfaces with brake system cleaner. Install a new rubber gasket, pressing it evenly into the grooves around the underside of the valve cover. Make sure the spark plug tube seals are in place on the underside of the valve cover before reinstalling it (see illustration). The mating surfaces of the timing chain cover, the cylinder head and valve cover must be perfectly clean when the valve cover is installed. If there's residue or oil on the mating surfaces when the valve cover is installed, oil leaks may develop.
6   Apply RTV sealant at the timing chain cover-to-cylinder head joint, then install the

valve cover and fasteners (see illustration).
7   Tighten the nuts/bolts to the torque listed in this Chapter's Specifications in two or three equal steps.
8   Reinstall the remaining parts, run the engine and check for oil leaks.

5   Variable Valve Timing (VVT) system - description

1   The VVT system varies intake and exhaust camshaft timing by directing oil pressure to advance or retard the camshaft sprocket/actuator assemblies. Changing the camshaft timing during certain engine conditions increases engine power output, fuel economy and reduces emissions.
2   System components include the Powertrain Control Module (PCM), the VVT oil control valves (OCV) and the intake and exhaust camshaft sprocket/actuator assemblies (see illustration).

4.6 Apply sealant at the timing chain cover-to-cylinder head joint before installing the valve cover

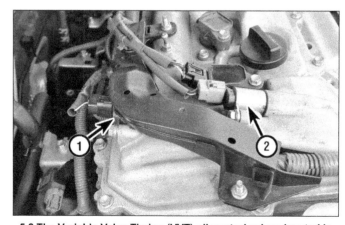

5.2 The Variable Valve Timing (VVT) oil control valves located in the valve cover

1   *Intake camshaft VVT oil control valve*
2   *Exhaust camshaft VVT oil control valve*

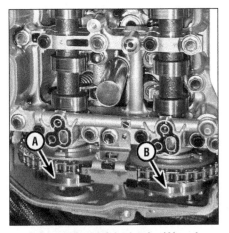

5.4 Locations of the intake (A) and exhaust (B) camshaft actuators

3   The PCM uses inputs from the following sensors to turn the oil control valve on or off:

a) Vehicle Speed Sensor (VSS)
b) Throttle Position Sensor (TPS)
c) Mass Airflow (MAF) sensor
d) Engine Coolant Temperature (ECT) sensor

4   Once the VVT oil control valves are actuated by the PCM, they direct the specified amount of oil pressure from the engine to advance or retard the camshaft sprocket/actuator assemblies (see illustration).

5   The camshaft sprocket/actuator assembly is equipped with an inner hub that is attached to the camshaft. The inner hub consists of a series of fixed vanes that use oil pressure as a wedge against the vanes to rotate the camshaft. The higher the oil pressure (or flow) the more the actuator assembly will rotate, thereby advancing or retarding the camshaft.

6   When oil is applied to the advance or retard side of the vanes, the actuator can advance or retard the camshaft timing. The PCM can also send a signal to the oil control valve to stop oil flow to both (advance and retard) passages to hold camshaft(s) in its current position.

7   For camshaft sprocket/actuator removal and installation, see Section 6. For VVT oil control valve removal and installation, see Chapter 6.

---

6   Timing chain and sprockets - removal, inspection and installation

---

Caution: *The timing system is complex, and severe engine damage will occur if you make any mistakes. Do not attempt this procedure unless you are highly experienced with this type of repair. If you are at all unsure of your abilities, be sure to consult an expert. Double-check all your work and be sure everything is correct before you attempt to start the engine.*

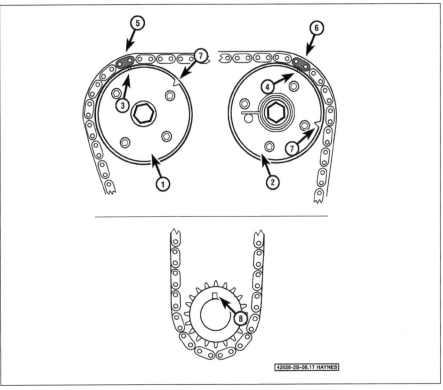

6.1 Camshafts and crankshaft TDC timing chain mark locations

1   Intake camshaft sprocket
2   Exhaust camshaft sprocket
3   Intake camshaft sprocket timing mark
4   Exhaust camshaft sprocket timing mark
5   Timing chain "yellow" colored link - intake

6   Timing chain "yellow" colored link - exhaust
7   Index mark, not a timing mark
8   Crankshaft "TDC" keyway at 12 o'clock position

Removal

1   Position the number one piston at TDC on the compression stroke (see Section 3). Confirm the engine is at TDC on the compression stroke by verifying that the timing mark on the crankshaft pulley/vibration damper is aligned with the "0" mark on the timing chain cover and the camshaft sprocket marks are aligned with the marks on the camshaft front bearing caps (see illustration 3.9 and the accompanying illustration).

Note: *There are two sets of marks on the camshaft sprockets. The marks that align at TDC are for TDC reference only; the other two marks are used to align the sprockets with the timing chain during installation.*

2   Remove the crankshaft pulley/vibration damper, being careful not to rotate the engine from TDC (see Section 11). If the engine rotates off TDC during this step, reposition the engine back to TDC before proceeding.

3   Remove the drivebelt tensioner and the crankshaft position sensor from the timing chain cover. Also remove the bolt securing the crankshaft position sensor wiring harness to the timing chain cover.

4   Remove the oil pan (see Section 13).

5   Remove the timing chain cover mounting

fasteners and carefully pry the cover off the engine from several locations.

Note: *The timing chain cover has several different size mounting bolts; take note of each bolt's location so they can be returned to the same locations on reassembly.*

6   Once the cover is off, remove the O-rings from the crankcase.

7   Remove the top chain guide mounting bolt and guide (see illustration).

6.7 Top chain guide, mounting bolt and locating pin

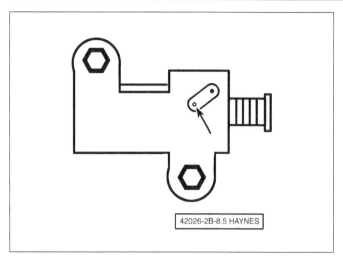

6.8 Align the hole on the lock with the hole in the tensioner, then insert a 0.06 inch (1.5 mm) pin or drill bit through both components to lock the tensioner

6.12 Hold the lug on the camshaft with a wrench to keep it from rotating as the sprocket bolts are loosened - when loosening the camshaft sprocket, loosen only the center bolt which secures the sprocket to the camshaft

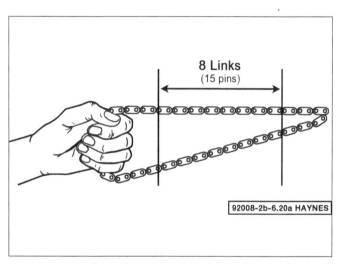

6.13a Timing chain stretch is measured by checking the length of the chain between 8 links (15 pins) at 3 or more places (selected randomly) around the chain - If chain stretch exceeds the specifications between any of the 15 pins, the chain must be replaced

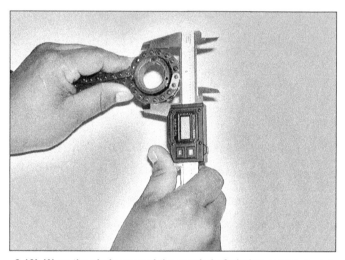

6.13b Wrap the chain around the crankshaft timing sprocket and measure the diameter of the sprocket across the chain rollers - if the measurement exceeds the minimum sprocket diameter, the chain and the timing sprockets must be replaced

8   Let the tensioner plunger extend until a 0.06 inch (1.5 mm) pin can be inserted into the alignment holes (see illustration) of the stopper plate and the tensioner. Release the tensioner and make sure the pin is secured.

9   Remove the mounting bolts and the timing chain tensioner.

10   Remove the timing chain tensioner pivot arm/chain guide and the stationary timing chain guide.

11   Lift the timing chain off the camshaft sprockets and remove the timing chain and crankshaft sprocket as an assembly from the engine. The crankshaft sprocket should slip off the crankshaft by hand. If not, carefully pry the sprocket off the crankshaft.

Note: *If you intend to reuse the timing chain, use white paint or chalk to make a mark indicating the front of the chain.If a used timing chain is reinstalled with the wear pattern in the* opposite direction, noise and increased wear may occur.

12   To remove the camshaft sprockets, loosen the bolts while holding the lug on the camshaft with a wrench (see illustration). Note the identification marks on the camshaft sprockets before removal, then remove the bolts. Pull on the sprockets by hand until they slip off the dowels. If necessary, use a small puller, with the legs inserted in the relief holes, to pull the sprockets off.

Note: *These models are equipped with variable valve timing, which consists of an actuator assembly attached to the camshaft sprocket. When removing the sprocket, only loosen and remove the center bolt, which fastens the sprocket to the camshaft. Do not loosen the outer four bolts that secure the VVT actuator to the sprocket.*

Inspection

13   Visually inspect all parts for wear and damage. Check the timing chain for loose pins, cracks, worn rollers and side plates. Check the sprockets for hook-shaped, chipped and broken teeth. Also check the timing chain for stretching and the diameter of the timing sprockets for wear with the chain assembled on the sprockets (see illustrations). Be sure to measure across the chain rollers when checking the sprocket diameter and to measure chain stretch at three or more places around the chain. Maximum chain elongation and minimum sprocket diameter (with chain) should not exceed the amount listed in this Chapter's Specifications. Replace the timing chain and sprockets as a set if the engine has high mileage or fails inspection.

14   Check the chain guides for excessive wear (see illustration). Replace the chain

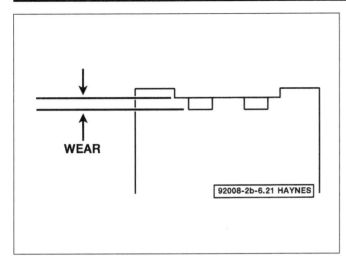

6.14 Timing chain guide wear is measured from the top of the chain contact surface to the bottom of the wear grooves

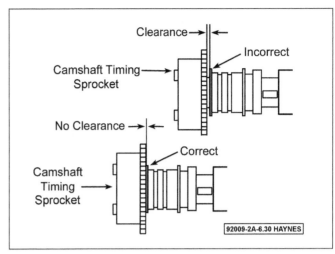

6.16 Camshaft timing sprocket installation details

guides if scoring or wear exceeds the amount listed in this Chapter's Specifications. Note that some scoring of the timing chain guide shoes is normal. If excessive wear is indicated, it will also be necessary to inspect the chain guide oil hole on the front of the block for clogging.

Installation

Caution: *Before starting the engine, carefully rotate the crankshaft by hand through at least two full revolutions (use a socket and breaker bar on the crankshaft pulley center bolt). If you feel any resistance, STOP! There is something wrong - most likely, valves are contacting the pistons. You must find the problem before proceeding. Check your work and see if any updated repair information is available.*

15   Remove all traces of old sealant from the timing chain cover and the mating surfaces of the engine block and cylinder head.

16   Make sure the camshafts are positioned with the dowel pins at the top in the 12 o'clock position, then install both camshaft sprockets in their original locations by aligning the dowel pin hole on the rear of the sprockets with the dowel pin on the camshaft. Apply medium strength thread locking compound to the camshaft sprocket bolt threads and make sure the washers are in place. Hold the camshaft from turning as described in Step 21 and tighten the bolts to the torque listed in this Chapter's Specifications. Make sure the camshaft sprocket is fully seated against the camshaft flange (see illustration).

17   Temporarily install the crankshaft pulley bolt and rotate the crankshaft until the keyway is pointing to the 10 o'clock position.

18   Rotate the camshafts as necessary to align the TDC marks on the camshaft sprockets (see illustration).

19   Install the stationary timing chain guide.

20   Loop the timing chain around the exhaust camshaft sprocket. Align the yellow colored link with the exhaust camshaft

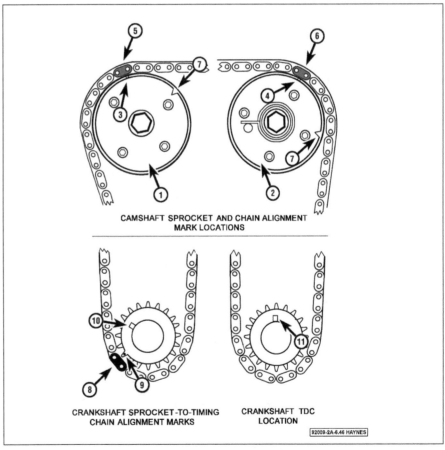

6.18 Timing chain alignment details

1 Intake camshaft sprocket	7 Index mark, not a timing mark
2 Exhaust camshaft sprocket	8 Timing chain "pink" colored
3 Intake camshaft sprocket timing mark	link - crankshaft
4 Exhaust camshaft sprocket	9 Crankshaft timing mark
timing mark	10 Crankshaft keyway 10 o'clock position
5 Timing chain "yellow" colored	11 Crankshaft "TDC" keyway at
link - intake	12 o'clock position
6 Timing chain "yellow" colored	
link - exhaust	

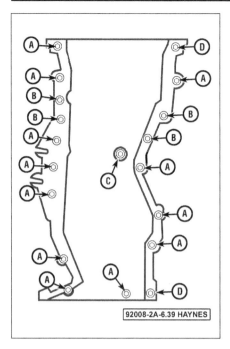

6.29a Timing chain cover fastener identification (see the Specifications for the torque settings)

Bolts A = 1.18 inch (30 mm)
Bolts B = 1.38 inch (35 mm)
Bolts C = 1.77 inch (45 mm)
Nuts D

sprocket timing mark.
Note: *There are three colored links on the timing chain. The pink colored link is the link farthest away from the two yellow colored links that are closest together.*
21   Guide the chain into the stationary chain guide and align the crankshaft sprocket with the pink colored link and the mark on the crankshaft sprocket.
Note: *Once the timing mark on the crankshaft sprocket and the colored link have been aligned, it may be necessary to tie the chain to the sprocket to prevent it from falling off during installation.*
22   Slip the timing chain around the intake

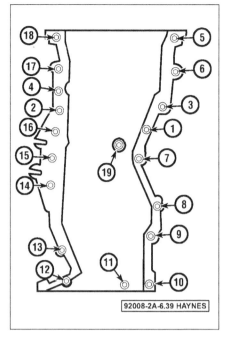

6.29b Timing chain cover fastener tightening sequence (see the Specifications for the torque settings)

camshaft sprocket but not onto the teeth of the sprocket. Using a wrench on the lug on the intake camshaft, rotate the camshaft counterclockwise to align the timing mark on the intake camshaft sprocket and the chain colored link. Once the camshaft sprocket and colored link are aligned, install the chain onto the teeth of the camshaft sprocket.
Caution: *Do not let go of the wrench holding the intake camshaft sprocket in place until the tensioner is installed.*
23   Using your other hand, remove the string holding the chain to the crankshaft sprocket. Rotate just the crankshaft clockwise until the keyway is pointing to the 12 o'clock position and all the slack is removed.
24   Install the tensioner guide and mounting bolt to the torque listed in this Chapter's Specifications.

25   Install a new gasket and the tensioner with the mounting bolts. Tighten the tensioner mounting bolts to the torque listed in this Chapter's Specifications, then remove the pin and allow the plunger to contact the tensioner guide.
26   Confirm that the number one piston is still at TDC on the compression stroke and that the timing marks on the crankshaft and camshaft sprockets are aligned with the colored links on the chain (see illustration 6.18).
27   Install the top chain guide mounting bolt and guide.
28   Apply a bead of RTV sealant to the timing chain cover sealing surfaces. Place the timing chain cover in position on the engine and install the bolts in their original locations.
29   Tighten the bolts and nut evenly in several steps to the torque listed in this Chapter's Specifications (see illustrations). Be sure to follow the sealant manufacturer's recommendations for assembly and sealant curing times.
30   Install the passenger's side mounting bracket (see illustration) to the front of the timing chain cover and tighten the bolts in sequence to the torque listed in this Chapter's Specifications.
31   Install the crankshaft pulley/vibration damper (see Section 11).
Caution: *Carefully rotate the crankshaft by hand through at least two full revolutions (use a socket and breaker bar on the crankshaft pulley center bolt). If you feel any resistance, STOP! There is something wrong - most likely, valves are contacting the pistons. You must find the problem before proceeding. Check your work and see if any updated repair information is available.*
32   Rotate the engine clockwise at least two revolutions and reposition the number one piston at TDC on the compression stroke (see Section 3). Visually confirm that the timing mark on the crankshaft pulley/vibration damper is aligned with the "0" mark on the timing chain cover and the camshaft sprocket marks are aligned and parallel with the top of the timing chain cover as shown in illustrations 6.1.
33   The remainder of installation is the reverse of removal.

---

7   Camshafts, camshaft housing, rocker arms and lash adjusters - removal, inspection and installation

---

Caution: *The camshafts should always be thoroughly inspected before installation, and camshaft endplay should always be checked prior to camshaft removal (see Step 20).*

Removal

1   Disconnect the cable from the negative battery terminal (see Chapter 5).
2   Remove the engine (see Chapter 2B).
3   Remove the valve cover (see Section 4). then refer to Section 3 and place the engine

6.30 Passenger's side mounting bracket bolt identifications

7.7 With the camshaft sprockets removed, hang the timing chain out of the way with a piece of wire and place a shop rag in the timing chain cover opening to prevent foreign objects from falling into the engine

7.15 Mount a dial indicator as shown to measure camshaft endplay - pry the camshaft forward and back and read the endplay on the dial

7.16 Inspect the No. 2 through No. 5 cam bearing surfaces in the camshaft housing for pits, score marks and abnormal wear - if wear or damage is noted, the camshaft housing must be replaced

at TDC for number 1 cylinder. Visually confirm the engine is at TDC on the compression stroke by verifying that the timing mark on the crankshaft pulley/vibration damper is aligned with the "0" mark on the timing chain cover and the camshaft sprocket TDC marks are aligned and parallel with the top of the timing chain cover (see illustrations 3.8 and 6.1).

4   With the TDC marks aligned, apply a dab of paint to the timing chain links where they meet the upper timing marks on the camshaft sprockets.

5   Using a wrench to hold the camshaft sprockets from turning, loosen the camshaft sprocket bolts several turns (see illustration 6.12).

6   Remove the timing chain cover and tensioner (see Section 6).

7   Remove the camshaft sprocket retaining bolts. Disengage the timing chain from the sprockets and remove the camshaft sprockets from the engine. Make sure to note that the Variable Valve Timing (VVT) actuator is installed on the intake camshaft and the exhaust camshaft. Also place a rag in the opening of the timing chain cover to prevent any foreign objects from falling into the engine (see illustration).

8   Loosen the camshaft housing bearing cap bolts in sequence in two or three steps, in the reverse order of the tightening sequence (see illustration 7.30).

9   Remove the camshaft housing by prying between the cylinder head and camshaft housing with a screwdriver.

Note: *Be careful not to damage the cylinder head and camshaft housing.*

10   Loosen the camshaft bearing cap bolts in sequence in two or three steps, in the reverse order of the tightening sequence (see illustrations 7.26).

11   Remove the bearing caps, if necessary the bearing inserts from the No.1 bearing cap, oil control filter and camshafts from the camshaft housing.

Note: *The Number 1 bearing cap has bearing inserts that can be replaced, make sure to keep them in order and install them into their original locations.*

12   Remove the rocker arms from the cylinder head.

Caution: *Keep the rocker arms in order. They must go back in the position from which they were removed.*

13   Remove the lash adjusters from the cylinder head.

Caution: *Keep the lash adjusters in order. They must go back in their original positions.*

14   Inspect the camshafts, camshaft bearings, camshaft housing, rockers and lash adjusters as described below. Also inspect the camshaft sprockets for wear on the teeth. Inspect the chains for cracks or excessive wear of the rollers, and for stretching (see Section 6). If any of the components show signs of excessive wear, they must be replaced.

Inspection

15   Before the camshafts are removed from the engine, check the camshaft endplay by placing a dial indicator with the stem in line with the camshaft and touching the snout (see illustration). Push the camshaft all the way to the rear and zero the dial indicator. Next, pry the camshaft to the front as far as possible and check the reading on the dial indicator. The distance it moves is the endplay. If the endplay for the intake camshaft is greater than the Specifications listed in this Chapter, check the thrust surfaces of the No.1 journal bearing for wear. If the thrust surface is worn, the bearings must be replaced. If the endplay for the exhaust camshaft is greater than the Specifications listed in this Chapter, the camshaft or the cylinder head (or both) may need to be replaced.

16   With the camshafts removed, visually check the camshaft bearing surfaces in the cylinder head for pitting, score marks, galling

7.17 Measure each journal diameter with a micrometer - if any journal measures less than the specified limit, replace the camshaft

and abnormal wear. If the camshaft housings or camshafts bearing surfaces are damaged, or the No.1 journal bearings may have to be replaced (see illustration).

17   Measure the outside diameter of each camshaft bearing journal and record your measurements (see illustration). Compare them to the journal outside diameter as specified in this Chapter's Specifications, then measure the inside diameter of each corresponding camshaft bearing and record the measurements. Subtract each cam journal outside diameter from its respective cam bearing bore inside diameter to determine the oil clearance for each bearing. Compare the results to the specified journal-to-bearing clearance. If any of the measurements fall outside the standard specified wear limits in this Chapter's Specifications, either the camshaft or the camshaft housing, or both, must be replaced.

Note: *If precision measuring tools are not available, Plastigage may be used to determine the bearing journal oil clearance.*

7.18 Measure the lobe heights on each camshaft - if any lobe height is less than the specified allowable minimum, replace that camshaft

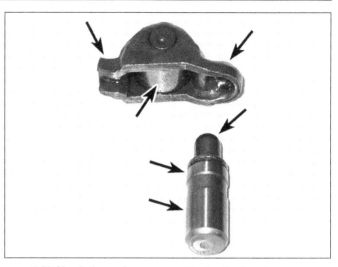

7.20 Check the rocker arms and lash adjusters for wear at the indicated points

18   Using a micrometer, measure the height of each camshaft lobe (see illustration). Compare your measurements with this Chapter's Specifications. If the height for any one lobe is less than the specified minimum, replace the camshaft.

19   Check the camshaft runout by placing the camshaft back into the cylinder head and set up a dial indicator on the center journal. Zero the dial indicator. Turn the camshaft slowly and note the dial indicator readings. Runout should not exceed 0.0012 inch (0.03 mm). If the measured runout exceeds the specified runout, replace the camshaft.

20   Inspect the rocker arms and lash adjusters for scuffing and score marks (see illustration).

Installation

21   Coat the lash adjusters with engine oil and install them into the cylinder head in the same locations from which they were removed.

22   Apply oil to the rocker arms and install them, making sure they are seated to their lash adjusters.

23   Install the camshaft bearings into the camshaft housing.

24   Apply camshaft installation lubricant to the camshaft lobes and journals and install the camshaft into the camshaft housing.

25   Install oil control valve filter into the No.1 bearing cap, then install the camshaft bearing caps and tighten the bearing cap bolts in several equal steps, to the torque listed in this Chapter's Specifications, using the proper tightening sequence (see illustration).

26   Check that the rocker arms are installed properly and have not moved.

27   Position the dowel pins facing upwards; the intake camshaft dowel pin should be at a 17-degree angle and the exhaust pin should be at a 2-degree angle (counterclockwise from the 12 o'clock position).

28   Clean the camshaft housing mounting surface. Apply a 1/4-inch wide bead of Toyota Genuine Seal Packing Black, Three Bond 1207B or equivalent to the camshaft housing-to-cylinder head contact surface.

Note: *The camshaft housing must be installed within 3 minutes and the bolts tightened*

within 10 minutes after applying seal packing.

29   Install the camshaft housing to the cylinder head, install the mounting bolts and tighten the bolts in several equal steps, to the torque listed in this Chapter's Specifications, using the proper tightening sequence (see illustration).

Caution: *Do not add oil or try to start the engine for at least 4 hours after installation or the seal packing will leak.*

30   Install the camshaft sprockets, timing chain and timing chain cover (see Section 6).

31   The remainder of installation is the reverse of removal.

8   Intake manifold - removal and installation

Warning: *Wait until the engine is completely cool before beginning this procedure.*

Removal

1   Relieve the fuel system pressure (see Chapter 4), then disconnect the cable from

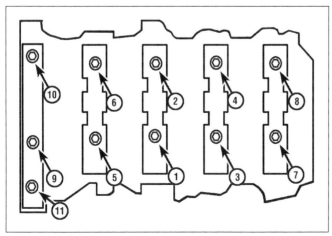

7.25 Camshaft bearing cap bolt TIGHTENING sequence

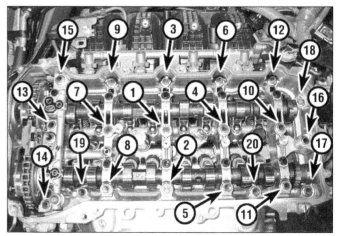

7.29 Camshaft housing bolt TIGHTENING sequence

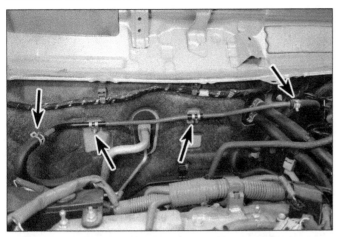

8.4 Disconnect the connection tube clamps and the tube retainers

8.7a Disconnect the electrical connector to the vacuum switching valve

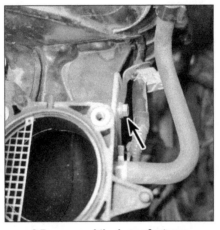

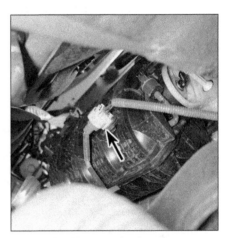

8.7b Remove the harness bracket upper fasteners . . .

8.7c . . . and the lower fastener

8.9 Tumble control valve actuator electrical connector

the negative battery terminal (see Chapter 5).

2   Remove the air intake duct and resonator (see Chapter 4). Remove the cowl cover and vent tray/sub-frame (see Chapter 11).
Note: *To remove the vent tray/sub-frame, the wiper motor and linkage must be completely removed (see Chapter 12).*

3   Remove the engine cover.

4   Disconnect the vacuum hose connection tube from the hoses and retainers at the firewall (see illustration).

5   Drain the cooling system (see Chapter 1) and remove the heater hoses going into the firewall (see Chapter 3).

6   Remove the throttle body from the intake manifold (see Chapter 4) and disconnect the vent tubes.

7   Disconnect the electrical connector to the vacuum switching valve (see illustration) then remove the harness bracket fasteners (see illustrations) and move the harness out of the way.

8   Disconnect the fuel inlet line to the fuel rail (see Chapter 4).

9   From under the intake manifold, disconnect the tumble control valve actuator electrical connector (see illustration).

10   Disconnect the electrical connectors from the fuel rail, remove the harness mounting fasteners and set the harness to the side. Disconnect and remove the main harness connector to the intake manifold and tumble control valves.

11   Using test leads, apply battery voltage to terminal no. 8 of the tumble control valve electrical connector and ground the no. 4 terminal

(see illustrations); this will close the tumble control valve to prevent it from being damaged when the manifold is removed.
Caution: *When the tumble control valves are in the open position, part of the valve is in the manifold inlet and the other half is inside the cylinder head. If the valves are not closed when the intake manifold is removed, the valves may be damaged. Also, don't apply*

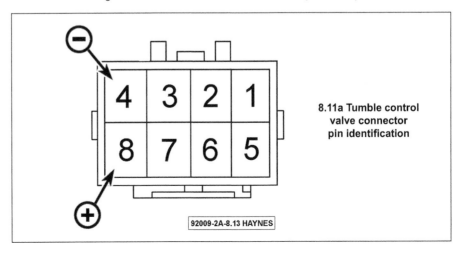

8.11a Tumble control valve connector pin identification

92009-2A-8.13 HAYNES

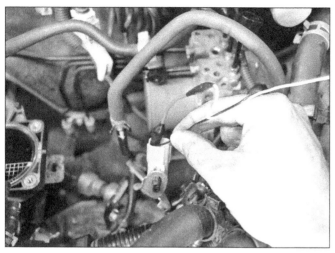

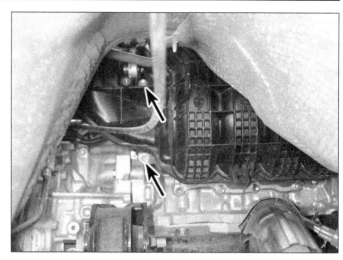

8.11b DO NOT apply voltage longer than three seconds or touch
any other terminals or the actuator may be damaged

8.12 Intake manifold lower bolt locations

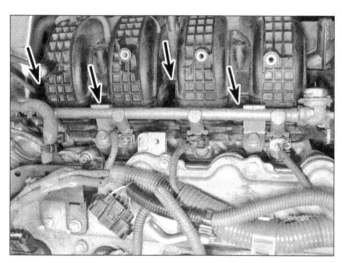

8.13 Intake manifold upper mounting bolt locations

8.15 Press the gasket into the groove on the intake manifold

*voltage longer than three seconds. If the volt-
age is applied longer than three seconds, or
any other terminals are touched, the actuator
may be damaged.*

12   Remove the intake manifold lower bolts
(see illustration).

13   Unscrew the intake manifold upper
mounting fasteners (see illustration) and
remove the manifold.

Installation

14   Clean the mating surfaces of the intake
manifold and the cylinder head mounting
surface with lacquer thinner or acetone. If
the gasket shows signs of leaking, check the
manifold for warpage with a straight edge. If
the manifold is warped it must be replaced.

15   Press a new gasket into the grooves on
the intake manifold (see illustration). Install
the manifold and gasket over the studs on the
cylinder head.

16   Tighten the manifold-to-cylinder head
nuts/bolts in three or four equal steps to the

torque listed in this Chapter's Specifications.
Work from the center out toward the ends to
avoid warping the manifold.

17   Install the remaining parts in the reverse
order of removal. Check the coolant level,
adding as necessary (see Chapter 1).

18   Run the engine and check for coolant
and vacuum leaks.

---

9   Exhaust manifold - removal and installation

Warning: *The engine must be completely cool
before beginning this procedure.*

Removal

1   Disconnect the cable from the negative
battery terminal (see Chapter 5).

2   Raise the front of the vehicle and support it securely on jackstands. Working below
the vehicle, remove the lower splash shields.

3   Apply penetrating oil to the bolts and

springs retaining the exhaust pipe to the manifold. After the bolts have soaked, remove the
bolts retaining the exhaust pipe to the manifold. Separate the front exhaust pipe from the
manifold, being careful not to damage the
oxygen sensor (see illustration).

4   Unbolt the lower exhaust manifold braces
and remove them from the engine. Also disconnect the oxygen sensor connectors.

5   Working in the engine compartment,
remove the upper heat shield from the manifold (see illustration).

Note: *There is also a lower heat shield, but it
is attached to the manifold from underneath
and does not need to be removed.*

6   Remove the oil dipstick tube mounting
bolt and remove the dipstick tube from the
crankcase stiffener.

7   Remove the exhaust manifold support
bracket bolt and bracket nut, then remove the
support bracket.

8   Remove the nuts and detach the manifold and gasket (see illustration).

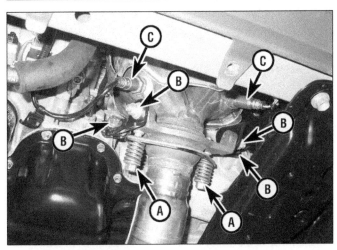

9.3 Working below the vehicle, remove the exhaust pipe-to-manifold mounting bolts (A) and lower the front exhaust pipe. Be careful not to damage the oxygen sensors (C) - (B) indicates the mounting bolts for the exhaust manifold lower brace

9.5 Working from the engine compartment, remove the upper heat shield mounting bolts . . .

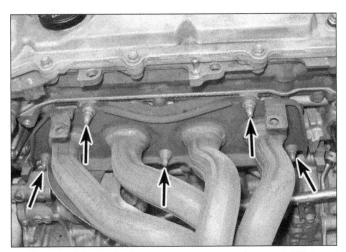

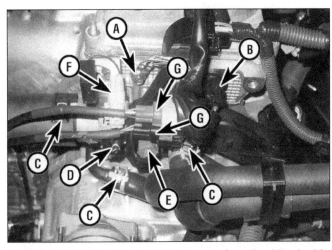

9.8 . . . and the exhaust manifold retaining nuts, then pull the manifold off the studs on the cylinder head and remove the manifold from above

10.11 Disconnect the following components from the driver's side of the cylinder head and position the electrical wiring harness aside

A	Ground straps	E	Coolant temperature sensor
B	Camshaft position sensor	F	Radio noise suppressor
C	Coolant hoses	G	Oxygen sensor connector
D	Oil pressure sending unit		

Installation

9   Use a scraper to remove all traces of old gasket material and carbon deposits from the manifold and cylinder head mating surfaces. If the gasket shows signs of leaking, check the manifold for warpage with a straight edge. If the manifold is warped it must be replaced.
Note: *If the manifold is being replaced with a new one it will be necessary to remove the lower heat shield and fasten it to the new manifold.*
10   Position a new gasket over the cylinder head studs, noting any directional marks or arrows on the gasket which may be present.
11   Install the manifold and thread the mounting nuts into place.
12   Working from the center out, tighten the nuts/bolts to the torque listed in this Chapter's Specifications in three or four equal steps.
13   Reinstall the remaining parts in the reverse order of removal.
14   Run the engine and check for exhaust leaks.

10   Cylinder head - removal and installation

Warning: *The engine must be completely cool before beginning this procedure.*
Caution: *New cylinder head bolts should be used when installing the cylinder head.*

Removal

1   Relieve the fuel system pressure (see Chapter 4), then disconnect the cable from the negative terminal of the battery (see Chapter 5).
2   Drain the engine coolant (see Chapter 1).
3   Remove the engine from the vehicle (see Chapter 2B).

4   Remove the drivebelt and the alternator (see Chapter 5).
5   Remove the valve cover (see Section 4).
6   Remove the throttle body, fuel injectors and fuel rail (see Chapter 4).
7   Remove the intake manifold (see Section 8) and the exhaust manifold (see Section 9).
8   Remove the timing chain and camshaft sprockets (see Section 6).
9   Remove the camshafts, lifters and the camshaft housing (see Section 7).
10   Remove the variable valve timing control valve (see illustration 5.2).
11   Label and remove the coolant hoses and electrical connections from the cylinder head (see illustration).
12   Using a 10 mm hex-head socket bit and a breaker bar, loosen the cylinder head

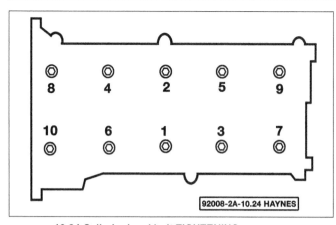

10.24 Cylinder head bolt TIGHTENING sequence

11.5 A puller base and several spacers can be mounted to the center hub of the pulley to keep the crankshaft from turning as the pulley retaining bolt is loosened - install the socket over the crankshaft bolt head before installing the puller, then insert the extension through the center hole of the puller

bolts in 1/4-turn increments until they can be removed by hand. Loosen the cylinder head bolts in the reverse order of the recommended tightening sequence (see illustration 10.24) to avoid warping or cracking the cylinder head.

13 Lift the cylinder head off the engine block. If it's stuck, very carefully pry up at the transaxle end, beyond the gasket surface.

14 Remove any remaining external components from the cylinder head to allow for thorough cleaning and inspection.

Installation

15 The mating surfaces of the cylinder head and block must be perfectly clean when the cylinder head is installed.

16 Use a gasket scraper to remove all traces of carbon and old gasket material, then clean the mating surfaces with brake system cleaner. If there's oil on the mating surfaces when the cylinder head is installed, the gasket may not seal correctly and leaks could develop. When working on the block, stuff

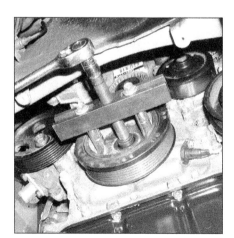

11.6 When using a puller on the crankshaft pulley, be sure to use the proper adapter between the puller screw and the nose of the crankshaft

the cylinders with clean shop rags to keep out debris.

17 Use a vacuum cleaner to remove material that falls into the cylinders.

18 Check the block and cylinder head mating surfaces for nicks, deep scratches and other damage. If damage is slight, it can be removed with a file; if it's excessive, machining may be the only alternative.

19 Use a tap of the correct size to chase the threads in the cylinder head bolt holes, then clean the holes with compressed air - make sure that nothing remains in the holes.

Warning: *Wear eye protection when using compressed air!*

20 Install the components that were removed from the cylinder head.

21 Position the new gasket over the dowel pins in the block. Apply RTV sealant to the ends of the cylinder head gasket on the timing chain cover side.

22 Carefully set the cylinder head on the block without disturbing the gasket.

23 Before installing the cylinder head bolts, apply a small amount of clean engine oil to the threads and under the bolt heads.

24 Install the new cylinder head bolts and tighten them finger-tight. Following the recommended sequence, tighten the bolts to the torque listed in this Chapter's Specifications (see illustration).

25 The remainder of installation is the reverse of removal.

Caution: *Wait at least four hours before adding oil and coolant and starting the engine.*

26 Change the engine oil and filter (see Chapter 1).

27 Refill the cooling system (see Chapter 1), run the engine and check for leaks.

11 Crankshaft pulley/vibration damper - removal and installation

1 Disconnect the cable from the negative battery terminal (see Chapter 5).

2 Remove the drivebelt (see Chapter 1).

3 Loosen the right front wheel lug nuts, then raise the front of the vehicle and support it securely on jackstands.

4 Remove the right front wheel and the splash shield from the wheelwell (see Chapter 11).

5 Remove the bolt from the front of the crankshaft. A breaker bar will probably be necessary, since the bolt is very tight (see illustration).

6 Using a puller that bolts to the crankshaft hub, remove the crankshaft pulley from the crankshaft (see illustration).

Caution: *Do not use a jaw-type puller - it will damage the pulley/damper assembly.*

Note: *Depending on the type of puller you have it may be necessary to support the engine from above, remove the right side engine mount and lower to engine to gain sufficient clearance to use the puller.*

7 Slide the pulley onto the crankshaft. Note that the slot (keyway) in the hub must be aligned with the Woodruff key in the end of the crankshaft and that the crankshaft bolt can also be used to press the crankshaft pulley into position.

8 Tighten the crankshaft bolt to the torque listed in this Chapter's Specifications.

9 The remainder of installation is the reverse of removal.

12 Crankshaft front oil seal - replacement

1 Remove the crankshaft pulley (see Section 11).

2 Note how the seal is installed - the new one must be installed to the same depth and facing the same way. Carefully pry the oil seal out of the cover with a seal puller or a large screwdriver (see illustration). Be very careful not to distort the cover or scratch the crankshaft! Wrap electrician's tape around the tip of the screwdriver to avoid damage to

12.2 Carefully pry the old seal out of the timing chain cover - don't damage the crankshaft in the process

12.3 Drive the new seal into place with a large socket and hammer

the crankshaft.

3 Apply clean engine oil or multi-purpose grease to the outer edge of the new seal, then install it in the cover with the lip (spring side) facing IN. Drive the seal into place with a seal driver or a large socket and a hammer (see illustration). Make sure the seal enters the bore squarely and stop when the front face is at the proper depth.

4 Check the surface on the pulley hub that the oil seal rides on. If the surface has been grooved from long-time contact with the seal, the pulley should be replaced.

5 Lubricate the pulley hub with clean engine oil and reinstall the crankshaft pulley (see Section 11).

6 Install the crankshaft pulley retaining bolt and tighten it to the torque listed in this Chapter's Specifications.

7 The remainder of installation is the reverse of the removal.

13 Oil pan - removal and installation

Removal

1 Disconnect the cable from the negative battery terminal (see Chapter 5).

2 Set the parking brake and block the rear wheels.

3 Raise the front of the vehicle and support it securely on jackstands.

4 Remove the two plastic splash shields under the engine, if equipped.

5 Drain the engine oil and remove the oil filter (see Chapter 1). Remove the oil dipstick.

6 Remove the bolts and detach the oil pan. If it's stuck, use a putty knife to break the bond caused by the sealant (see illustrations). Don't damage the mating surfaces of the pan and block or oil leaks could develop.

Installation

7 Use a scraper to remove all traces of old sealant from the block and oil pan. Clean the mating surfaces with lacquer thinner or acetone.

8 Make sure the threaded bolt holes in the block are clean.

9 Check the oil pan flange for distortion, particularly around the bolt holes. Remove any nicks or burrs as necessary.

10 Inspect the oil pump pick-up tube assembly for cracks and a blocked strainer. If the pick-up was removed, clean it thoroughly and install it now, using a new gasket. Tighten the nuts/bolts to the torque listed in this Chapter's Specifications.

11 Apply a 3/16-inch wide bead of RTV sealant to the mating surface of the oil pan.

Note: *Be sure follow the sealant manufacturers' recommendations for assembly and seal-*

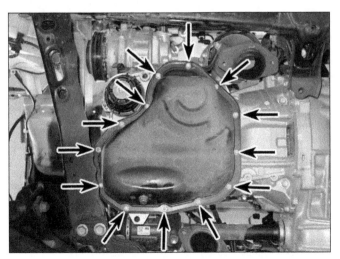

13.6a Oil pan mounting bolts

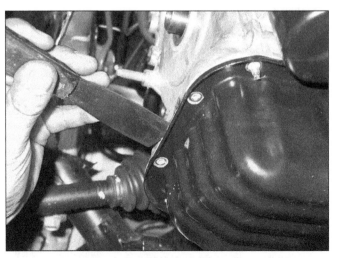

13.6b Use a putty knife, working around the pan to cut the sealant bond - be careful not to damage the mating surfaces of the pan and block or oil leaks may develop

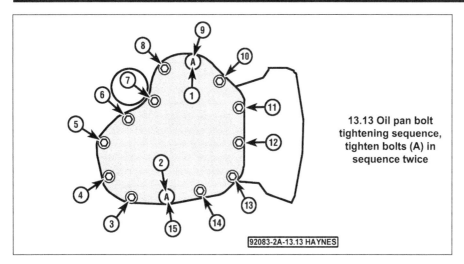

13.13 Oil pan bolt tightening sequence, tighten bolts (A) in sequence twice

92083-2A-13.13 HAYNES

ant curing times.

12　Carefully position the oil pan on the engine block and install the oil pan-to-engine block bolts loosely.

13　Tighten the oil pan-to-engine block bolts in sequence (see illustration), to the torque listed in this Chapter's Specifications in three or four steps. Make sure to tighten bolts marked "A" twice.

14　The remainder of installation is the reverse of removal. Be sure to wait at least one hour before adding oil to allow the sealant to properly cure.

15　Run the engine and check for oil pressure and leaks.

14　Oil pump - removal and installation

Note: *The oil pump is not serviceable separately from the timing chain cover. If the oil pump is defective, the timing chain cover and pump must be replaced as a unit (see Section 6).*

Removal

1　Remove the engine (see Chapter 2B).
2　Remove the valve cover (see Section 4).
3　Remove the crankshaft pulley (see Section 11).
4　Remove the engine mount and mounting bracket.
5　Remove the timing chain cover (see Section 6).

Installation

6　Install new oil pump gaskets and oil hole cover gasket to the lower crankcase.
7　Lubricate the pump cavity by pouring clean engine oil into the inlet side of the oil pump and turning the drive gear shaft.
8　Align the flats on the crankshaft timing sprocket with the flats on the oil pump drive rotor.
9　Install the timing chain cover (see Section 6) and tighten the bolts to the torque listed in this Chapter's Specifications in several steps (see illustration 6.38b).

10　Reinstall the remaining parts in the reverse order of removal.
11　Add oil to the proper level, start the engine and check for oil pressure and leaks.

15　Driveplate - removal and installation

Removal

1　Refer to Chapter 7A and remove the transaxle. If it's leaking, now would be a very good time to replace the front pump seal/O-ring (automatic transaxle only).
2　Use a center punch or paint to make alignment marks on the driveplate and crankshaft to ensure correct alignment during installation.
3　Remove the bolts that secure the driveplate to the crankshaft. If the crankshaft turns, wedge a screwdriver in the ring gear teeth to jam the driveplate.
4　Remove the driveplate from the crankshaft. Also, there are spacers on both sides of the driveplate; don't lose them.

Installation

5　Check for cracked and broken ring gear teeth.
6　Clean and inspect the mating surfaces of the driveplate and the crankshaft. If the crankshaft rear seal is leaking, replace it (see Section 16) before reinstalling the driveplate.
7　Position the driveplate against the crankshaft. Be sure to align the marks made during removal. Note that some engines have an alignment dowel or staggered bolt holes to ensure correct installation. Before installing the bolts, apply thread locking compound to the threads.
8　Wedge a screwdriver in the ring gear teeth to keep the driveplate from turning and tighten the bolts to the torque listed in this Chapter's Specifications. Follow a criss-cross pattern and work up to the final torque in three or four steps.
9　The remainder of installation is the reverse of the removal procedure.

16　Rear main oil seal - replacement

1　Remove the engine/transaxle assembly (see Chapter 2B).
2　Remove the driveplate (see Section 15).
3　Pry the oil seal from the rear of the engine with a screwdriver. Be careful not to nick or scratch the crankshaft or the seal bore. Be sure to note how far it's recessed into the bore before removal so the new seal can be installed to the same depth. Thoroughly clean the seal bore in the block with a shop towel. Remove all traces of oil and dirt.
4　Lubricate the outside diameter of the seal and install the seal over the end of the crankshaft. Make sure the lip of the seal points toward the engine. Preferably, a seal installation tool (available at most auto parts stores) is needed to press the new seal back into place. If the proper seal installation tool is unavailable, use a large socket, section of pipe or a blunt tool and carefully drive the new seal squarely into the seal bore and flush with the edge of the engine block.
5　Install the driveplate (see Section 16).
6　Install the transaxle to the engine (see Chapter 7A), then install the engine/transaxle assembly (see Chapter 2B).

17　Powertrain mounts - check and replacement

1　The powertrain mounts seldom require attention, but broken or deteriorated mounts should be replaced immediately or the added strain placed on the driveline components may cause damage or wear.

Check

2　During the check, the engine must be raised slightly to remove the weight from the mounts.
3　Raise the vehicle and support it securely on jackstands. Remove the splash shields under the engine and position a jack under the engine oil pan. Place a large block of wood between the jack head and the oil pan, then carefully raise the engine just enough to take the weight off the mounts. Do not position the wood block under the drain plug.
Warning: *DO NOT place any part of your body under the engine when it's supported only by a jack!*
4　Check the mounts to see if the rubber is cracked, hardened or separated from the bushing in the center of the mount.
5　Check for relative movement between the mount plates and the engine or frame (use a large screwdriver or pry bar to attempt to move the mounts). If movement is noted, lower the engine and tighten the mount fasteners.
6　Rubber preservative should be applied to the mounts to slow deterioration.

Replacement

7　Disconnect the negative battery cable from the battery (see Chapter 5), then

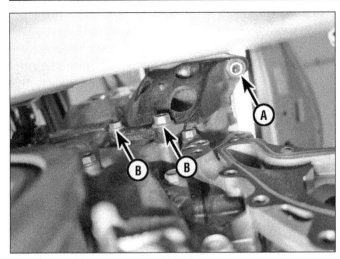

17.8 Working from below, remove the passenger's side engine mount-to-engine bracket retaining nut (A). (B) are the lower bracket-to-engine block bolts

17.9 Working from above, remove the passenger's side engine mount fasteners

raise the vehicle and support it securely on jackstands (if not already done). Support the engine as described in Step 3.

Passenger's side engine mount

8   Working below the vehicle, remove the nut securing the right side mount to the engine bracket (see illustration).

9   Working above in the engine compartment, remove the remaining nut and bolts securing the right side mount to the engine bracket (see illustration). If the vehicle is equipped with ABS brakes, it will be necessary to remove the ABS actuator and bracket first to allow access to the mount (see Chapter 9).

10   Detach the mount from the vehicle.

11   Installation is the reverse of removal. Use thread locking compound on the mount bolts/nuts and be sure to tighten them securely.

Caution: *Be sure to bleed the brakes prop-*

erly if the vehicle is equipped with an Antilock Brake System (see Chapter 9).

Front and rear transaxle mounts

12   Working up through the holes in the crossmember, remove the mount-to-crossmember retaining bolts (see illustration).

13   Remove the through-bolt securing the mount to the transaxle bracket and detach the mount from the vehicle (see illustration).

14   Installation is the reverse of removal. Use thread locking compound on the mount bolts/nuts and be sure to tighten them securely.

Driver's side transaxle mount

15   Remove the air filter housing (see Chapter 4).

16   Remove the nuts and bolts securing the mount to the transaxle (see illustration).

17   Remove the through-bolt securing the mount to the chassis bracket and remove the

17.12 Working from below, remove the transaxle mount fasteners

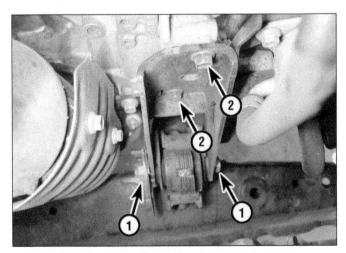

17.13 Remove the transaxle mount through-bolt - front mount shown, rear mount similar

1   *Through-bolt and flag nut (hold the nut and turn the bolt)*
2   *Mount bracket-to-transaxle bolts*

17.16 Driver's side transaxle mount fasteners

1   *Mount-to-transaxle bolts*
2   *Through-bolt and nut*
3   *Mount bracket-to-chassis bolts*

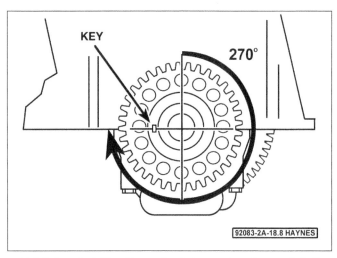

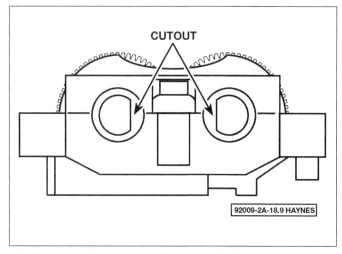

18.8 Before installing the engine balancer, turn the crankshaft 270-degrees clockwise from its TDC position

18.9 Make sure the cutouts at the rear ends of the balancer shafts are facing each other like this

mount from the vehicle

18   Installation is the reverse of removal. Use thread locking compound on the mount bolts/ nuts and be sure to tighten them securely.

18   Engine balancer assembly - removal and installation

1   Remove the engine (see Chapter 2B).

2   Position the engine at TDC compression for cylinder No. 1 (see Section 3). Disconnect the negative battery cable (see Chapter 5).

3   Remove the engine oil pan (see Section 13), then remove the oil strainer mounting bolts and remove the strainer.

4   Remove the oil baffle plate mounting bolts and remove the baffle plate. Then remove the balancer assembly mounting bolts and lift the assembly off of the lower crankcase.

5   Before installing the balancer assembly, check that the alignment line mark of the balance shaft damper cover and dot on the balance shaft driven gear are aligned.

6   If the alignment marks are not aligned, place a wrench on the rear cutout of the No. 2 balance shaft and hold the assembly.

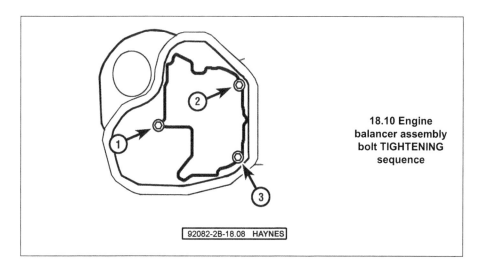

18.10 Engine balancer assembly bolt TIGHTENING sequence

7   Rotate the balance shaft gear of the No. 1 balance shaft counterclockwise to align the alignment dot of the balance shaft gear with the alignment mark line of the balance shaft damper cover. Make sure the balance shafts stay in this position until the assembly is installed.

8   Turn the crankshaft 270-degrees clockwise so the keyway is aligned with the bottom of the cylinder block (see illustration).

9   Check to make sure the cutouts at the rear ends of the shafts are facing each other (see illustration).

10   Install the engine balancer and tighten the bolts in the proper sequence (see illustration) to the torque listed in this Chapter's Specifications.

Notes

Notes

Chapter 2 Part B
General engine overhaul procedures

Contents

Specifications

General

Engine designation	2AR-FE
Displacement	152.2 cubic inches (2.5 liters)
Bore and stroke	3.54 x 3.86 inches (90.0 x 98.0 mm)
Cylinder compression pressure @ 250 rpm	
Standard	210 psi (14.7 kg/cm2) or more
Minimum	142 psi (10 kg/cm2)
Difference between cylinders	29 psi (2.0 kg/cm2) or less
Oil pressure (engine warm)	
At idle	4.4 psi (0.3 kg/cm2) minimum
At 4000 rpm	38 or more (2.7 kg/cm2) minimum

Torque specifications

	Ft-lbs (unless otherwise indicated)	Nm
Connecting rod bearing cap bolts		
Step 1	30	40
Step 2	Tighten an additional 90-degrees	
Main bearing cap bolts		
Step 1	15	20
Step 2	30	40
Step 3	Tighten an additional 90-degrees	
Lower crankcase bolts (see illustration 11.2b)		
Bolts "A"	18	24
All others	32	43

1.10a An engine block being bored. An engine rebuilder will use special machinery to recondition the cylinder bores

1.10b If the cylinders are bored, the machine shop will normally hone the engine on a machine like this

1   General information - engine overhaul

1   Included in this portion of Chapter 2 are general information and diagnostic testing procedures for determining the overall mechanical condition of your engine.

2   The information ranges from advice concerning preparation for an overhaul and the purchase of replacement parts and/or components to detailed, step-by-step procedures covering removal and installation.

3   The following Sections have been written to help you determine whether your engine needs to be overhauled and how to remove and install it once you've determined it needs to be rebuilt. For information concerning in-vehicle engine repair, see Chapter 2A.

4   The Specifications included in this Part are general in nature and include only those necessary for testing the oil pressure and checking the engine compression. Refer to Chapter 2A for additional engine Specifications.

5   It's not always easy to determine when, or if, an engine should be completely overhauled, because a number of factors must be considered.

6   High mileage is not necessarily an indication that an overhaul is needed, while low mileage doesn't preclude the need for an overhaul. Frequency of servicing is probably the most important consideration. An engine that's had regular and frequent oil and filter changes, as well as other required maintenance, will most likely give many thousands of miles of reliable service. Conversely, a neglected engine may require an overhaul very early in its service life.

7   Excessive oil consumption is an indication that piston rings, valve seals and/or valve guides are in need of attention. Make sure that oil leaks aren't responsible before deciding that the rings and/or guides are bad. Perform a cylinder compression check to determine the extent of the work required (see Section 3). Also check the vacuum readings under various conditions (see Section 4).

8   Check the oil pressure with a gauge installed in place of the oil pressure sending unit and compare it to this Chapter's Specifications (see Section 2). If it's extremely low, the bearings and/or oil pump are probably worn out.

9   Loss of power, rough running, knocking or metallic engine noises, excessive valve train noise and high fuel consumption rates may also point to the need for an overhaul, especially if they're all present at the same time. If a complete tune-up doesn't remedy the situation, major mechanical work is the only solution.

10   An engine overhaul involves restoring the internal parts to the specifications of a new engine. During an overhaul, the piston rings are replaced and the cylinder walls are reconditioned (rebored and/or honed) (see illustrations 1.10a and 1.10b). If a rebore is done by an automotive machine shop, new oversize pistons will also be installed. The main bearings, connecting rod bearings and camshaft bearings are generally replaced with new ones and, if necessary, the crankshaft may be reground to restore the journals (see illustration 1.10c). Generally, the valves are serviced as well, since they're usually in less-than-perfect condition at this point. While the engine is being overhauled, other components, such as the starter and alternator, can be rebuilt as well. The end result should be similar to a new engine that will give many trouble free miles.

Caution: *Critical cooling system components such as the hoses, drivebelts, thermostat and water pump should be replaced with new parts when an engine is overhauled. The radiator should be checked carefully to ensure that it isn't clogged or leaking (see Chapter 3). If you purchase a rebuilt engine or short block, some rebuilders will not warranty their engines unless the radiator has been professionally flushed. Also, we don't recommend overhauling the oil pump - always install a new one when an engine is rebuilt.*

Caution: *These engines have a removable cylinder block water jacket and should not be machined. In the event of a warped block deck, misalignment of the main bearing bores, or worn cylinders, the manufacturer states that the cylinder block must be replaced.*

11   Overhauling the internal components on today's engines is a difficult and time-consuming task which requires a significant amount of specialty tools and is best left to a professional engine rebuilder (see illustrations 1.11a, 1.11b and 1.11c). A competent engine rebuilder will handle the inspection

1.10c A crankshaft having a main bearing journal ground

1.11a A machinist checks for a bent connecting rod, using specialized equipment

of your old parts and offer advice concerning the reconditioning or replacement of the original engine; never purchase parts or have machine work done on other components until the block has been thoroughly inspected by a professional machine shop. As a general rule, time is the primary cost of an overhaul, especially since the vehicle may be tied up for a minimum of two weeks or more. Be aware that some engine builders only have the capability to rebuild the engine you bring them while other rebuilders have a large inventory of rebuilt exchange engines in stock. Also be aware that many machine shops could take as much as two weeks time to completely rebuild your engine depending on shop workload. Sometimes it makes more sense to simply exchange your engine for another engine that's already rebuilt to save time.

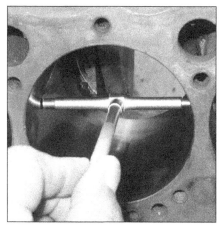

1.11b A bore gauge being used to check a cylinder bore

1.11c Uneven piston wear like this indicates a bent connecting rod

2   Oil pressure check

1   Low engine oil pressure can be a sign of an engine in need of rebuilding. A low oil pressure indicator (often called an "idiot light") is not a test of the oiling system. Such indicators only come on when the oil pressure is dangerously low. Even a factory oil pressure gauge in the instrument panel is only a relative indication, although much better for driver information than a warning light. A better test is with a mechanical (not electrical) oil pressure gauge.
2   Locate the oil pressure sending unit: The oil pressure sending unit is located on the side of the cylinder head close to the drivebelt end and just above the exhaust manifold (see illustration).
3   Remove the oil pressure sending unit and install a fitting which will allow you to directly connect your hand-held, mechanical oil pressure gauge. Use Teflon tape or sealant on the threads of the adapter and the fitting on the end of your gauge's hose.
4   Connect an accurate tachometer to the engine, according to the tachometer manu-

facturer's instructions.
5   Check the oil pressure with the engine running (normal operating temperature) at the specified engine speed, and compare it to this Chapter's Specifications. If it's extremely low, the bearings and/or oil pump are probably worn out.

3   Cylinder compression check

1   A compression check will tell you what mechanical condition the upper end of your engine (pistons, rings, valves, head gaskets) is in. Specifically, it can tell you if the compression is down due to leakage caused by worn piston rings, defective valves and seats or a blown head gasket.
Note: *The engine must be at normal operating temperature and the battery must be fully charged for this check.*
2   Disable the fuel pump circuit (see Chapter 4, Section 3).
3   Clean the area around the spark plugs before you remove them to prevent dirt from

getting into the cylinders as the compression check is being done (compressed air should be used, if available).
4   Remove all of the spark plugs from the engine (see Chapter 1).
5   Install the compression gauge in the spark plug hole (see illustration).
6   Have an assistant depress the accelerator pedal to the floor and crank the engine over at least seven compression strokes while you watch the gauge. The compression should build up quickly in a healthy engine. Low compression on the first stroke, followed by gradually increasing pressure on successive strokes, indicates worn piston rings. A low compression reading on the first stroke, which doesn't build up during successive strokes, indicates leaking valves or a blown head gasket (a cracked head could also be the cause). Deposits on the undersides of the valve heads can also cause low compression. Record the highest gauge reading obtained.
7   Repeat the procedure for the remaining cylinders and compare the results to this Chapter's Specifications.
8   Add some engine oil (about three squirts

2.2 The oil pressure sending unit is located at the front side of the cylinder head on the drivebelt end, above the exhaust manifold

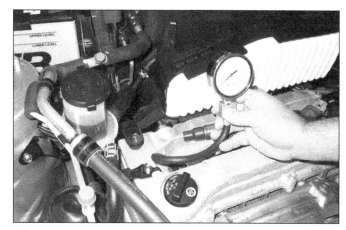

3.5 A compression gauge with a threaded fitting for the spark plug hole is preferred over the type that requires hand pressure to maintain the seal - be sure to block open the throttle valve as far as possible during the compression check

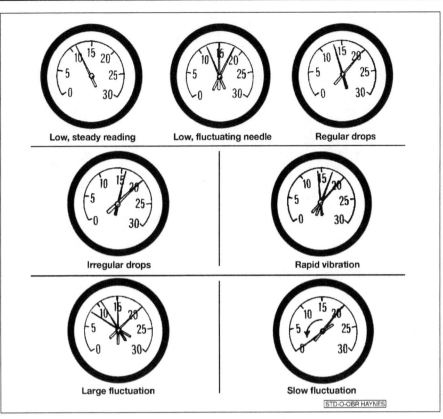

4.6 Typical vacuum gauge readings

4.4 A simple vacuum gauge can be handy in diagnosing engine condition and performance

from a plunger-type oil can) to each cylinder, through the spark plug hole, and repeat the test.

9   If the compression increases after the oil is added, the piston rings are definitely worn. If the compression doesn't increase significantly, the leakage is occurring at the valves or head gasket. Leakage past the valves may be caused by burned valve seats and/or faces or warped, cracked or bent valves.

10   If two adjacent cylinders have equally low compression, there's a strong possibility that the head gasket between them is blown. The appearance of coolant in the combustion chambers or the crankcase would verify this condition.

11   If one cylinder is slightly lower than the others, and the engine has a slightly rough idle, a worn lobe on the camshaft could be the cause.

12   If the compression is unusually high, the combustion chambers are probably coated with carbon deposits. If that's the case, the cylinder head(s) should be removed and decarbonized.

13   If compression is way down or varies greatly between cylinders, it would be a good idea to have a leak-down test performed by an automotive repair shop. This test will pinpoint exactly where the leakage is occurring and how severe it is.

14   After all of the cylinders have been checked, install the spark plugs and ignition coils and reconnect the fuel pump electrical connector.

---

4   Vacuum gauge diagnostic checks

---

1   A vacuum gauge provides valuable information about what is going on in the engine at a low cost. You can check for worn rings or cylinder walls, leaking head or intake manifold gaskets, fuel injection system problems, restricted exhaust, stuck or burned valves, weak valve springs, improper ignition or valve timing and ignition problems.

2   Unfortunately, vacuum gauge readings are easy to misinterpret, so they should be used in conjunction with other tests to confirm the diagnosis.

3   Both the absolute readings and the rate of needle movement are important for accurate interpretation. Most gauges measure vacuum in inches of mercury (in-Hg). The following references to vacuum assume the diagnosis is being performed at sea level. As elevation increases (or atmospheric pressure decreases), the reading will decrease. For every 1,000 foot increase in elevation above approximately 2000 feet, the gauge readings will decrease about one inch of mercury.

4   Connect the vacuum gauge directly to intake manifold vacuum, not to ported (throttle body) vacuum (see illustration). Be sure no hoses are left disconnected during the test or false readings will result.

5   Before you begin the test, allow the engine to warm up completely. Block the wheels and set the parking brake. With the transmission in Park, start the engine and allow it to run at normal idle speed.

Warning: *Keep your hands and the vacuum gauge clear of the fans.*

6   Read the vacuum gauge; an average, healthy engine should normally produce about 17 to 22 in-Hg with a fairly steady needle (see illustration). Refer to the following vacuum gauge readings and what they indicate about the engine's condition.

7   A low steady reading usually indicates a

leaking gasket between the intake manifold and cylinder head(s) or throttle body, a leaky vacuum hose, late ignition timing or incorrect camshaft timing. Check ignition timing with a timing light and eliminate all other possible causes, utilizing the tests provided in this Chapter before you remove the timing chain cover to check the timing marks.

8   If the reading is three to eight inches below normal and it fluctuates at that low reading, suspect an intake manifold gasket leak at an intake port or a faulty fuel injector.

9   If the needle has regular drops of about two-to-four inches at a steady rate, the valves are probably leaking. Perform a compression check or leak-down test to confirm this.

10   An irregular drop or down-flick of the needle can be caused by a sticking valve or an ignition misfire. Perform a compression check or leak-down test and read the spark plugs.

11   A rapid vibration of about four in-Hg vibration at idle combined with exhaust smoke indicates worn valve guides. Perform a leak-down test to confirm this. If the rapid vibration occurs with an increase in engine speed, check for a leaking intake manifold gasket or head gasket, weak valve springs, burned valves or ignition misfire.

12   A slight fluctuation, say one inch up and down, may mean ignition problems. Check all the usual tune-up items and, if necessary, run the engine on an ignition analyzer.

13   If there is a large fluctuation, perform a

6.3a After tightly wrapping water-vulnerable components, use a spray cleaner on everything, with particular concentration on the greasiest areas, usually around the valve cover and lower edges of the block. If one section dries out, apply more cleaner

6.3b Depending on how dirty the engine is, let the cleaner soak in according to the directions and hose off the grime and cleaner. Get the rinse water down into every area you can get at; then dry important components with a hair dryer or paper towels

compression or leak-down test to look for a weak or dead cylinder or a blown head gasket.

14   If the needle moves slowly through a wide range, check for a clogged PCV system, or throttle body or intake manifold gasket leaks.

15   Check for a slow return after revving the engine by quickly snapping the throttle open until the engine reaches about 2,500 rpm and let it shut. Normally the reading should drop to near zero, rise above normal idle reading (about 5 in-Hg over) and then return to the previous idle reading. If the vacuum returns slowly and doesn't peak when the throttle is snapped shut, the rings may be worn. If there is a long delay, look for a restricted exhaust system (often the muffler or catalytic converter). An easy way to check this is to temporarily disconnect the exhaust ahead of the suspected part and redo the test.

5   Engine rebuilding alternatives

1   The do-it-yourselfer is faced with a number of options when purchasing a rebuilt engine. The major considerations are cost, warranty, parts availability and the time required for the rebuilder to complete the project. The decision to replace the engine block, piston/connecting rod assemblies and crankshaft depends on the final inspection results of your engine. Only then can you make a cost effective decision whether to have your engine overhauled or simply purchase an exchange engine for your vehicle.

2   Some of the rebuilding alternatives include:

3   **Individual parts** - If the inspection procedures reveal that the engine block and most engine components are in reusable condition, purchasing individual parts and having a rebuilder rebuild your engine may be the most

economical alternative. The block, crankshaft and piston/connecting rod assemblies should all be inspected carefully by a machine shop first.

4   **Short block** - A short block consists of an engine block with a crankshaft and piston/connecting rod assemblies already installed. All new bearings are incorporated and all clearances will be correct. The existing camshafts, valve train components, cylinder head and external parts can be bolted to the short block with little or no machine shop work necessary.

5   **Long block** - A long block consists of a short block plus an oil pump, oil pan, cylinder head, valve cover, camshaft and valve train components, timing sprockets and chain or gears and timing cover. All components are installed with new bearings, seals and gaskets incorporated throughout. The installation of manifolds and external parts is all that's necessary.

6   **Low mileage used engines** - Some companies now offer low mileage used engines which is a very cost effective way to get your vehicle up and running again. These engines often come from vehicles which have been totaled in accidents or come from other countries which have a higher vehicle turn over rate. A low mileage used engine also usually has a similar warranty like the newly remanufactured engines.

7   Give careful thought to which alternative is best for you and discuss the situation with local automotive machine shops, auto parts dealers and experienced rebuilders before ordering or purchasing replacement parts.

6   Engine removal - methods and precautions

1   If you've decided that an engine must be removed for overhaul or major repair work,

several preliminary steps should be taken. Read all removal and installation procedures carefully prior to committing to this job.

2   Locating a suitable place to work is extremely important. Adequate work space, along with storage space for the vehicle, will be needed. If a shop or garage isn't available, at the very least a flat, level, clean work surface made of concrete or asphalt is required.

3   Cleaning the engine compartment and engine before beginning the removal procedure will help keep tools clean and organized (see illustrations).

4   An engine hoist will also be necessary. Make sure the hoist is rated in excess of the combined weight of the engine and transaxle. Safety is of primary importance, considering the potential hazards involved in removing the engine from the vehicle.

5   A vehicle hoist will be necessary for engine removal, since on these models the subframe must be removed and the engine/transaxle assembly must be lowered from the engine compartment, then the vehicle is raised and the powertrain unit is removed from under the vehicle. If the necessary equipment is not available, the engine will have to be removed by a qualified automotive repair facility.

6   If you're a novice at engine removal, get at least one helper. One person cannot easily do all the things you need to do to remove a big heavy engine and transaxle assembly from the engine compartment. Also helpful is to seek advice and assistance from someone who's experienced in engine removal.

7   Plan the operation ahead of time. Arrange for or obtain all of the tools and equipment you'll need prior to beginning the job (see illustration). Some of the equipment necessary to perform engine removal and installation safely and with relative ease are (in addition to a vehicle hoist and an engine hoist) a heavy duty floor jack (preferably fitted with a transaxle jack head adapter), complete sets of wrenches and sockets as described

6.7 Get an engine stand sturdy enough to firmly support the engine while you're working on it. Stay away from three-wheeled models; they have a tendency to tip over more easily, so get a four-wheeled unit

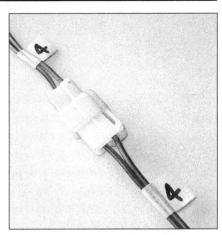

7.11 Label both ends of each wire and hose before disconnecting it

in the front of this manual, wooden blocks, plenty of rags and cleaning solvent for mopping up spilled oil, coolant and gasoline.

8   Plan for the vehicle to be out of use for quite a while. A machine shop can do the work that is beyond the scope of the home mechanic. Machine shops often have a busy schedule, so before removing the engine, consult the shop for an estimate of how long it will take to rebuild or repair the components that may need work.

---

7   Engine - removal and installation

Warning: *The models covered by this manual are equipped with a Supplemental Restraint System (SRS), more commonly known as airbags. Always disable the airbag system before working in the vicinity of airbag system components to avoid the possibility of accidental deployment of the airbag, which could cause personal injury (see Chapter 12).*
Warning: *Gasoline is extremely flammable, so take extra precautions when you work on any part of the fuel system. Don't smoke or allow open flames or bare light bulbs near the work area, and don't work in a garage where a gas-type appliance (such as a water heater or a clothes dryer) is present. Since gasoline is carcinogenic, wear fuel-resistant gloves when there's a possibility of being exposed to fuel, and, if you spill any fuel on your skin, rinse it off immediately with soap and water. Mop up any spills immediately and do not store fuel-soaked rags where they could ignite. The fuel system is under constant pressure, so, if any fuel lines are to be disconnected, the fuel pressure in the system must be relieved first (see Chapter 4 for more information). When you perform any kind of work on the fuel system, wear safety glasses and have a Class B type fire extinguisher on hand.*
Warning: *The air conditioning system is under high pressure. Do not loosen any hose fittings or remove any components until the system has been discharged. Air conditioning refrigerant should be properly discharged into an*

EPA-approved recovery/recycling unit by a dealer service department or an automotive air conditioning repair facility. Always wear eye protection when disconnecting air conditioning system fittings.*
Warning: *Wait until the engine is completely cool before beginning this procedure.*
Note: *Engine removal on these models is a difficult job, especially for the do-it-yourself mechanic working at home. Because of the vehicle's design, the manufacturer states that the engine and transaxle have to be removed as a unit from the bottom of the vehicle, not the top. With a floor jack and jackstands the vehicle can't be raised high enough and supported safely enough for the engine/transaxle assembly to slide out from underneath. The manufacturer recommends that removal of the engine/transaxle assembly only be performed on a vehicle hoist.*
Note: *The TRANSAXLE COMPENSATION CODE and RESET MEMORY procedures (A/T initialization) must be performed when replacing the automatic transaxle assembly, engine assembly or ECM. The TRANSAXLE COMPENSATION CODE and RESET MEMORY can be performed only with the intelligent tester and should be done by a dealer service department or transmission repair shop.*

Removal

1   Have the air conditioning system discharged by an automotive air conditioning technician.
2   Park the vehicle on a frame-contact type vehicle hoist. The pads of the hoist arms must contact the body welt along each side of the vehicle (see *Jacking and towing* at the front of this manual).
3   Relieve the fuel system pressure (see Chapter 4).
4   Place protective covers on the fenders and cowl and remove the hood (see Chapter 11).
5   Remove the air filter housing (see Chapter 4).
6   Remove the cowl cover and the vent tray (see Chapter 11).

7   Disconnect the battery cables and remove the battery and battery tray (see Chapter 5). Also remove the battery tray reinforcement brackets.
8   Remove the Powertrain Control Module (PCM) (see Chapter 6).
9   Loosen the front wheel lug nuts and the driveaxle/hub nuts, then raise the vehicle on the hoist. Drain the cooling system and engine oil and remove the drivebelts (see Chapter 1).
10   Remove the coolant expansion tank (see Chapter 3).
11   Clearly label, then disconnect all vacuum lines, coolant and emissions hoses, wiring harness connectors, ground straps and fuel lines. Masking tape and/or a touch up paint applicator work well for marking items (see illustration). Take photos or sketch the locations of components and brackets.
12   Remove all fuel and/or emission control components that might be damaged during engine removal (see Chapter 4 and Chapter 6).
13   Remove the engine wiring harness connectors from the engine compartment junction box.
14   Remove the alternator and the starter (see Chapter 5).
15   Remove the engine splash shield fasteners and splash shield from the sides and bottom of the vehicle.
16   Remove the cooling fan(s), shroud(s) and radiator (see Chapter 3).
Note: *This step is not absolutely necessary, but it will help avoid damage to the cooling fans and radiator as the engine is lowered out of the vehicle. If the radiator is not removed, it will still be necessary to detach the transaxle oil cooler lines from the bottom of the radiator on automatic transaxle equipped vehicles.*
17   Disconnect and cap the refrigerant lines to the air conditioning compressor, then unbolt the air conditioning compressor (see Chapter 3) and set it aside. Make sure to cover the openings on the compressor.
18   Remove the front section of the exhaust system (see Chapter 4).
19   Disconnect the shift cable from the transaxle (see Chapter 7A). Disconnect any

7.23 With the chain or sling attached securely to the engine, take up the slack until there is slight tension on the chain - typical procedure shown

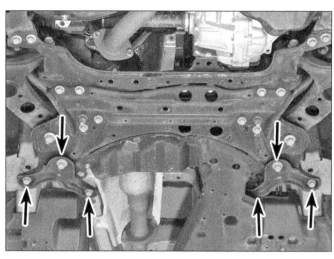

7.26 Subframe left and right rear brace mounting bolt locations

wiring harness connectors from the transaxle.

20   Disconnect the intermediate shaft from the steering gear (see Chapter 10).

21   Remove both front driveaxles (see Chapter 8). On AWD models, remove the rear driveshaft.

22   Detach the stabilizer bar links from the bar (see Chapter 10). Remove the two bolts securing the steering gear to the front crossmember and the fasteners securing the lower balljoints to the lower control arms (see Chapter 10). Support the steering gear from above with rope.

23   Attach a lifting sling or chain to the engine (see illustration). If no lifting hooks or brackets are present, you'll have to fasten the chains or slings to some substantial part of the engine - ones that are strong enough to take the weight, but in locations that will provide balance. Position an engine hoist and connect the sling to it. Take up the slack until there is slight tension on the sling or chain. Remember that the transaxle end of the engine will be heavier, so position the chain on the hoist so it balances the engine and the transaxle level

with the vehicle.

24   Recheck to be sure nothing except the mounts are still connecting the engine to the vehicle or to the transaxle. Disconnect and label anything still remaining.

25   Remove the front crossmember/lower radiator support mounting bolts and lower the crossmember from the vehicle.

26   Remove the subframe left and right rear brace mounting bolts and braces (see illustration).

27   Remove the subrfame left and right reinforcement mounting bolts and reinforcements (see illustration).

28   Remove the through-bolt on the driver's side transaxle mount and the passenger's side engine mount (see Chapter 2A).

29   Remove the subframe (see Chapter 10).

30   Slowly lower the engine/transaxle from the vehicle.

Note: *Placing a sheet of hardboard or paneling between the engine and the floor makes moving the powertrain easier.*

31   Once the powertrain is on the floor, disconnect the engine lifting hoist and raise the

vehicle hoist until the powertrain can be slid out from underneath.

Note: *An assistant will be needed to move the powertrain.*

32   Reconnect the chain or sling and raise the engine with the hoist several inches off the ground. Remove the through-bolts on the front and rear mounts and slide the front crossmember out from under the engine. Lower the engine back to the ground and separate the engine from the transaxle (see Chapter 7A). Disregard the steps that do not apply since the transaxle is already removed from the vehicle.

33   Remove the driveplate/flywheel and mount the engine on a stand.

Installation

34   Check the engine/transaxle mounts. If they're worn or damaged, replace them.

35   Inspect the converter seal and bushing. If the seal and bushing are okay, apply a dab of grease to the nose of the torque converter.

36   Carefully guide the transaxle into place, following the procedure outlined in Chapter 7A.

Caution: *Do not use the bolts to force the engine and transaxle into alignment. It may crack or damage major components.*

37   Install the engine-to-transaxle bolts and tighten them securely.

38   Slide the engine/transaxle over a sheet of hardboard or paneling until it is in the appropriate position under the vehicle, then lower the vehicle over the engine.

39   Roll the engine hoist into position, attach the chain or sling in a position that will allow a good balance and slowly raise the powertrain until the mount at the transaxle end can be attached.

40   Reinstall the subframe and attach the passenger side engine mount. Tighten the subframe mounting bolts to the torque values listed in the Chapter 10 Specifications.

41   Reinstall the remaining components and fasteners in the reverse order of removal.

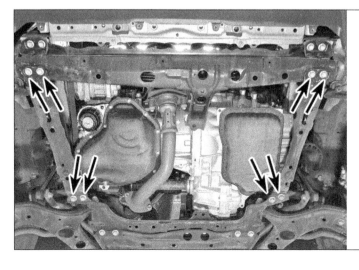

7.27 Subframe reinforcement mounting bolt locations

9.1 Before you try to remove the pistons from engines with very worn cylinders, use a ridge reamer to remove the raised material (ridge) from the top of the cylinders

9.3 Checking the connecting rod endplay (side clearance)

9.5 If the connecting rods and caps are not marked, use permanent ink or paint to mark the caps to the rods by cylinder number (for example, this would be the No. 4 connecting rod)

When installing the air conditioning compressor line, replace any O-rings with new ones specifically made for the purpose and lubricate them with refrigerant oil.

42   Add coolant, oil and transmission fluids as needed (see Chapter 1).

43   Run the engine and check for proper operation and leaks. Shut off the engine and recheck the fluid levels.

44   Have the air conditioning system evacuated, recharged and leak tested by the shop that discharged it.

8   Engine overhaul - disassembly sequence

1   It's much easier to remove the external components if the engine is mounted on a portable engine stand. A stand can often be rented quite cheaply from an equipment rental yard. Before the engine is mounted on a stand, the flywheel/driveplate should be removed from the engine.

2   If a stand isn't available, it's possible to remove the external engine components with it blocked up on the floor. Be extra careful not to tip or drop the engine when working without a stand.

3   If you're going to obtain a rebuilt engine, all external components must come off first, to be transferred to the replacement engine. These components include:

Emissions control components
Ignition coils
Spark plugs
Thermostat and housing cover
Water pump
Water bypass tube
EFI components
Intake/exhaust manifolds
Oil filter
Engine mount brackets
Driveplate

Note: *When removing the external compnents from the engine, pay close attention to details that may be helpful or important during installation. Note the installed position of gaskets, seals, spacers, pins, brackets, washers, bolts and other small items.*

4   If you're obtaining a short block, which consists of the engine block, crankshaft, pistons and connecting rods all assembled, then the timing belt or chain, the cylinder head, the oil pan, the oil pump, the lower crankcase will have to be removed as well from your engine so that your short block can be turned in to the rebuilder as a core (see Chapter 2A). See *Engine rebuilding alternatives* for additional information regarding the different possibilities to be considered.

9   Pistons and connecting rods - removal and installation

Caution: *New connecting rod cap bolts should be used when reinstalling the pistons/connecting rods.*

Removal

Note: *Prior to removing the piston/connecting rod assemblies, remove the cylinder head, oil pan, balancer assembly (see Chapter 2A) and lower crankcase.*

1   Use your fingernail to feel if a ridge has formed at the upper limit of ring travel (about 1/4-inch down from the top of each cylinder). If carbon deposits or cylinder wear have produced ridges, they must be completely removed with a special tool (see illustration). Follow the manufacturer's instructions provided with the tool. Failure to remove the ridges before attempting to remove the piston/connecting rod assemblies may result in piston breakage.

Caution: *These engines have a removable cylinder block water jacket and should not*

be machined. In the event of a warped block deck, misalignment of the main bearing bores, or worn cylinders, the manufacturer states that the cylinder block must be replaced.

2   After the cylinder ridges have been removed, turn the engine so the crankshaft is facing up. Remove the engine balancer assembly (see Chapter 2A).

3   Before the connecting rods are removed, check the connecting rod endplay with feeler gauges. Slide them between the first connecting rod and the crankshaft throw until the play is removed (see illustration). Repeat this procedure for each connecting rod. The endplay is equal to the thickness of the feeler gauge(s).

4   Check with an automotive machine shop for the endplay service limit (a typical end play limit should measure between 0.005 to 0.015 inch [0.127 to 0.369 mm]). If the play exceeds the service limit, new connecting rods will be required. If new rods (or a new crankshaft) are installed, the endplay may fall under the minimum allowable. If it does, the rods will have to be machined to restore it. If necessary, consult an automotive machine shop for advice.

5   Check the connecting rods and caps for identification marks (see illustration). If they aren't plainly marked, use paint or permanent marker to clearly identify each rod and cap (1, 2, 3, etc., depending on the cylinder they're associated with).

6   Loosen each of the connecting rod cap fasteners 1/2-turn at a time until they can be removed by hand. Remove the number one connecting rod cap and bearing insert. Don't drop the bearing insert out of the cap.

7   Remove the bearing insert and push the connecting rod/piston assembly out through the top of the engine. Use a wooden or plastic hammer handle to push on the upper bearing surface in the connecting rod. If resistance is felt, double-check to make sure that all of the ridge was removed from the cylinder.

8   Repeat the procedure for the remaining cylinders.

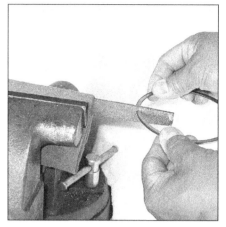

9.13 Install the piston ring into the cylinder, then push it down into position using a piston so the ring will be square in the cylinder

9.14 With the ring square in the cylinder, measure the ring end gap with a feeler gauge

9.15 If the ring end gap is too small, clamp a file in a vise and file the piston ring ends - file from the outside of the ring inward only

9   Reassemble the connecting rod caps and bearing inserts in their respective connecting rods and install the cap fasteners finger-tight. Leaving the old bearing inserts in place until reassembly will help prevent the connecting rod bearing surfaces from being accidentally nicked or gouged.

10   The pistons and connecting rods are now ready for inspection and overhaul at an automotive machine shop.

Piston ring installation

11   Before installing the new piston rings, the ring end gaps must be checked. It's assumed that the piston ring side clearance has been checked and verified correct.

12   Lay out the piston/connecting rod assemblies and the new ring sets so the ring sets will be matched with the same piston and cylinder during the end gap measurement and engine assembly.

13   Insert the top (number one) ring into the first cylinder and square it up with the cylinder walls by pushing it in with the top of the piston (see illustration). The ring should be near the bottom of the cylinder, at the lower limit of ring travel.

14   To measure the end gap, slip feeler gauges between the ends of the ring until a gauge equal to the gap width is found (see illustration). The feeler gauge should slide between the ring ends with a slight amount of drag. Check with an automotive machine shop for the correct end gap for your engine. If the gap is larger or smaller than specified, double-check to make sure you have the correct rings before proceeding.

15   If the gap is too small, it must be enlarged or the ring ends may come in contact with each other during engine operation, which can cause serious damage to the engine. The end gap can be increased by filing the ring ends very carefully with a fine file. Mount the file in a vise equipped with soft jaws, slip the ring over the file with the ends contacting the file face and slowly move the ring to remove

material from the ends. When performing this operation, file only by pushing the ring from the outside end of the file towards the vise (see illustration). Be sure to remove all raised material.

16   Excess end gap isn't critical unless it's greater than approximately 0.040-inch. Again, double-check to make sure you have the correct ring type and that you are referencing the correct section and category of specifications.

17   Repeat the procedure for each ring that will be installed in the first cylinder and for each ring in the remaining cylinders. Remember to keep rings, pistons and cylinders matched up.

18   Once the ring end gaps have been checked/corrected, the rings can be installed on the pistons.

19   The oil control ring (lowest one on the piston) is usually installed first. It's composed of three separate components. Slip the spacer/expander into the groove (see illustration). If an anti-rotation tang is used, make sure it's inserted into the drilled hole in the ring groove. Next, install the upper side rail in the same manner (see illustration). Don't use a piston ring installation tool on the oil ring side rails, as they may be damaged. Instead, place one end of the side rail into the groove between the spacer/expander and the ring land, hold it firmly in place and slide a finger around the piston while pushing the rail into the groove. Finally, install the lower side rail.

20   After the three oil ring components have been installed, check to make sure that both the upper and lower side rails can be rotated smoothly inside the ring grooves.

21   The number two (middle) ring is installed next. It's usually stamped with a mark which must face up, toward the top of the piston. Do not mix up the top and middle rings, as they have different profiles.

Note: *Always follow the instructions printed on the ring package or box - different manufacturers may require different approaches.*

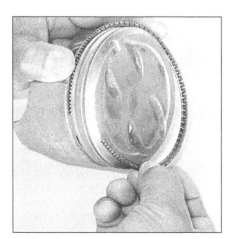

9.19a Installing the spacer/expander in the oil ring groove

9.19b DO NOT use a piston ring installation tool when installing the oil control side rails

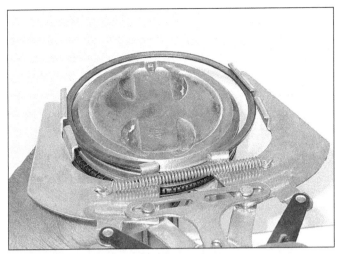

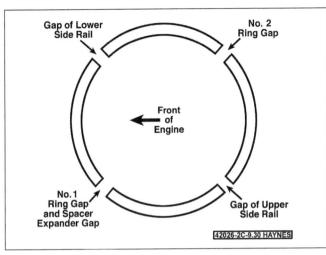

9.22 Use a piston ring installation tool to install the number 2 and the number 1 (top) rings - be sure the directional mark on the piston ring(s) is facing toward the top of the piston

9.29 Position the piston ring end gaps as shown

22   Use a piston ring installation tool and make sure the identification mark is facing the top of the piston, then slip the ring into the middle groove on the piston (see illustration). Don't expand the ring any more than necessary to slide it over the piston.

23   Install the number one (top) ring in the same manner. Make sure the mark is facing up. Be careful not to confuse the number one and number two rings.

24   Repeat the procedure for the remaining pistons and rings.

Installation

25   Before installing the piston/connecting rod assemblies, the cylinder walls must be perfectly clean, the top edge of each cylinder bore must be chamfered, and the crankshaft must be in place.

26   Remove the cap from the end of the number one connecting rod (refer to the marks made during removal). Remove the original bearing inserts and wipe the bearing surfaces of the connecting rod and cap with a clean, lint-free cloth. They must be kept spotlessly clean.

Connecting rod bearing oil clearance check

27   Clean the back side of the new upper bearing insert, then lay it in place in the connecting rod. Make sure the tab on the bearing fits into the recess in the rod. Don't hammer the bearing insert into place and be very careful not to nick or gouge the bearing face. Don't lubricate the bearing at this time.

28   Clean the back side of the other bearing insert and install it in the rod cap. Again, make sure the tab on the bearing fits into the recess in the cap, and don't apply any lubricant. It's critically important that the mating surfaces of the bearing and connecting rod are perfectly clean and oil free when they're assembled.

29   Position the piston ring gaps at 90-degree intervals around the piston as shown

(see illustration).

30   Lubricate the piston and rings with clean engine oil and attach a piston ring compressor to the piston. Leave the skirt protruding about 1/4-inch to guide the piston into the cylinder. The rings must be compressed until they're flush with the piston.

31   Rotate the crankshaft until the number one connecting rod journal is at BDC (bottom dead center) and apply a liberal coat of engine oil to the cylinder walls.

32   With the mark (cavity) on top of the piston facing the front of the engine, gently insert the piston/connecting rod assembly into the number one cylinder bore and rest the bottom edge of the ring compressor on the engine block.

33   Tap the top edge of the ring compressor to make sure it's contacting the block around its entire circumference.

34   Gently tap on the top of the piston with the end of a wooden or plastic hammer handle (see illustration) while guiding the end of the connecting rod into place on the crank-

shaft journal.

35   The piston rings may try to pop out of the ring compressor just before entering the cylinder bore, so keep some downward pressure on the ring compressor. Work slowly, and if any resistance is felt as the piston enters the cylinder, stop immediately. Find out what's hanging up and fix it before proceeding. Do not, for any reason, force the piston into the cylinder - you might break a ring and/or the piston.

36   Once the piston/connecting rod assembly is installed, the connecting rod bearing oil clearance must be checked before the rod cap is permanently installed.

37   Cut a piece of the appropriate size Plastigage slightly shorter than the width of the connecting rod bearing and lay it in place on the number one connecting rod journal, parallel with the journal axis (see illustration).

38   Clean the connecting rod cap bearing face and install the rod cap. Make sure the mating mark on the cap is on the same side as the mark on the connecting rod.

9.34 Use a plastic or wooden hammer handle to push the piston into the cylinder

9.37 Place Plastigage on each connecting rod bearing journal parallel to the crankshaft centerline

ENGINE BEARING ANALYSIS

Debris

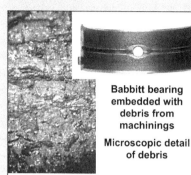

Babbitt bearing embedded with debris from machinings

Microscopic detail of debris

Microscopic detail of gouges

Overplated copper alloy bearing gouged by cast iron debris

Aluminum bearing embedded with glass beads

Microscopic detail of glass beads

Damaged lining caused by dirt left on the bearing back

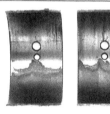

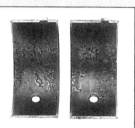

Misassembly

Result of a lower half assembled as an upper - blocking the oil flow

Excessive oil clearance is indicated by a short contact arc

Polished and oil-stained backs are a result of a poor fit in the housing bore

Result of a wrong, reversed, or shifted cap

Overloading

Damage from excessive idling which resulted in an oil film unable to support the load imposed

Damaged upper connecting rod bearings caused by engine lugging; the lower main bearings (not shown) were similarly affected

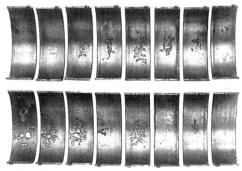

The damage shown in these upper and lower connecting rod bearings was caused by engine operation at a higher-than-rated speed under load

Misalignment

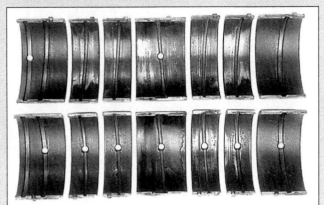

A warped crankshaft caused this pattern of severe wear in the center, diminishing toward the ends

A poorly finished crankshaft caused the equally spaced scoring shown

A tapered housing bore caused the damage along one edge of this pair

A bent connecting rod led to the damage in the "V" pattern

Lubrication

Result of dry start: The bearings on the left, farthest from the oil pump, show more damage

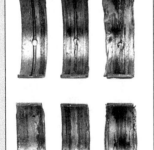

Result of a low oil supply or oil starvation

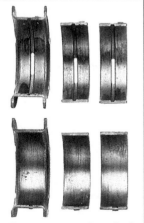

Severe wear as a result of inadequate oil clearance

Corrosion

Microscopic detail of corrosion

Corrosion is an acid attack on the bearing lining generally caused by inadequate maintenance, extremely hot or cold operation, or inferior oils or fuels

Microscopic detail of cavitation

Example of cavitation - a surface erosion caused by pressure changes in the oil film

Damage from excessive thrust or insufficient axial clearance

Bearing affected by oil dilution caused by excessive blow-by or a rich mixture

9.41 Use the scale on the Plastigage package to determine the bearing oil clearance - be sure to measure the widest part of the Plastigage and use the correct scale; it comes with both standard and metric scales

10.1 Checking crankshaft endplay with a dial indicator

10.3 Checking crankshaft endplay with feeler gauges at the thrust bearing journal

39   Install the old rod bolts and tighten them to the torque listed in this Chapter's Specifications.

Note: *Use a thin-wall socket to avoid erroneous torque readings that can result if the socket is wedged between the rod cap and the fastener. If the socket tends to wedge itself between the fastener and the cap, lift up on it slightly until it no longer contacts the cap. DO NOT rotate the crankshaft at any time during this operation.*

40   Remove the fasteners and detach the rod cap, being very careful not to disturb the Plastigage.

41   Compare the width of the crushed Plastigage to the scale printed on the Plastigage envelope to obtain the oil clearance (see illustration). The connecting rod oil clearance is usually about 0.002 inch. Consult an automotive machine shop for the clearance specified for the rod bearings on your engine.

42   If the clearance is not as specified, the bearing inserts may be the wrong size (which means different ones will be required). Before deciding that different inserts are needed, make sure that no dirt or oil was between the bearing inserts and the connecting rod or cap when the clearance was measured. Also, recheck the journal diameter. If the Plastigage was wider at one end than the other, the journal may be tapered. If the clearance still exceeds the limit specified, the bearing will have to be replaced with an undersize bearing.

Caution: *When installing a new crankshaft, always use a standard size bearing unless the crankshaft is marked differently.*

Final installation

43   Carefully scrape all traces of the Plastigage material off the rod journal and/or bearing face. Be very careful not to scratch the

bearing - use your fingernail or the edge of a plastic card.

44   Make sure the bearing faces are perfectly clean, then apply a uniform layer of clean moly-base grease or engine assembly lube to both of them. You'll have to push the piston into the cylinder to expose the face of the bearing insert in the connecting rod.

45   Slide the connecting rod back into place on the journal, install the rod cap, install the new bolts and tighten them to the torque listed in this Chapter's Specifications. Again, work up to the torque in three steps.

46   Repeat the entire procedure for the remaining pistons/connecting rods.

47   The important points to remember are:

a) *Keep the back sides of the bearing inserts and the insides of the connecting rods and caps perfectly clean when assembling them.*

b) *Make sure you have the correct piston/rod assembly for each cylinder.*

c) *The mark on the piston must face the front of the engine.*

d) *Lubricate the cylinder walls liberally with clean oil.*

e) *Lubricate the bearing faces when installing the rod caps after the oil clearance has been checked.*

48   After all the piston/connecting rod assemblies have been correctly installed, rotate the crankshaft a number of times by hand to check for any obvious binding.

49   As a final step, check the connecting rod endplay again. If it was correct before disassembly and the original crankshaft and rods were reinstalled, it should still be correct. If new rods or a new crankshaft were installed, the endplay may be inadequate. If so, the rods will have to be removed and taken to an automotive machine shop for resizing.

10   Crankshaft - removal and installation

Caution: *New main bearing cap bolts should be used when reinstalling the crankshaft.*

Removal

Note: *The crankshaft can be removed only after the engine has been removed from the vehicle. It's assumed that the flywheel driveplate, crankshaft pulley, timing chain, oil pan, oil pump body, balance shaft assembly, lower crankcase, piston oil jet nozzles, and piston/connecting rod assemblies have already been removed. The rear main oil seal retainer must be unbolted and separated from the block before proceeding with crankshaft removal.*

1   Before the crankshaft is removed, measure the endplay. Mount a dial indicator with the indicator in line with the crankshaft and touching the end of the crankshaft (see illustration).

2   Pry the crankshaft all the way to the rear and zero the dial indicator. Next, pry the crankshaft to the front as far as possible and check the reading on the dial indicator. The distance traveled is the endplay. A typical crankshaft endplay will fall between 0.003 to 0.010 inch (0.076 to 0.254 mm). If it's greater than that, check the crankshaft thrust surfaces for wear after it's removed. If no wear is evident, new main bearings should correct the endplay.

3   If a dial indicator isn't available, feeler gauges can be used. Gently pry the crankshaft all the way to the front of the engine. Slip feeler gauges between the crankshaft and the front face of the thrust bearing or washer to determine the clearance (see illustration).

4   Loosen the main bearing cap bolts 1/4-turn at a time each, until they can be removed by hand.

10.17 Place the Plastigage (arrow) onto the crankshaft bearing journal as shown

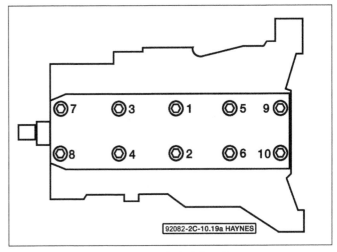

92082-2C-10.19a HAYNES

10.19 Main bearing cap bolt tightening sequence

5   Gently tap the main bearing cap(s) with a soft-faced hammer. Pull the main bearing cap(s) straight up and off the cylinder block. Try not to drop the bearing inserts if they come out with the assembly.

6   Carefully lift the crankshaft out of the engine. It may be a good idea to have an assistant available, since the crankshaft is quite heavy and awkward to handle. With the bearing inserts in place inside the engine block and main bearing caps, reinstall the main bearing cap assembly onto the engine block and tighten the bolts finger-tight. Make sure you install the main bearing cap(s) with the arrow facing the front end of the engine.

Installation

7   Crankshaft installation is the first step in engine reassembly. It's assumed at this point that the engine block and crankshaft have been cleaned, inspected and repaired or reconditioned.

8   Position the engine block with the bottom facing up.

9   Remove the mounting bolts and lift off the main bearing caps.

10   If they're still in place, remove the original bearing inserts from the block and from the main bearing cap(s). Wipe the bearing surfaces of the block and main bearing cap(s) with a clean, lint-free cloth. They must be kept spotlessly clean. This is critical for determining the correct bearing oil clearance.

Main bearing oil clearance check

11   Without mixing them up, clean the back sides of the new upper main bearing inserts (with grooves and oil holes) and lay one in each main bearing saddle in the block. Each upper bearing has an oil groove and oil hole in it. The thrust washer is located on the number 3 crankshaft journal. Install the thrust washers with the grooved side facing out, with one set located in the block and the

other set with the main bearing cap. Clean the back sides of the lower main bearing inserts (without grooves) and lay them in the corresponding main bearing caps. Make sure the tab on the bearing insert fits into the recess in the block or main bearing cap.

Caution: *The oil holes in the block must line up with the oil holes in the upper bearing inserts.*

Caution: *Do not hammer the bearing insert into place and don't nick or gouge the bearing faces. DO NOT apply any lubrication at this time.*

12   Clean the faces of the bearing inserts in the block and the crankshaft main bearing journals with a clean, lint-free cloth.

13   Check or clean the oil holes in the crankshaft, as any dirt here can go only one way - straight through the new bearings.

14   Once you're certain the crankshaft is clean, carefully lay it in position in the cylinder block.

15   Before the crankshaft can be permanently installed, the main bearing oil clearance must be checked.

16   Cut several strips of the appropriate size of Plastigage (they must be slightly shorter than the width of the main bearing journal).

17   Place one piece on each crankshaft main bearing journal, parallel with the journal axis (see illustration).

18   Clean the faces of the bearing inserts in the main bearing caps. Hold the bearing inserts in place and install the caps onto the crankshaft and cylinder block. DO NOT disturb the Plastigage. Make sure you install the main bearing cap assembly with the arrow facing the front of the engine.

19   Apply clean engine oil to all bolt threads prior to installation, then install all bolts finger-tight. Tighten the main bearing cap bolts in the proper sequence (see illustration) to the torque listed in this Chapter's Specifications. DO NOT rotate the crankshaft at any time during this operation.

Note: *Use the old main bearing cap bolts for*

the oil clearance checking procedure.

20   Remove the bolts a little at a time (and in the reverse order of the tightening sequence) and carefully lift the main bearing caps straight up and off the block. Do not disturb the Plastigage or rotate the crankshaft. If a main bearing cap is difficult to remove, tap it gently from side-to-side with a soft-face hammer to loosen it.

21   Compare the width of the crushed Plastigage on each journal to the scale printed on the Plastigage envelope to determine the main bearing oil clearance (see illustration). A typical main bearing oil clearance should fall between 0.0015 to 0.0023-inch. Check with an automotive machine shop for the clearance specified for your engine.

22   If the clearance is not as specified, the bearing inserts may be the wrong size (which means different ones will be required). Before deciding if different inserts are needed, make sure that no dirt or oil was between the bearing inserts and the cap assembly or block when

10.21 Use the scale on the Plastigage package to determine the bearing oil clearance - be sure to measure the widest part of the Plastigage and use the correct scale; it comes with both standard and metric scales

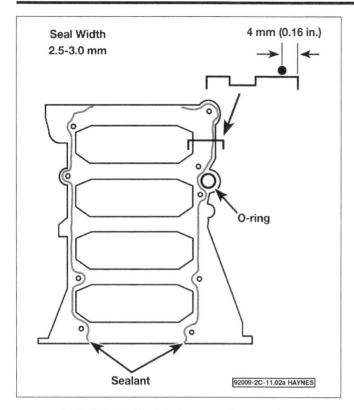

11.2a Cylinder block-to-lower crankcase sealant installation details

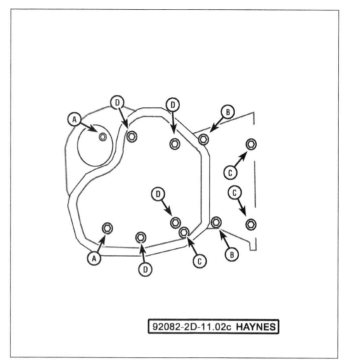

11.2b Lower crankcase bolt locations - four different length bolts are used

A	2.56 inches (65 mm) long	C	4.92 inches (125 mm) long
B	1.38 inches (35 mm) long	D	6.50 inches (165 mm) long

the clearance was measured. If the Plasti-gage was wider at one end than the other, the crankshaft journal may be tapered. If the clearance still exceeds the limit specified, the bearing insert(s) will have to be replaced with an undersize bearing insert(s).

Caution: *When installing a new crankshaft, always install a standard bearing insert set unless the crankshaft is marked otherwise.*

23   Carefully scrape all traces of the Plasti-gage material off the main bearing journals and/or the bearing insert faces. Be sure to remove all residue from the oil holes. Use your fingernail or the edge of a plastic card - don't nick or scratch the bearing faces.

Final installation

24   Carefully lift the crankshaft out of the cylinder block.
25   Clean the bearing insert faces in the cylinder block, then apply a thin, uniform layer of moly-base grease or engine assembly lube to each of the bearing surfaces. Be sure to coat the thrust faces as well as the journal face of the thrust bearing.
26   Make sure the crankshaft journals are clean, then lay the crankshaft back in place in the cylinder block.
27   Clean the remaining bearing insert faces and apply the same lubricant to them.
28   Hold the bearing inserts in place and install the main bearing caps on the crank-

shaft and cylinder block. Tap the bearing caps into place with a brass punch or a soft-faced hammer.
29   Apply clean engine oil to the new bolt threads, wipe off any excess oil and install the bolts finger-tight.
30   Tighten the main bearing cap bolts to 10 or 12 foot-pounds (see illustration 10.19).
31   Push the crankshaft forward using a screwdriver or prybar to seat the thrust bearing. Once the crankshaft is pushed fully forward to seat the thrust bearing, leave the screwdriver in position so that pressure stays on the crankshaft until after all main bearing cap bolts have been tightened.
32   Tighten the main bearing cap bolts in the indicated sequence to the torque and angle of rotation listed in this Chapter's Specifications (see illustration 10.19).
33   Recheck crankshaft endplay with a feeler gauge or a dial indicator. The endplay should be correct if the crankshaft thrust faces aren't worn or damaged and if new bearings have been installed.
34   Rotate the crankshaft a number of times by hand to check for any obvious binding. It should rotate with a running torque of 50 in-lbs or less. If the running torque is too high, correct the problem at this time.
35   Install the new rear main oil seal (see Chapter 2A).

11   Engine overhaul - reassembly sequence

1   Before beginning engine reassembly, make sure you have all the necessary new parts, gaskets and seals as well as the following items on hand:

Common hand tools
A 1/2-inch drive torque wrench
New engine oil
Thread locking compound

2   If you obtained a short block it will be necessary to install the cylinder head, the oil pump, the timing belt or chain, the lower crankcase (see illustrations), the oil pan and the valve cover (see Chapter 2A). In order to save time and avoid problems, the external components must be installed in the following general order:

Thermostat and housing cover
Water pump
Water bypass tube
Intake/exhaust manifolds
EFI components
Emissions control components
Spark plugs
Ignition coils
Oil filter
Engine mount brackets
Driveplate

12 Initial start-up and break-in after installation

Warning: *Have a fire extinguisher handy when starting the engine for the first time.*

1 Once the engine has been installed in the vehicle, double-check the engine oil and coolant levels.

2 With the spark plugs out of the engine and the ignition system and fuel pump disabled (see Chapter 4, Section 3), crank the engine until oil pressure registers on the gauge or the light goes out.

3 Install the spark plugs and ignition coils and reconnect the fuel pump electrical connector.

4 Start the engine. It may take a few moments for the fuel system to build up pressure, but the engine should start without a great deal of effort.

5 After the engine starts, it should be allowed to warm up to normal operating temperature. While the engine is warming up, make a thorough check for fuel, oil and coolant leaks.

6 Shut the engine off and recheck the engine oil and coolant levels.

7 Drive the vehicle to an area with minimum traffic, accelerate from 30 to 50 mph, then allow the vehicle to slow to 30 mph with the throttle closed. Repeat the procedure 10 or 12 times. This will load the piston rings and cause them to seat properly against the cylinder walls. Check again for oil and coolant leaks.

8 Drive the vehicle gently for the first 500 miles (no sustained high speeds) and keep a constant check on the oil level. It is not unusual for an engine to use oil during the break-in period.

9 At approximately 500 to 600 miles, change the oil and filter.

10 For the next few hundred miles, drive the vehicle normally. Do not pamper it or abuse it.

11 After 2000 miles, change the oil and filter again and consider the engine broken in.

B

Backlash - The amount of play between two parts. Usually refers to how much one gear can be moved back and forth without moving the gear with which it's meshed.

Bearing Caps - The caps held in place by nuts or bolts which, in turn, hold the bearing surface. This space is for lubricating oil to enter.

Bearing clearance - The amount of space left between shaft and bearing surface. This space is for lubricating oil to enter.

Bearing crush - The additional height which is purposely manufactured into each bearing half to ensure complete contact of the bearing back with the housing bore when the engine is assembled.

Bearing knock - The noise created by movement of a part in a loose or worn bearing.

Blueprinting - Dismantling an engine and reassembling it to EXACT specifications.

Bore - An engine cylinder, or any cylindrical hole; also used to describe the process of enlarging or accurately refinishing a hole with a cutting tool, as to bore an engine cylinder. The bore size is the diameter of the hole.

Boring - Renewing the cylinders by cutting them out to a specified size. A boring bar is used to make the cut.

Bottom end - A term which refers collectively to the engine block, crankshaft, main bearings and the big ends of the connecting rods.

Break-in - The period of operation between installation of new or rebuilt parts and time in which parts are worn to the correct fit. Driving at reduced and varying speed for a specified mileage to permit parts to wear to the correct fit.

Bushing - A one-piece sleeve placed in a bore to serve as a bearing surface for shaft, piston pin, etc. Usually replaceable.

C

Camshaft - The shaft in the engine, on which a series of lobes are located for operating the valve mechanisms. The camshaft is driven by gears or sprockets and a timing chain. Usually referred to simply as the cam.

Carbon - Hard, or soft, black deposits found in combustion chamber, on plugs, under rings, on and under valve heads.

Cast iron - An alloy of iron and more than two percent carbon, used for engine blocks and heads because it's relatively inexpensive and easy to mold into complex shapes.

Chamfer - To bevel across (or a bevel on) the sharp edge of an object.

Chase - To repair damaged threads with a tap or die.

Combustion chamber - The space between the piston and the cylinder head, with the piston at top dead center, in which air-fuel mixture is burned.

Compression ratio - The relationship between cylinder volume (clearance volume) when the piston is at top dead center and cylinder volume when the piston is at bottom dead center.

Connecting rod - The rod that connects the crank on the crankshaft with the piston. Sometimes called a con rod.

Connecting rod cap - The part of the connecting rod assembly that attaches the rod to the crankpin.

Core plug - Soft metal plug used to plug the casting holes for the coolant passages in the block.

Crankcase - The lower part of the engine in which the crankshaft rotates; includes the lower section of the cylinder block and the oil pan.

Crank kit - A reground or reconditioned crankshaft and new main and connecting rod bearings.

Crankpin - The part of a crankshaft to which a connecting rod is attached.

Crankshaft - The main rotating member, or shaft, running the length of the crankcase, with offset throws to which the connecting rods are attached; changes the reciprocating motion of the pistons into rotating motion.

Cylinder sleeve - A replaceable sleeve, or liner, pressed into the cylinder block to form the cylinder bore.

D

Deburring - Removing the burrs (rough edges or areas) from a bearing.

Deglazer - A tool, rotated by an electric motor, used to remove glaze from cylinder walls so a new set of rings will seat.

E

Endplay - The amount of lengthwise movement between two parts. As applied to a crankshaft, the distance that the crankshaft can move forward and back in the cylinder block.

F

Face - A machinist's term that refers to removing metal from the end of a shaft or the face of a larger part, such as a flywheel.

Fatigue - A breakdown of material through a large number of loading and unloading cycles. The first signs are cracks followed shortly by breaks.

Feeler gauge - A thin strip of hardened steel, ground to an exact thickness, used to check clearances between parts.

Free height - The unloaded length or height of a spring.

Freeplay - The looseness in a linkage, or an assembly of parts, between the initial application of force and actual movement. Usually perceived as slop or slight delay.

Freeze plug - See Core plug.

G

Gallery - A large passage in the block that forms a reservoir for engine oil pressure.

Glaze - The very smooth, glassy finish that develops on cylinder walls while an engine is in service.

H

Heli-Coil - A rethreading device used when threads are worn or damaged. The device is installed in a retapped hole to reduce the thread size to the original size.

I

Installed height - The spring's measured length or height, as installed on the cylinder head. Installed height is measured from the spring seat to the underside of the spring retainer.

J

Journal - The surface of a rotating shaft which turns in a bearing.

K

Keeper - The split lock that holds the valve spring retainer in position on the valve stem.

Key - A small piece of metal inserted into matching grooves machined into two parts fitted together - such as a gear pressed onto a shaft - which prevents slippage between the two parts.

Knock - The heavy metallic engine sound, produced in the combustion chamber as a result of abnormal combustion - usually detonation. Knock is usually caused by a loose or worn bearing. Also referred to as detonation, pinging and spark knock. Connecting rod or main bearing knocks are created by too much oil clearance or insufficient lubrication.

L

Lands - The portions of metal between the piston ring grooves.

Lapping the valves - Grinding a valve face and its seat together with lapping compound.

Lash - The amount of free motion in a gear train, between gears, or in a mechanical assembly, that occurs before movement can begin. Usually refers to the lash in a valve train.

Lifter - The part that rides against the cam to transfer motion to the rest of the valve train.

M

Machining - The process of using a machine to remove metal from a metal part.

Main bearings - The plain, or babbit, bearings that support the crankshaft.

Main bearing caps - The cast iron caps, bolted to the bottom of the block, that support the main bearings.

O

O.D. - Outside diameter.

Oil gallery - A pipe or drilled passageway in the engine used to carry engine oil from one area to another.

Oil ring - The lower ring, or rings, of a piston; designed to prevent excessive amounts of oil from working up the cylinder walls and into the combustion chamber. Also called an oil-control ring.

Oil seal - A seal which keeps oil from leaking out of a compartment. Usually refers to a dynamic seal around a rotating shaft or other moving part.

O-ring - A type of sealing ring made of a special rubberlike material; in use, the O-ring is compressed into a groove to provide the sealing action.

Overhaul - To completely disassemble a unit, clean and inspect all parts, reassemble it with the original or new parts and make all adjustments necessary for proper operation.

P

Pilot bearing - A small bearing installed in the center of the flywheel (or the rear end of the crankshaft) to support the front end of the input shaft of the transmission.

Pip mark - A little dot or indentation which indicates the top side of a compression ring.

Piston - The cylindrical part, attached to the connecting rod, that moves up and down in the cylinder as the crankshaft rotates. When the fuel charge is fired, the piston transfers the force of the explosion to the connecting rod, then to the crankshaft.

Piston pin (or wrist pin) - The cylindrical and usually hollow steel pin that passes through the piston. The piston pin fastens the piston to the upper end of the connecting rod.

Piston ring - The split ring fitted to the groove in a piston. The ring contacts the sides of the ring groove and also rubs against the cylinder wall, thus sealing space between piston and wall. There are two types of rings: Compression rings seal the compression pressure in the combustion chamber; oil rings scrape excessive oil off the cylinder wall.

Piston ring groove - The slots or grooves cut in piston heads to hold piston rings in position.

Piston skirt - The portion of the piston below the rings and the piston pin hole.

Plastigage - A thin strip of plastic thread, available in different sizes, used for measuring clearances. For example, a strip of plastigage is laid across a bearing journal and mashed as parts are assembled. Then parts are disassembled and the width of the strip is measured to determine clearance between journal and bearing. Commonly used to measure crankshaft main-bearing and connecting rod bearing clearances.

Press-fit - A tight fit between two parts that requires pressure to force the parts together. Also referred to as drive, or force, fit.

Prussian blue - A blue pigment; in solution, useful in determining the area of contact between two surfaces. Prussian blue is commonly used to determine the width and location of the contact area between the valve face and the valve seat.

R

Race (bearing) - The inner or outer ring that provides a contact surface for balls or rollers in bearing.

Ream - To size, enlarge or smooth a hole by using a round cutting tool with fluted edges.

Ring job - The process of reconditioning the cylinders and installing new rings.

Runout - Wobble. The amount a shaft rotates out-of-true.

S

Saddle - The upper main bearing seat.

Scored - Scratched or grooved, as a cylinder wall may be scored by abrasive particles moved up and down by the piston rings.

Scuffing - A type of wear in which there's a transfer of material between parts moving against each other; shows up as pits or grooves in the mating surfaces.

Seat - The surface upon which another part rests or seats. For example, the valve seat is the matched surface upon which the valve face rests. Also used to refer to wearing into a good fit; for example, piston rings seat after a few miles of driving.

Short block - An engine block complete with crankshaft and piston and, usually, camshaft assemblies.

Static balance - The balance of an object while it's stationary.

Step - The wear on the lower portion of a ring land caused by excessive side and back-clearance. The height of the step indicates the ring's extra side clearance and the length of the step projecting from the back wall of the groove represents the ring's back clearance.

Stroke - The distance the piston moves when traveling from top dead center to bottom dead center, or from bottom dead center to top dead center.

Stud - A metal rod with threads on both ends.

T

Tang - A lip on the end of a plain bearing used to align the bearing during assembly.

Tap - To cut threads in a hole. Also refers to the fluted tool used to cut threads.

Taper - A gradual reduction in the width of a shaft or hole; in an engine cylinder, taper usually takes the form of uneven wear, more pronounced at the top than at the bottom.

Throws - The offset portions of the crankshaft to which the connecting rods are affixed.

Thrust bearing - The main bearing that has thrust faces to prevent excessive endplay, or forward and backward movement of the crankshaft.

Thrust washer - A bronze or hardened steel washer placed between two moving parts. The washer prevents longitudinal movement and provides a bearing surface for thrust surfaces of parts.

Tolerance - The amount of variation permitted from an exact size of measurement. Actual amount from smallest acceptable dimension to largest acceptable dimension.

U

Umbrella - An oil deflector placed near the valve tip to throw oil from the valve stem area.

Undercut - A machined groove below the normal surface.

Undersize bearings - Smaller diameter bearings used with re-ground crankshaft journals.

V

Valve grinding - Refacing a valve in a valve-refacing machine.

Valve train - The valve-operating mechanism of an engine; includes all components from the camshaft to the valve.

Vibration damper - A cylindrical weight attached to the front of the crankshaft to minimize torsional vibration (the twist-untwist actions of the crankshaft caused by the cylinder firing impulses). Also called a harmonic balancer.

W

Water jacket - The spaces around the cylinders, between the inner and outer shells of the cylinder block or head, through which coolant circulates.

Web - A supporting structure across a cavity.

Woodruff key - A key with a radiused backside (viewed from the side).

Notes

Chapter 3
Cooling, heating and air conditioning systems

Contents

Specifications

General

Expansion tank cap pressure rating	13.5 to 18 psi (93 to 123 kPa)
Thermostat rating	
Opens	176 to 183-degrees F (80 to 84-degrees C)
Fully open...	203-degrees F (95-degrees C)
Refrigerant type...	R-134a
Refrigerant capacity...	15.9 +/- 1.1 ounces (450 +/- 30 grams)

Torque specifications

Note: *One foot-pound (ft-lb) of torque is equivalent to 12 inch-pounds (in-lbs) of torque. Torque values below approximately 15 foot-pounds are expressed in inch-pounds, because most foot-pound torque wrenches are not accurate at these smaller values.*

	Ft-lbs (unless otherwise indicated)	Nm
Thermostat housing bolts	88 in-lbs	10
Water pump bolts	15	21
Water inlet housing bolts	88 in-lbs	10
Upper and lower radiator support bolts	96 in-lbs	10.5
Receiver/drier end plug	35 in-lbs	3.9

1 General information

Warning: *Do not allow antifreeze to come in contact with your skin or painted surfaces of the vehicle. Rinse off spills immediately with plenty of water. Antifreeze is highly toxic if ingested. Never leave antifreeze lying around in an open container or in puddles on the floor; children and pets are attracted by its sweet smell and may drink it. Check with local authorities about disposing of used antifreeze. Many communities have collection centers which will see that antifreeze is disposed of safely. Never dump used antifreeze on the ground or pour it into drains.*

Engine cooling system

1   All modern vehicles employ a pressurized engine cooling system with thermostatically controlled coolant circulation. The cooling system consists of a radiator, an expansion tank, a pressure cap (located on the expansion tank), a thermostat, a cooling fan, and a water pump.

2   The water pump circulates coolant through the engine. The coolant flows around each cylinder and around the intake and exhaust ports, near the spark plug areas and in close proximity to the exhaust valve guides.

3   A thermostat controls engine coolant temperature. During warm up, the closed thermostat prevents coolant from circulating through the radiator. As the engine nears normal operating temperature, the thermostat opens and allows hot coolant to travel through the radiator, where it's cooled before returning to the engine.

Heating system

4   The heating system consists of a blower fan and heater core located in a housing under the dash, the hoses connecting the heater core to the engine cooling system and the heater/air conditioning control head on the dashboard. Hot engine coolant is circulated through the heater core. When the heater mode is activated, a flap door in the housing opens to expose the heater core to the passenger's compartment through air ducts. A fan switch on the control head activates the blower motor, which forces air through the core, heating the air.

Air conditioning system

5   The air conditioning system consists of a condenser mounted in front of the radiator, an evaporator mounted adjacent to the heater core, a compressor mounted on the engine, a receiver-drier or accumulator and the plumbing connecting all of the above components.

6   A blower fan forces the warmer air of the passenger's compartment through the evaporator core (sort of a radiator-in-reverse), transferring the heat from the air to the refrigerant. The liquid refrigerant boils off into low pressure vapor, taking the heat with it when it leaves the evaporator.

2 Troubleshooting

Coolant leaks

1   A coolant leak can develop anywhere in the cooling system, but the most common causes are:

a) A loose or weak hose clamp
b) A defective hose
c) A faulty pressure cap
d) A damaged radiator
e) A bad heater core
f) A faulty water pump
g) A leaking gasket at any joint that carries coolant

2   Coolant leaks aren't always easy to find. Sometimes they can only be detected when the cooling system is under pressure. Here's where a cooling system pressure tester comes in handy. After the engine has cooled completely, the tester is attached in place of the pressure cap, then pumped up to the pressure value equal to that of the pressure cap rating (see illustration). Now, leaks that only exist when the engine is fully warmed up will become apparent. The tester can be left connected to locate a nagging slow leak.

Coolant level drops, but no external leaks

3   If you find it necessary to keep adding coolant, but there are no external leaks, the probable causes include:

a) A blown head gasket
b) A leaking intake manifold gasket (only on engines that have coolant passages in the manifold)
c) A cracked cylinder head or cylinder block

4   Any of the above problems will also usually result in contamination of the engine oil, which will cause it to take on a milkshake-like appearance. A bad head gasket or cracked head or block can also result in engine oil contaminating the cooling system.

5   Combustion leak detectors (also known as block testers) are available at most auto parts stores. These work by detecting exhaust gases in the cooling system, which indicates a compression leak from a cylinder into the coolant. The tester consists of a large bulb-type syringe and bottle of test fluid (see illustration). A measured amount of the fluid is added to the syringe. The syringe is placed over the cooling system filler neck and, with the engine running, the bulb is squeezed and a sample of the gases present in the cooling system are drawn up through the test fluid (see illustration). If any combustion gases are present in the sample taken, the test fluid will change color.

6   If the test indicates combustion gas is present in the cooling system, you can be sure that the engine has a blown head gasket or a crack in the cylinder head or block, and will require disassembly to repair.

2.2 The cooling system pressure tester is connected in place of the pressure cap, then pumped up to pressurize the system

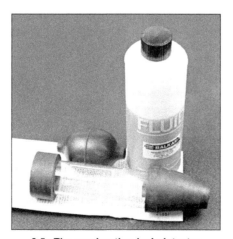

2.5a The combustion leak detector consists of a bulb, syringe and test fluid

2.5b Place the tester over the cooling system filler neck and use the bulb to draw a sample into the tester

Pressure cap

Warning: *Wait until the engine is completely cool before beginning this check.*

7   The cooling system is sealed by a spring-loaded cap, which raises the boiling point of the coolant. If the cap's seal or spring are worn out, the coolant can boil and escape past the cap. With the engine completely cool, remove the cap and check the seal; if it's cracked, hardened or deteriorated in any way, replace it with a new one.

8   Even if the seal is good, the spring might not be; this can be checked with a cooling system pressure tester (see illustration). If the cap can't hold a pressure within approximately 1-1/2 lbs of its rated pressure (which is marked on the cap), replace it with a new one.

9   The cap is also equipped with a vacuum relief spring. When the engine cools off, a vacuum is created in the cooling system. The vacuum relief spring allows air back into the system, which will equalize the pressure and prevent damage to the radiator (the radiator tanks could collapse if the vacuum is great enough). If, after turning the engine off and allowing it to cool down you notice any of the cooling system hoses collapsing, replace the pressure cap with a new one.

Thermostat

10   Before assuming the thermostat (see illustration) is responsible for a cooling system problem, check the coolant level (see Chapter 1), drivebelt tension (see Chapter 1) and temperature gauge (or light) operation.

11   If the engine takes a long time to warm up (as indicated by the temperature gauge or heater operation), the thermostat is probably stuck open. Replace the thermostat with a new one.

12   If the engine runs hot or overheats, a thorough test of the thermostat should be performed.

13   Definitive testing of the thermostat can only be made when it is removed from the vehicle. If the thermostat is stuck in the open position at room temperature, it is faulty and must be replaced.

Caution: *Do not drive the vehicle without a thermostat. The computer may stay in open loop and emissions and fuel economy will suffer.*

14   To test a thermostat, suspend the (closed) thermostat on a length of string or wire in a pot of cold water.

15   Heat the water on a stove while observing thermostat. The thermostat should fully open before the water boils.

16   If the thermostat doesn't open and close as specified, or sticks in any position, replace it.

Cooling fan

Electric cooling fan

17   If the engine is overheating and the cooling fan is not coming on when the engine temperature rises to an excessive level, unplug the fan motor electrical connector(s) and connect the motor directly to the battery with fused jumper wires. If the fan motor doesn't come on, replace the motor.

18   If the radiator fan motor is okay, but it isn't coming on when the engine gets hot, the fan relay might be defective. A relay is used to control a circuit by turning it on and off in response to a control decision by the Powertrain Control Module (PCM). These control circuits are fairly complex, and checking them should be left to a qualified automotive technician. Sometimes, the control system can be fixed by simply identifying and replacing a bad relay.

19   Locate the fan relays in the engine compartment fuse/relay box.

20   Test the relay (see Chapter 12).

21   If the relay is okay, check all wiring and connections to the fan motor. Refer to the wiring diagrams at the end of Chapter 12. If no obvious problems are found, the problem could be the Engine Coolant Temperature (ECT) sensor or the Powertrain Control Module (PCM). Have the cooling fan system and circuit diagnosed by a dealer service department or repair shop with the proper diagnostic equipment.

Belt-driven cooling fan

22   Disconnect the cable from the negative terminal of the battery and rock the fan back and forth by hand to check for excessive bearing play.

23   With the engine cold (and not running), turn the fan blades by hand. The fan should turn freely.

24   Visually inspect for substantial fluid leakage from the clutch assembly. If problems are noted, replace the clutch assembly.

25   With the engine completely warmed up, turn off the ignition switch and disconnect the negative battery cable from the battery. Turn the fan by hand. Some drag should be evident. If the fan turns easily, replace the fan clutch.

Water pump

26   A failure in the water pump can cause serious engine damage due to overheating.

Drivebelt-driven water pump

27   There are two ways to check the operation of the water pump while it's installed on the engine. If the pump is found to be defective, it should be replaced with a new or rebuilt unit.

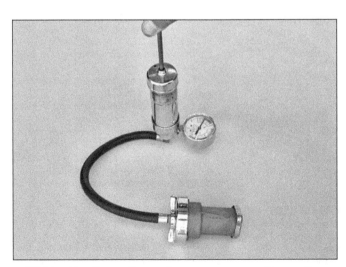

2.8 Checking the cooling system pressure cap with a cooling system pressure tester

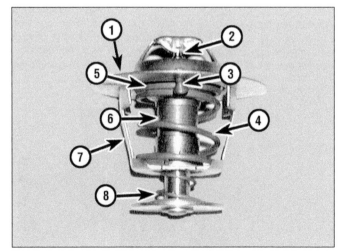

2.10 Typical thermostat:

1	Flange	5	Valve seat
2	Piston	6	Valve
3	Jiggle valve	7	Frame
4	Main coil spring	8	Secondary coil spring

28 Water pumps are equipped with weep (or vent) holes (see illustration). If a failure occurs in the pump seal, coolant will leak from the hole.

29 If the water pump shaft bearings fail, there may be a howling sound at the pump while it's running. Shaft wear can be felt with the drivebelt removed if the water pump pulley is rocked up and down (with the engine off). Don't mistake drivebelt slippage, which causes a squealing sound, for water pump bearing failure.

Timing chain or timing belt-driven water pump

30 Water pumps driven by the timing chain or timing belt are located underneath the timing chain or timing belt cover.

31 Checking the water pump is limited because of where it is located. However, some basic checks can be made before deciding to remove the water pump. If the pump is found to be defective, it should be replaced with a new or rebuilt unit.

32 One sign that the water pump may be failing is that the heater (climate control) may not work well. Warm the engine to normal operating temperature, confirm that the coolant level is correct, then run the heater and check for hot air coming from the ducts.

33 Check for noises coming from the water pump area. If the water pump impeller shaft or bearings are failing, there may be a howling sound at the pump while the engine is running.

Note: *Be careful not to mistake drivebelt noise (squealing) for water pump bearing or shaft failure.*

34 It you suspect water pump failure due to noise, wear can be confirmed by feeling for play at the pump shaft. This can be done by rocking the drive sprocket on the pump shaft up and down. To do this you will need to remove the tension on the timing chain or belt as well as access the water pump.

All water pumps

35 In rare cases or on high-mileage vehicles, another sign of water pump failure may be the presence of coolant in the engine oil. This condition will adversely affect the engine in varying degrees.

Note: *Finding coolant in the engine oil could indicate other serious issues besides a failed water pump, such as a blown head gasket or a cracked cylinder head or block.*

36 Even a pump that exhibits no outward signs of a problem, such as noise or leakage, can still be due for replacement. Removal for close examination is the only sure way to tell. Sometimes the fins on the back of the impeller can corrode to the point that cooling efficiency is diminished significantly.

Heater system

37 Little can go wrong with a heater. If the fan motor will run at all speeds, the electrical part of the system is okay. The three basic heater problems fall into the following general categories:

  a) *Not enough heat*
  b) *Heat all the time*
  c) *No heat*

38 If there's not enough heat, the control valve or door is stuck in a partially open position, the coolant coming from the engine isn't hot enough, or the heater core is restricted. If the coolant isn't hot enough, the thermostat in the engine cooling system is stuck open, allowing coolant to pass through the engine so rapidly that it doesn't heat up quickly enough. If the vehicle is equipped with a temperature gauge instead of a warning light, watch to see if the engine temperature rises to the normal operating range after driving for a reasonable distance.

39 If there's heat all the time, the control valve or the door is stuck wide open.

40 If there's no heat, coolant is probably not reaching the heater core, or the heater core is plugged. The likely cause is a collapsed or plugged hose, core, or a frozen heater control valve. If the heater is the type that flows coolant all the time, the cause is a stuck door or a broken or kinked control cable.

Air conditioning system

41 If the cool air output is inadequate: Inspect the condenser coils and fins to make sure they're clear.

  a) *Check the compressor clutch for slippage.*

  b) *Check the blower motor for proper operation.*
  c) *Inspect the blower discharge passage for obstructions.*
  d) *Check the system air intake filter for clogging.*

42 If the system provides intermittent cooling air:

  a) *Check the circuit breaker, blower switch and blower motor for a malfunction.*
  b) *Make sure the compressor clutch isn't slipping.*
  c) *Inspect the plenum door to make sure it's operating properly.*
  d) *Inspect the evaporator to make sure it isn't clogged.*
  e) *If the unit is icing up, it may be caused by excessive moisture in the system, incorrect super heat switch adjustment or low thermostat adjustment.*

43 If the system provides no cooling air:

  a) *Inspect the compressor drivebelt. Make sure it's not loose or broken.*
  b) *Make sure the compressor clutch engages. If it doesn't, check for a blown fuse.*
  c) *Inspect the wire harness for broken or disconnected wires.*
  d) *If the compressor clutch doesn't engage, bridge the terminals of the A/C pressure switch(es) with a jumper wire; if the clutch now engages, and the system is properly charged, the pressure switch is bad.*
  e) *Make sure the blower motor is not disconnected or burned out.*
  f) *Make sure the compressor isn't partially or completely seized.*
  g) *Inspect the refrigerant lines for leaks.*
  h) *Check the components for leaks.*
  i) *Inspect the receiver-drier/accumulator or expansion valve/tube for clogged screens.*

44 If the system is noisy:

  a) *Look for loose panels in the passenger's compartment.*
  b) *Inspect the compressor drivebelt. It may be loose or worn.*
  c) *Check the compressor mounting bolts. They should be tight.*
  d) *Listen carefully to the compressor. It may be worn out.*
  e) *Listen to the idler pulley and bearing and the clutch. Either may be defective.*
  f) *The winding in the compressor clutch coil or solenoid may be defective.*
  g) *The compressor oil level may be low.*
  h) *The blower motor fan bushing or the motor itself may be worn out.*
  i) *If there is an excessive charge in the system, you'll hear a rumbling noise in the high pressure line, a thumping noise in the compressor, or see bubbles or cloudiness in the sight glass.*
  j) *If there's a low charge in the system, you might hear hissing in the evaporator case at the expansion valve, or see bubbles or cloudiness in the sight glass.*

2.28 The water pump weep hole is generally located on the underside of the pump

3 Air conditioning and heating system - check and maintenance

Air conditioning system

Warning: *The air conditioning system is under high pressure. Do not loosen any hose fittings or remove any components until after the system has been discharged. Air conditioning refrigerant should be properly discharged into an EPA-approved recovery/recycling unit at a dealer service department or an automotive air conditioning repair facility. Always wear eye protection when disconnecting air conditioning system fittings.*

Caution: *All models covered by this manual use environmentally friendly R-134a. This refrigerant (and its appropriate refrigerant oils) are not compatible with R-12 refrigerant system components and must never be mixed or the components will be damaged.*

Caution: *When replacing entire components, additional refrigerant oil should be added equal to the amount that is removed with the component being replaced. Be sure to read the can before adding any oil to the system, to make sure it is compatible with the R-134a system.*

1 The following maintenance checks should be performed on a regular basis to ensure that the air conditioning continues to operate at peak efficiency.

 a) *Inspect the condition of the compressor drivebelt. If it is worn or deteriorated, replace it (see Chapter 1).*
 b) *Check the drivebelt tension (see Chapter 1).*
 c) *Inspect the system hoses. Look for cracks, bubbles, hardening and deterioration. Inspect the hoses and all fittings for oil bubbles or seepage. If there is any evidence of wear, damage or leakage, replace the hose(s).*

 d) *Inspect the condenser fins for leaves, bugs and any other foreign material that may have embedded itself in the fins. Use a fin comb or compressed air to remove debris from the condenser.*
 e) *Make sure the system has the correct refrigerant charge.*
 f) *If you hear water sloshing around in the dash area or have water dripping on the carpet, check the evaporator housing drain tube (see illustration) and insert a piece of wire into the opening to check for blockage.*

2 It's a good idea to operate the system for about ten minutes at least once a month. This is particularly important during the winter months because long term non-use can cause hardening, and subsequent failure, of the seals. Note that using the Defrost function operates the compressor.

3 If the air conditioning system is not working properly, proceed to Step 6 and perform the general checks outlined below.

4 Because of the complexity of the air conditioning system and the special equipment necessary to service it, in-depth troubleshooting and repairs beyond checking the refrigerant charge and the compressor clutch operation are not included in this manual. However, simple checks and component replacement procedures are provided in this Chapter. For more complete information on the air conditioning system, refer to the *Haynes Automotive Heating and Air Conditioning Manual*.

5 The most common cause of poor cooling is simply a low system refrigerant charge. If a noticeable drop in system cooling ability occurs, one of the following quick checks will help you determine if the refrigerant level is low.

Checking the refrigerant charge

6 Warm the engine up to normal operating temperature.

7 Place the air conditioning temperature selector at the coldest setting and put the blower at the highest setting.

8 After the system reaches operating temperature, feel the larger pipe exiting the evaporator at the firewall. The outlet pipe should be cold (the tubing that leads back to the compressor). If the evaporator outlet pipe is warm, the system probably needs a charge.

9 Insert a thermometer in the center air distribution duct (see illustration) while operating the air conditioning system at its maximum setting - the temperature of the output air should be 35 to 40 degrees F below the ambient air temperature (down to approximately 40 degrees F). If the ambient (outside) air temperature is very high, say 110 degrees F, the duct air temperature may be as high as 60 degrees F, but generally the air conditioning is 30 to 40 degrees F cooler than the ambient air.

10 Further inspection or testing of the system requires special tools and techniques and is beyond the scope of the home mechanic.

Adding refrigerant

Caution: *Make sure any refrigerant, refrigerant oil or replacement component you purchase is designated as compatible with R-134a systems.*

11 Purchase an R-134a automotive charging kit at an auto parts store (see illustration). A charging kit includes a can of refrigerant, a tap valve and a short section of hose that can be attached between the tap valve and the system low side service valve.

Caution: *Never add more than one can of refrigerant to the system. If more refrigerant than that is required, the system should be evacuated and leak tested.*

12 Back off the valve handle on the charging kit and screw the kit onto the refrigerant can, making sure first that the O-ring or rubber seal inside the threaded portion of the kit

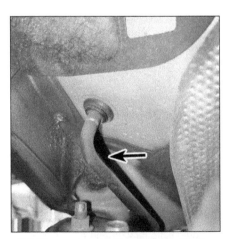

3.1 The evaporator drain hose exits the floorpan under the passenger's footwell

3.9 Insert a thermometer in the center vent, turn on the air conditioning system and wait for it to cool down; depending on the humidity, the output air should be 30 to 40 degrees cooler than the ambient air temperature

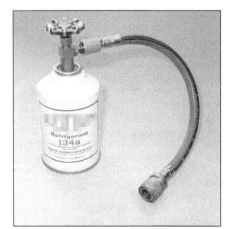

3.11 R-134a automotive air conditioning charging kit

3.13 Location of the low-side charging port

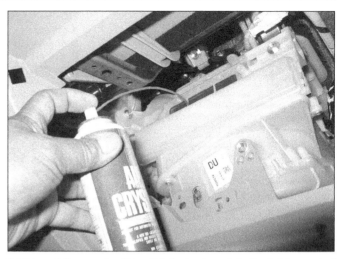

3.24 Insert the nozzle of the disinfectant can into the intake port of the HVAC housing

is in place.

Warning: *Wear protective eyewear when dealing with pressurized refrigerant cans.*

13　Remove the dust cap from the low-side charging port and attach the hose's quick-connect fitting to the port (see illustration). The fittings on the charging kit are designed to fit only on the low side of the system.

Warning: *DO NOT hook the charging kit hose to the system high side!*

14　Warm up the engine and turn On the air conditioning. Keep the charging kit hose away from the fan and other moving parts.

Note: *The charging process requires the compressor to be running. If the clutch cycles off, you can put the air conditioning switch on High and leave the car doors open to keep the clutch on and compressor working. The compressor can be kept on during the charging by removing the connector from the pressure switch and bridging it with a paper clip or jumper wire during the procedure.*

15　Turn the valve handle on the kit until the stem pierces the can, then back the handle out to release the refrigerant. You should be able to hear the rush of gas. Keep the can upright at all times, but shake it occasionally. Allow stabilization time between each addition.

Note: *The charging process will go faster if you wrap the can with a hot-water-soaked rag to keep the can from freezing up.*

16　If you have an accurate thermometer, you can place it in the center air conditioning duct inside the vehicle and keep track of the output air temperature. A charged system that is working properly should cool down to approximately 40 degrees F. If the ambient (outside) air temperature is very high, say 110 degrees F, the duct air temperature may be as high as 60 degrees F, but generally the air conditioning is 35 to 40 degrees F cooler than the ambient air.

17　When the can is empty, turn the valve handle to the closed position and release the connection from the low-side port. Reinstall the dust cap.

18　Remove the charging kit from the can and store the kit for future use with the piercing valve in the UP position, to prevent inadvertently piercing the can on the next use.

Heating systems

19　If the carpet under the heater core is damp, or if antifreeze vapor or steam is coming through the vents, the heater core is leaking. Remove it (see Section 11) and install a new unit (most radiator shops will not repair a leaking heater core).

20　If the air coming out of the heater vents isn't hot, the problem could stem from any of the following causes:

a) *The thermostat is stuck open, preventing the engine coolant from warming up enough to carry heat to the heater core. Replace the thermostat (see Section 4).*

b) *There is a blockage in the system, preventing the flow of coolant through the heater core. Feel both heater hoses at the firewall. They should be hot. If one of them is cold, there is an obstruction in one of the hoses or in the heater core, or the heater control valve is shut. Detach the hoses and back flush the heater core with a water hose. If the heater core is clear but circulation is impeded, remove the two hoses and flush them out with a water hose.*

c) *If flushing fails to remove the blockage from the heater core, the core must be replaced (see Section 11).*

Eliminating air conditioning odors

21　Unpleasant odors that often develop in air conditioning systems are caused by the growth of a fungus, usually on the surface of the evaporator core. The warm, humid environment there is a perfect breeding ground for mildew to develop.

22　The evaporator core on most vehicles is difficult to access, and factory dealerships have a lengthy, expensive process for eliminating the fungus by opening up the evaporator case and using a powerful disinfectant and rinse on the core until the fungus is gone. You can service your own system at home, but it takes something much stronger than basic household germ-killers or deodorizers.

23　Aerosol disinfectants for automotive air conditioning systems are available in most auto parts stores, but remember when shopping for them that the most effective treatments are also the most expensive. The basic procedure for using these sprays is to start by running the system in the RECIRC mode for ten minutes with the blower on its highest speed. Use the highest heat mode to dry out the system and keep the compressor from engaging by disconnecting the wiring connector at the compressor.

24　The disinfectant can usually comes with a long spray hose. Insert the nozzle into an intake port inside the cabin, and spray according to the manufacturer's recommendations (see illustration).

25　Once the evaporator has been cleaned, the best way to prevent the mildew from coming back again is to make sure your evaporator housing drain tube is clear (see illustration 3.1).

Automatic heating and air conditioning systems

26　Some vehicles are equipped with an optional automatic climate control system. This system has its own computer that receives inputs from various sensors in the heating and air conditioning system. This computer, like the PCM, has self-diagnostic capabilities to help pinpoint problems or faults within the system. Vehicles equipped with automatic heating and air conditioning systems are very complex and considered beyond the scope of the home mechanic. Vehicles equipped with automatic heating and air conditioning systems should be taken to a dealer service department or other qualified facility for repair.

4.3 With the alternator removed, remove the nuts securing the thermostat housing to the side of the engine block

4.5 The thermostat gasket fits around the edge of the thermostat like a grooved sealing ring

4.6 Position the jiggle valve straight up

4   Thermostat - replacement

Warning: *Do not start this procedure until the engine is completely cool.*

1   Disconnect the cable from the negative battery terminal (see Chapter 5).
2   Drain the cooling system (see Chapter 1). Disconnect the radiator hose from the thermostat housing or water inlet pipe.
3   Remove the alternator (see Chapter 5), then unbolt the thermostat housing from the engine block (see illustration).
4   Remove the thermostat, noting the direction in which it was installed in the housing, and thoroughly clean the sealing surfaces.
5   Fit a new rubber gasket onto the thermostat (see illustration). Make sure it is evenly installed all the way around.
6   Install the thermostat into the engine, then install the thermostat housing. Be sure to position the jiggle valve upward at the highest point (see illustration).
7   Tighten the housing fasteners to the

torque listed in this Chapter's Specifications and reinstall the remaining components in the reverse order of removal.
8   Refill the cooling system (see Chapter 1). Run the engine and check for leaks and proper operation.

5   Engine cooling fans - replacement

Warning: *To avoid possible injury, keep clear of the fan blades, as they may start turning whenever the battery is connected!*

1   Disconnect the cable from the negative battery terminal (see Chapter 5).
2   Disconnect the electrical connector at the fan motors and free the harness from the clips on the fan shroud (see illustration).
3   Drain the cooling system (see Chapter 1).
4   Remove the radiator and fan shroud as an assembly (see Section 6).
5   Hold the fan blades and remove the fan

retaining fastener(s) (see illustration).
6   Unbolt the fan motor(s) from the shroud (see illustration).
7   Installation is the reverse of removal.

6   Radiator and coolant expansion tank - removal and installation

Warning: *Do not start this procedure until the engine is completely cool.*
Warning: *Do not allow antifreeze to come in contact with your skin or painted surfaces of the vehicle. Rinse off spills immediately with plenty of water. Antifreeze is highly toxic if ingested. Never leave antifreeze lying around in an open container or in puddles on the floor; children and pets are attracted by its sweet smell and may drink it. Check with local authorities about disposing of used antifreeze. Many communities have collection centers which will see that antifreeze is disposed of safely. Never dump used antifreeze on the ground or into drains.*

5.2 Fan motor electrical connectors

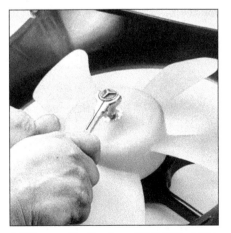

5.5 Remove the fan from the motor - some fan blades are retained by a single nut, others by three screws

5.6 Remove the screws and separate the motor from the shroud

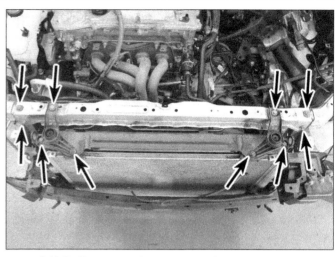

6.9 Unclip the cooling fan wiring harness and disconnect the electrical connectors to the cooling fans

6.10 Radiator mounting bracket and radiator support fastener locations

Warning: *The air conditioning system is under high pressure. Do not loosen any hose fittings or remove any components until the system has been discharged. Air conditioning refrigerant should be properly discharged into an EPA-approved recovery/recycling unit by a dealer service department or an automotive air conditioning repair facility. Always wear eye protection when disconnecting air conditioning system fittings.*

Note: *Non-toxic coolant is available at local auto parts stores. Although the coolant is non-toxic when fresh, proper disposal of used coolant is still required.*

Radiator

1   Have the refrigerant discharged by an automotive air conditioning technician.
2   Disconnect the cable from the negative battery terminal, then remove the battery hold-down clamp (see Chapter 5).
3   Remove the front bumper cover (see Chapter 11) and the headlight housings (see Chapter 12).
4   Drain the cooling system (see Chapter 1).
5   Remove the hood latch, then detach the

hood release cable from the clips along the radiator support (see Chapter 11).
6   Disengage the radiator air deflector clips and remove the air deflector from the top of the radiator.
7   Disconnect the electrical connectors to the horns (see Chapter 12) then remove the front bumper upper support reinforcement bolts and remove the support along with the horns as an assembly.
8   Disconnect the inlet and outlet lines to the condenser (see Section 14). Cap the open fittings immediately to keep moisture and dirt out of the system.
9   Unclip the cooling fan wiring harness along the fan housing, then disconnect the electrical connectors from the cooling fans (see illustration).
10   Remove the bolts from the radiator mounting brackets and remove the brackets (see illustration).
11   Remove the four upper radiator support mounting bolts and remove the upper radiator support.
12   Depress the locking tabs and disconnect the by-pass hoses from the radiator

hose assembly.
13   Squeeze the clamps and slide them back on the hoses, then disconnect the upper and lower radiator hoses from the radiator.
14   Disconnect the expansion tank hoses from the radiator (see illustration).
15   Lift out the radiator and fan and condenser assembly. Be aware of dripping fluids and the sharp fins.
16   Remove the oil cooler tube clips and bolts from the fan shroud.
17   Remove the fan shroud support mounting bolts and separate the support from the top of the fan shroud.
18   Remove the air conditioning condenser, then remove the radiator from the shroud.
19   With the radiator removed, it can be inspected for leaks, damage and internal blockage.
20   If in need of repairs, have a professional radiator shop or dealer service department perform the work, as special techniques are required.
21   Bugs and dirt can be cleaned from the radiator with compressed air and a soft brush. Don't bend the cooling fins as this is done.
Warning: *Wear eye protection.*
22   Installation is the reverse of removal. Be sure the rubber mounts are in place.
23   After installation, fill the cooling system with the proper mixture of antifreeze and water (see Chapter 1).
24   Have the air conditioning system evacuated, charged and leak tested by the shop that discharged it. If the condenser or receiver/drier was replaced, have them add the proper amount of refrigeration oil to the compressor. Use only compressor oil compatible with R-134a refrigerant.

Coolant reservoir/expansion tank

25   Drain the cooling system until the level has dropped below the expansion tank (see Chapter 1), then disconnect the hoses from

6.14 Disconnect the reservoir hoses from the radiator

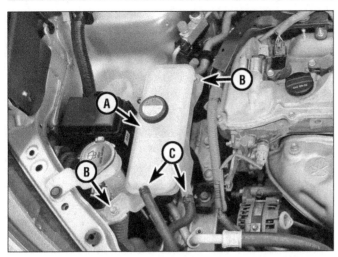

6.25 Coolant expansion tank (A), mounting fasteners (B) and hoses (C)

7.7 Water pump mounting fasteners

the tank and remove the tank mounting fasteners (see illustration).

26   Installation is the reverse of removal. Refill the cooling system (see Chapter 1).

7   Water pump - replacement

Warning: *Do not start this procedure until the engine is completely cool.*

Warning: *Do not allow antifreeze to come in contact with your skin or painted surfaces of the vehicle. Rinse off spills immediately with plenty of water. Antifreeze is highly toxic if ingested. Never leave antifreeze lying around in an open container or in puddles on the floor; children and pets are attracted by its sweet smell and may drink it. Check with local authorities about disposing of used antifreeze. Many communities have collection centers which will see that antifreeze is disposed of safely. Never dump used antifreeze on the ground or into drains.*

Note: *Non-toxic coolant is available at local auto parts stores. Although the coolant is non-toxic when fresh, proper disposal of used coolant is still required.*

1   Disconnect the cable from the negative battery terminal (see Chapter 5) and drain the cooling system (see Chapter 1).

2   Raise the vehicle and support it securely on jackstands. Remove the engine cover by pulling it straight up. Also remove the splash shields from under the engine and the right side inner fender splash shield.

3   Drain the engine coolant (see Chapter 1).

4   Remove the drivebelt and tensioner (see Chapter 1).

5   Remove the alternator (see Chapter 5).

6   Remove the wiring harness from the clamp on the water pump and harness bracket.

7   Remove the water pump mounting fasteners and detach the water pump (see illustration).

Caution: *If the water pump sticks to the housing, use a soft-face hammer to tap it free. Don't pry between the sealing surfaces.*

8   Thoroughly clean all sealing surfaces, removing all traces of old gasket.

9   Installation is the reverse of removal. Be sure to use a new gasket and tighten the fasteners to the torque listed in this Chapter's Specifications.

10   Refill the cooling system (see Chapter 1), run the engine and check for leaks and proper operation.

8   Coolant temperature sending - replacement

1   The coolant temperature gauge sending unit is an integral part of the Engine Coolant Temperature (ECT) sensor. Refer to Chapter 6 for the engine coolant temperature sensor replacement procedure.

9   Blower motor - removal and installation

Warning: *The models covered by this manual are equipped with a Supplemental Restraint System (SRS), more commonly known as airbags. Always disable the airbag system before working in the vicinity of the impact sensors, steering column or instrument panel to avoid the possibility of accidental deployment of the airbag, which could cause personal injury (see Chapter 12).*

1   Disconnect the cable from the negative battery terminal (see Chapter 5).

Note: *The blower unit is located in the passenger's compartment above the right front footwell.*

2   Remove the right side lower instrument panel fasteners, then remove the panel (see Chapter 11).

3   Disconnect the electrical connector(s) from the blower motor (see illustration).

4   Remove the screws and lower the unit from the housing.

5   If the motor is being replaced, transfer the fan to the new motor prior to installation.

6   Installation is the reverse of removal.

10   Heater and air conditioning control assembly - removal, installation and cable adjustment

Warning: *The models covered by this manual are equipped with a Supplemental Restraint System (SRS), more commonly known as airbags. Always disable the airbag system before working in the vicinity of any airbag system components to avoid the possibility of accidental deployment of the airbag, which could cause personal injury (see Chapter 12).*

9.3 The blower motor is located under the right end of the instrument panel

A   *Electrical connector*
B   *Mounting screws*

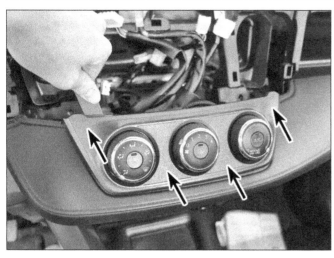

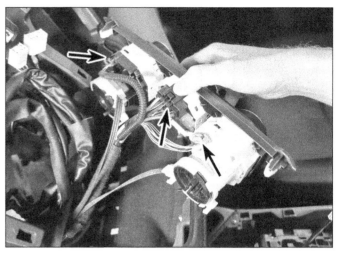

10.5 Heater/air conditioning control assembly - arrows indicate the area of two plastic retaining clips, be careful not to break them

10.6a Disconnect the electrical connectors from the control assembly . . .

Removal and installation

1   Disconnect the cable from the negative battery terminal (see Chapter 5).

2   Remove the center instrument trim panel (see Chapter 11).

3   Remove the center trim (see Chapter 11), then disengace the clips for the center vent registers on each side of the radio and remove the vents.

4   Remove the radio (see Chapter 12).

5   Using a trim tool, disengage the control assembly retaining clips and carefully pry the assembly out of the instrument panel (see illustration).

6   Pull the control assembly out of the instrument panel and disconnect the electrical connectors (see illustrations).

7   Installation is the reverse of removal.

---

11   Heater core - replacement

Warning: *The models covered by this manual*

10.6b . . . then squeeze the connector to the control nobs and disconnect the electrical connectors from the knobs

are equipped with a Supplemental Restraint System (SRS), more commonly known as airbags. Always disable the airbag system before working in the vicinity of any airbag system components to avoid the possibility of accidental deployment of the airbag, which could cause personal injury (see Chapter 12).

Warning: *Do not allow antifreeze to come in contact with your skin or painted surfaces of the vehicle. Rinse off spills immediately with plenty of water. Antifreeze is highly toxic if ingested. Never leave antifreeze lying around in an open container or in puddles on the floor; children and pets are attracted by its sweet smell and may drink it. Check with local authorities about disposing of used antifreeze. Many communities have collection centers which will see that antifreeze is disposed of safely. Never dump used antifreeze on the ground or into drains.*

Warning: *The air conditioning system is under high pressure. DO NOT loosen any fittings or remove any components until after the system has been discharged. Air conditioning refrig-*

11.4 Squeeze the clamps and disconnect the heater hoses at the firewall

erant should be properly discharged into an EPA-approved container at a dealer service department or an automotive air conditioning repair facility. Always wear eye protection when disconnecting air conditioning system fittings.

Warning: *Wait until the engine is completely cool before beginning this procedure.*

Note: *Replacement of the heater core is a difficult procedure for the home mechanic, involving removal of the entire dashboard, floor console and many wiring connectors. If you attempt this procedure at home, keep track of the assemblies by taking notes and keeping screws and other hardware in small, marked plastic bags for reassembly.*

1   If the vehicle is equipped with air conditioning, have the air conditioning system discharged at a dealer service department or service station.

2   Turn the heater control setting to HOT. Drain the cooling system (see Chapter 1). If the coolant is relatively new, or is in good condition, save it and re-use it.

3   Disconnect the cable from the negative battery terminal (see Chapter 5).

4   Working in the engine compartment, disconnect the heater hoses from the heater core tubes at the firewall (see illustration).

5   Remove the rubber grommets where the heater core tubes go through the firewall.

6   Remove the air conditioning refrigerant lines manifold bolt to the expansion valve (see illustration), then remove the expansion valve mounting bolt and remove the valve from the evaporator core. Cap the open fittings immediately to keep moisture and dirt out of the system.

Warning: *Always wear eye protection when disconnecting air conditioning system fittings and always cap the fitting ends to prevent moisture from entering the refrigerant lines.*

7   Remove the steering column (see Chapter 10).

8   Remove the entire instrument panel (see Chapter 11).

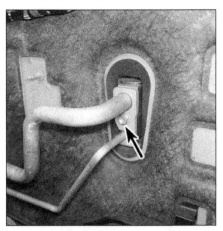

11.6 Air conditioning refrigerant lines manifold bolt-to-expansion valve location

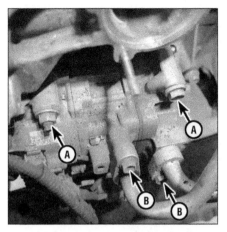

12.4 Compressor mounting details

A   Upper mounting fasteners
B   Refrigerant line retaining bolts

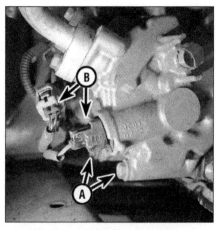

12.5 Air conditioning compressor lower mounting bolts (A) and electrical connectors (B) locations

9   Disconnect the electrical connectors from the blower motor and the blower resistor, then remove the blower unit.

10   Remove the rear floor duct, then remove the upper air duct mounting nuts and remove the duct.

11   Disengage the defroster duct clips and remove the defroster duct assembly.

12   Remove the vertical instrument panel brace fasteners and remove the two braces.

13   Remove the instrument panel reinforcement beam (see Chapter 11).

14   Remove the remaining air conditioning unit fasteners and remove the assembly from the vehicle.

15   Disconnect the wiring harness from the clips along the blower motor housing, then remove the blower motor housing-to-air conditioning unit fasteners and separate the two housings.

16   Remove the quick heater fasteners, then disconnect the electrical connectors and remove the quick heater from the air conditioning unit housing.

17   Detach the retaining clamp screw and remove the clamp, then remove the heater core from the air conditioning unit housing.

Note: The heater core is located on the passenger's side of the A/C unit.

18   Reinstall the remaining parts in the reverse order of removal.

19   Refill the cooling system (see Chapter 1), reconnect the battery and run the engine. Check for leaks and proper operation of the system. Have the air conditioning system recharged if equipped.

---

12   Air conditioning compressor - removal and installation

Warning: The air conditioning system is under high pressure. Do not loosen any hose fittings or remove any components until the system

has been discharged. Air conditioning refrigerant should be properly discharged into an EPA-approved recovery/recycling unit by a dealer service department or an automotive air conditioning repair facility. Always wear eye protection when disconnecting air conditioning system fittings.

Caution: The receiver/drier should be replaced whenever the air conditioning compressor is replaced (see Section 13).

1   Have the refrigerant discharged by an automotive air conditioning technician.

2   Disconnect the cable from the negative battery terminal (see Chapter 5). Loosen the right front wheel lug nuts. Raise the front of the vehicle and support it securely on jackstands, then remove the lower splash shield(s) and the wheel.

3   Remove the drivebelt (see Chapter 1).

4   Remove the upper mounting fasteners and detach the refrigerant lines (see illustration).

5   Detach the wiring connector, remove the compressor lower mounting fasteners (see illustration) and lower it from the vehicle.

6   If a new or rebuilt compressor is being installed, the oil level inside must be adjusted:

a) Remove the drain bolt from the old compressor, pour the oil into a graduated container and record the amount.

b) Subtract the measured amount from 70 cc (2.37 oz.), which is the amount that is present in the new compressor.

c) Remove the drain bolt from the new compressor and drain off the amount calculated in the previous step.

d) Reinstall the drain bolt and tighten it securely

7   Installation is the reverse of removal. Replace any O-rings with new ones specifically made for the purpose and lubricate them with refrigerant oil.

8   Have the system evacuated, recharged and leak tested by the shop that discharged it.

13.4 The receiver/drier is located in a tube on the side of the condenser

---

13   Air conditioning receiver/drier - removal and installation

---

Warning: The air conditioning system is under high pressure. Do not loosen any hose fittings or remove any components until the system has been discharged. Air conditioning refrigerant should be properly discharged into an EPA-approved recovery/recycling unit by a dealer service department or an automotive air conditioning repair facility. Always wear eye protection when disconnecting air conditioning system fittings.

1   Have the refrigerant discharged by an air conditioning technician.

2   Disconnect the cable from the negative battery terminal (see Chapter 5).

3   Remove the bumper cover (see Chapter 11).

4   Use a hex bit or an Allen wrench to unscrew the end plug (see illustration) and remove the receiver/drier from the condenser.

14.8 Condenser inlet and outlet line bolt locations

5   Installation is the reverse of removal. Be sure to tighten the end plug to the torque listed in this Chapter's Specifications.

6   Have the system evacuated, charged and leak tested by the shop that discharged it. If the receiver/drier was replaced, have them add the proper amount of refrigeration oil to the compressor. Use only compressor oil compatible with R-134a refrigerant.

14   Air conditioning condenser - removal and installation

Warning: *The air conditioning system is under high pressure. Do not loosen any hose fittings or remove any components until the system has been discharged. Air conditioning refrigerant should be properly discharged into an EPA-approved recovery/recycling unit by a dealer service department or an automotive air conditioning repair facility. Always wear eye protection when disconnecting air conditioning system fittings.*

Warning: *Wait until the engine is completely cool before beginning this procedure.*

Caution: *The receiver/drier should be replaced whenever the air conditioning condenser is replaced (see Section 13).*

1   Have the refrigerant discharged by an automotive air conditioning technician.

2   Disconnect the cable from the negative battery terminal, then remove the battery hold-down clamp (see Chapter 5).

3   Remove the front bumper cover (see Chapter 11) and the headlight housings (see Chapter 12).

4   Drain the cooling system (see Chapter 1).

5   Remove the hood latch, then detach the hood release cable from the clips along the radiator support (see Chapter 11).

6   Disengage the radiator air deflector clips and remove the air deflector from the top of the radiator.

7   Disconnect the electrical connectors to the horns (see Chapter 12), then remove the front bumper upper support reinforcement bolts and remove the support along with the horns as an assembly.

8   Disconnect the inlet and outlet line bolts to the condenser (see illustration). Cap the open fittings immediately to keep moisture and dirt out of the system.

9   Detach the water by-pass pipe from the fan shroud.

10   Remove the fan shroud support mounting bolts and separate the support from the top of the fan shroud.

11   Carefully lift the air conditioning condenser up and out making sure not to damage the cooling fins on the condenser or the radiator as the condenser is lifted out.

12   Install the condenser to the fan shroud, making sure the rubber cushions fit on the mounting points properly.

13   Reconnect the refrigerant lines, using new O-rings where needed.

14   If a new condenser is installed, add 1.35 ounces (40 cc) of refrigerant oil to the condenser before installing it.

15   Reinstall the remaining parts in the reverse order of removal. Have the system evacuated, charged and leak tested by the shop that discharged it.

Notes

Notes

Chapter 4
Fuel and exhaust systems

Contents

Specifications

Fuel system

Fuel system pressure (idle)	44 to 50 psi (304 to 345 kPa)
Fuel system hold pressure (after five minutes)	21 psi minimum (145 kPa)
Injector resistance (approximate)	11.6 to 12.4 ohms

Torque specifications

Note: *One foot-pound (ft-lb) of torque is equivalent to 12 inch-pounds (in-lbs) of torque. Torque values below approximately 15 foot-pounds are expressed in inch-pounds, because most foot-pound torque wrenches are not accurate at these smaller values.*

	Ft-lbs (unless otherwise indicated)	Nm
Fuel rail mounting bolts	15	21
Throttle body mounting bolts	84 in-lbs	10
Fuel pulsation damper bolts	84 in-lbs	10
Fuel tank strap bolts	33	45
Fuel pump module bolts	53 in-lbs	6

1   General information

Fuel system warnings

Warning: *Gasoline is extremely flammable and repairing fuel system components can be dangerous. Consider your automotive repair knowledge and experience before attempting repairs which may be better suited for a professional mechanic.*

a) *Don't smoke or allow open flames or bare light bulbs near the work area*
b) *Don't work in a garage with a gas-type appliance (water heater, clothes dryer)*
c) *Use fuel-resistant gloves. If any fuel spills on your skin, wash it off immediately with soap and water*
d) *Clean up spills immediately*
e) *Do not store fuel-soaked rags where they could ignite*
f) *Prior to disconnecting any fuel line, you must relieve the fuel pressure (see Section 3)*
g) *Wear safety glasses*
h) *Have a proper fire extinguisher on hand*

Fuel system

1   The fuel system consists of the fuel tank, electric fuel pump/fuel level sending unit (located in the fuel tank), fuel rail and fuel injectors. The fuel injection system is a multi-port system; multi-port fuel injection uses timed impulses to inject the fuel directly into the intake port of each cylinder. The Powertrain Control Module (PCM) controls the injectors. The PCM monitors various engine parameters and delivers the exact amount of fuel required into the intake ports.

2   Fuel is circulated from the fuel pump to the fuel rail through fuel lines running along the underside of the vehicle. Various sections of the fuel line are either rigid metal or nylon, or flexible fuel hose. The various sections of the fuel hose are connected either by quick-connect fittings or threaded metal fittings.

3   These models are NOT equipped with a separately serviceable fuel filter. The fuel filter is located within the fuel pump module.

Exhaust system

4   The exhaust system consists of the exhaust manifold(s), catalytic converter(s), muffler(s), tailpipe and all connecting pipes, flanges and clamps. The catalytic converters are an emission control device added to the exhaust system to reduce pollutants.

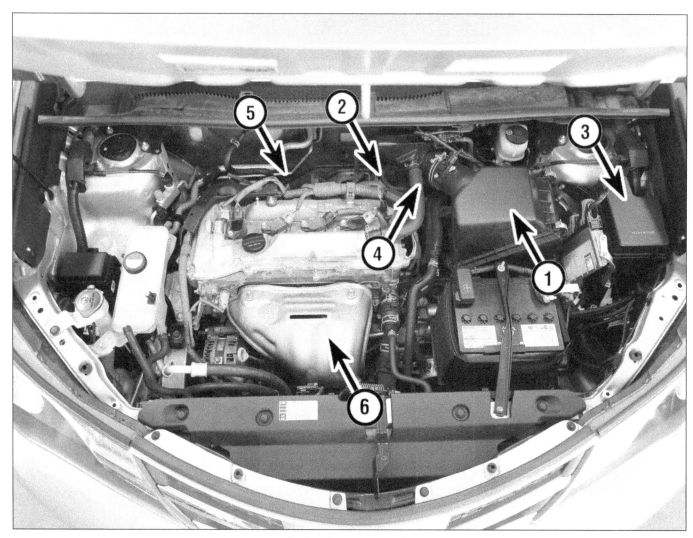

Fuel system components

1   Air filter housing
2   Throttle body
3   No. 1 relay and fuse box

4   Air intake duct
5   Fuel injectors (mounted on the back side of the cylinder head)

6   Exhaust manifold

2 Troubleshooting

Fuel pump

1 The fuel pump is located inside the fuel tank. Sit inside the vehicle with the windows closed, turn the ignition key to On (not Start) and listen for the sound of the fuel pump as it's briefly activated. You will only hear the sound for a second or two, but that sound tells you that the pump is working. Alternatively, have an assistant listen at the fuel filler cap.

2 If the pump does not come on, check the fuel pump-related fuses and relays (see illustrations). If the fuses and relays are okay, check the wiring back to the fuel pump. If the fuses, relays and wiring are okay, the fuel pump is probably defective. If the pump runs continuously with the ignition key in the On position, the Powertrain Control Module (PCM) is probably defective. Have the PCM checked by a professional mechanic.

Fuel injection system

Note: *The following procedure is based on the assumption that the fuel pump is working and the fuel pressure is adequate (see Section 4).*

3 Check all electrical connectors that are related to the system. Check the ground wire connections for tightness.

4 Verify that the battery is fully charged (see Chapter 5).

5 Inspect the air filter element (see Chapter 1).

6 Check all fuses related to the fuel system (see Chapter 12).

7 Check the air induction system between the throttle body and the intake manifold for air leaks. Also inspect the condition of all vacuum hoses connected to the intake manifold and to the throttle body.

8 Remove the air intake duct from the throttle body and look for dirt, carbon, varnish, or other residue in the throttle body, particularly around the throttle plate. If it's dirty, clean it with carb cleaner, a toothbrush and a clean shop towel.

9 With the engine running, place an automotive stethoscope against each injector, one at a time, and listen for a clicking sound that indicates operation.

Warning: *Stay clear of the drivebelt and any rotating or hot components.*

10 If you can hear the injectors operating, but the engine is misfiring, the electrical circuits are functioning correctly, but the injectors might be dirty or clogged. Try a commercial injector cleaning product (available at auto parts stores). If cleaning the injectors doesn't

help, replace the injectors.

11 If an injector is not operating (it makes no sound), disconnect the injector electrical connector and measure the resistance across the injector terminals with an ohmmeter. Compare this measurement to the other injectors. If the resistance of the non-operational injector is quite different from the other injectors, replace it.

12 If the injector is not operating, but the resistance reading is within the range of resistance of the other injectors, the PCM or the circuit between the PCM and the injector might be faulty.

3 Fuel pressure relief procedure

Warning: *Gasoline is extremely flammable. See* Fuel system warnings *in Section 1.*

1 Remove the fuel filler cap - this will relieve any pressure built up in the tank.

2 Remove the center console box (see Chapter 11, Section 22).

3 Locate the fuel pump/sending unit electrical connector and unplug it (the harness will exit through a grommet in the vehicle's floor) (see illustration).

4 Start the engine and allow it to run until it stops (it should only run for a few seconds, or it might not even start). Turn the ignition switch Off and disconnect the cable from the negative terminal of the battery before working on the fuel system (see Chapter 5).

5 The fuel system pressure is now relieved. Place a shop rag around any fitting to be disconnected to catch the residual fuel as it bleeds off. Dispose of the fuel soaked rag in an approved safety container.

6 When you're finished working on the fuel system, reconnect the fuel pump/sending unit harness connector and connect the negative cable to the battery.

7 Performing this procedure may cause a Diagnostic Trouble Code to set (fuel system too lean). If this happens, refer to Chapter 6 and erase the trouble code.

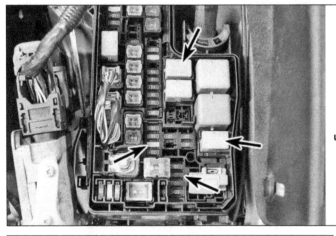

2.2a Fuel pump related fuses and relays in the underhood fuse block (models up to 9/15)

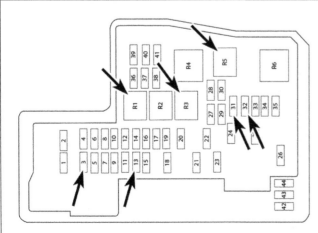

2.2b Fuel pump related fuses and relays in the underhood fuse block (models from 10/15)

3.3 Disconnect the fuel pump electrical connector

4   Fuel pressure - check

Warning: *Gasoline is extremely flammable. See* Fuel system warnings *in Section 1.*
Note: *The following procedure assumes that the fuel pump is receiving voltage and runs.*

General checks

1   If you suspect insufficient fuel delivery check the following items first:

a) *Check the battery and make sure it's fully charged (see Chapter 5).*
b) *Check the fuel pump fuse.*
c) *Inspect all fuel lines to ensure that the problem is not simply a leak in a line.*

2   Verify the fuel pump actually runs. Have an assistant turn the ignition switch to On - you should hear a brief whirring noise (for approximately two seconds) as the pump comes on and pressurizes the system.
Note: *The fuel pump is easily heard through the fuel tank filler neck. If there is no response from the fuel pump (makes no sound), check the fuel pump electrical circuit. If the fuel pump runs, but a fuel system problem is suspected, continue with the fuel pump pressure check.*

Fuel pump pressure check

Note: *In order to perform the fuel pressure test, you will need to obtain a fuel pressure gauge capable of measuring high fuel pressure and the proper adapter set for the specific fuel injection system.*
Note: *The fuel filter is mounted in the fuel pump assembly.*
3   Relieve the fuel system pressure (see Section 2).
4   Disconnect the fuel tube connector at the fuel rail and connect the fuel pressure gauge using the proper adapters (see illustrations).
5   Turn all the accessories Off and switch the ignition key On. The fuel pump should run for about two seconds; note the reading on the gauge. If the fuel pressure is higher than

specified, replace the fuel pressure regulator. If the pressure is too low, the fuel filter (or in-tank strainer) could be clogged, the lines could be restricted or leaking, a fuel injector could be leaking, or the fuel pressure regulator and/or the fuel pump could be defective.
6   Start the engine and let it idle at normal operating temperature. The pressure should fall within the range listed in this Chapter's Specifications. If the pressure is lower than specified, check the items listed above.
Note: *If no obvious problems are found, most likely the fuel pressure regulator and/or the fuel pump is defective. In this situation, it is recommended that both the fuel pressure regulator and fuel pump are replaced to prevent any future fuel pressure problems.*
7   Turn the engine off and check the gauge - the pressure should hold steady. After five minutes it should not drop below the minimum listed in this Chapter's Specifications. If it does drop, the fuel pump or pressure regulator could be defective, or a fuel injector could be leaking.
8   After the testing is done, relieve the fuel pressure (see Section 2) and remove the fuel pressure gauge.

5   Fuel lines and fittings - general information and disconnection

Warning: *Gasoline is extremely flammable. See* Fuel system warnings *in Section 1.*
1   Relieve the fuel pressure before servicing fuel lines or fittings (see Section 3), then disconnect the cable from the negative battery terminal (see Chapter 5) before proceeding.
2   The fuel supply line connects the fuel pump in the fuel tank to the fuel rail on the engine. The Evaporative Emission (EVAP) system lines connect the fuel tank to the EVAP canister and connect the canister to the intake manifold.
3   Whenever you're working under the

vehicle, be sure to inspect all fuel and evaporative emission lines for leaks, kinks, dents and other damage. Always replace a damaged fuel or EVAP line immediately.
4   If you find signs of dirt in the lines during disassembly, disconnect all lines and blow them out with compressed air. Inspect the fuel strainer on the fuel pump pick-up unit for damage and deterioration.

Steel tubing

5   It is critical that the fuel lines be replaced with lines of equivalent type and specification.
6   Some steel fuel lines have threaded fittings. When loosening these fittings, hold the stationary fitting with a wrench while turning the tube nut.

Plastic tubing

7   When replacing fuel system plastic tubing, use only original equipment replacement plastic tubing.
Caution: *When removing or installing plastic fuel line tubing, be careful not to bend or twist it too much, which can damage it. Also, plastic fuel tubing is NOT heat resistant, so keep it away from excessive heat.*

Flexible hoses

8   When replacing fuel system flexible hoses, use only original equipment replacements.
9   Don't route fuel hoses (or metal lines) within four inches of the exhaust system or within ten inches of the catalytic converter. Make sure that no rubber hoses are installed directly against the vehicle, particularly in places where there is any vibration. If allowed to touch some vibrating part of the vehicle, a hose can easily become chafed and it might start leaking. A good rule of thumb is to maintain a minimum of 1/4-inch clearance around a hose (or metal line) to prevent contact with the vehicle underbody.

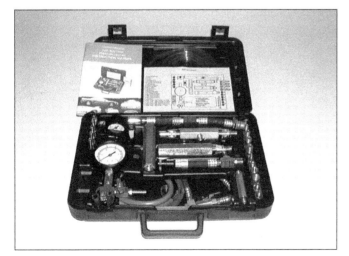

4.4a This fuel pressure testing kit contains all the necessary fittings and adapters, along with the fuel pressure gauge, to test most automotive systems

4.4b Fuel feed line fitting at the fuel rail

Disconnecting Fuel Line Fittings

Two-tab type fitting; depress both tabs with your fingers, then pull the fuel line and the fitting apart

On this type of fitting, depress the two buttons on opposite sides of the fitting, then pull it off the fuel line

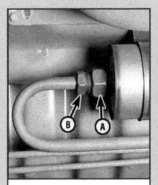

Threaded fuel line fitting; hold the stationary portion of the line or component (A) while loosening the tube nut (B) with a flare-nut wrench

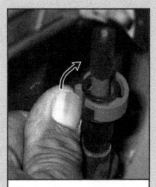

Plastic collar-type fitting; rotate the outer part of the fitting

Metal collar quick-connect fitting; pull the end of the retainer off the fuel line and disengage the other end from the female side of the fitting . . .

. . . insert a fuel line separator tool into the female side of the fitting, push it into the fitting and pull the fuel line off the pipe

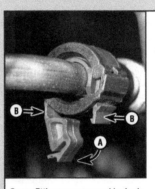

Some fittings are secured by lock tabs. Release the lock tab (A) and rotate it to the fully-opened position, squeeze the two smaller lock tabs (B) . . .

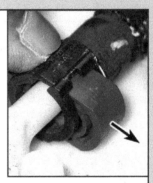

. . . then push the retainer out and pull the fuel line off the pipe

Spring-lock coupling; remove the safety cover, install a coupling release tool and close the tool around the coupling . . .

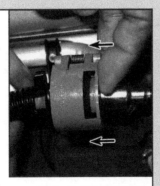

. . . push the tool into the fitting, then pull the two lines apart

Hairpin clip type fitting: push the legs of the retainer clip together, then push the clip down all the way until it stops and pull the fuel line off the pipe

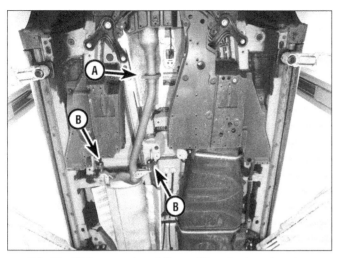

6.1a Identifying an exhaust flange (A) and exhaust system hangers (B)

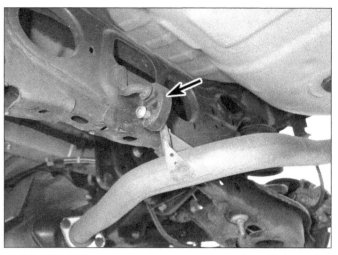

6.1b A typical exhaust system hanger. Inspect regularly and replace at the first sign of damage or deterioration

6   Exhaust system servicing - general information

Warning: *Allow exhaust system components to cool before inspection or repair. Also, when working under the vehicle, make sure it is securely supported on jackstands.*

1   The exhaust system consists of the exhaust manifolds, catalytic converter, muffler, tailpipe and all connecting pipes, flanges and clamps (see illustration). The exhaust system is isolated from the vehicle body and from chassis components by a series of rubber hangers (see illustration). Periodically inspect these hangers for cracks or other signs of deterioration, replacing them as necessary.

2   Conduct regular inspections of the exhaust system to keep it safe and quiet. Look for any damaged or bent parts, open seams, holes, loose connections, excessive corrosion or other defects which could allow exhaust fumes to enter the vehicle. Do not repair deteriorated exhaust system components; replace them with new parts.

3   If the exhaust system components are extremely corroded, or rusted together, a cutting torch is the most convenient tool for removal. Consult a properly-equipped repair shop. If a cutting torch is not available, you can use a hacksaw, or if you have compressed air, there are special pneumatic cutting chisels that can also be used. Wear safety goggles to protect your eyes from metal chips and wear work gloves to protect your hands.

4   Here are some simple guidelines to follow when repairing the exhaust system:

  a) *Work from the back to the front when removing exhaust system components.*

  b) *Apply penetrating oil to the exhaust system component fasteners to make them easier to remove.*

  c) *Use new gaskets, hangers and clamps.*

  d) *Apply anti-seize compound to the threads of all exhaust system fasteners during reassembly.*

  e) *Allow sufficient clearance between newly installed parts and all points on the underbody to avoid overheating the floor pan and possibly damaging the interior*

carpet and insulation. Pay particularly close attention to the catalytic converter and heat shield.

7   Fuel pump module/fuel level sending unit - removal and installation

Warning: *Gasoline is extremely flammable, so take extra precautions when you work on any part of the fuel system. See* Fuel system warnings *in Section 1.*
Note: *All models are equipped with a separate fuel level sending unit located in the front half of the tank.*

Removal

1   Relieve the fuel system pressure (see Section 3) and remove the fuel tank cap.

2   Disconnect the cable from the negative battery terminal (see Chapter 5).

3   Remove the fuel tank (see Section 10).

4   Clean the top of the fuel tank, fuel pump module and fuel sending unit to prevent contaminants from getting into the tank once the module is removed.

Fuel pump module

5   Remove the clip and disconnect the fuel supply line and connector at the fuel pump module (see illustration).

6   Remove the eight fuel pump module retaining bolts.

7   Carefully withdraw the fuel pump module assembly from the fuel tank, free the clamp and disconnect the fuel sub-suction hose (see illustration).

8   Remove the fuel pump module from the tank and drain the fuel into a suitable container.

Fuel pump and fuel filter

9   Disconnect the electrical connector from the fuel pump main body (see illustration).

7.7 Disconnect the fuel sub-suction hose

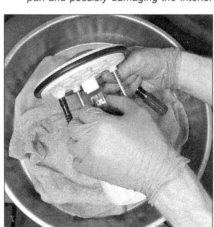

7.9 Disconnect the fuel pump connector from the main body

7.10 Remove the clips using needle-nosed pliers

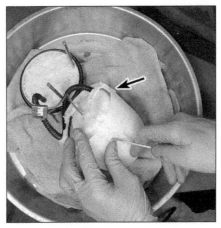

7.11a Disengage the retaining tabs . . .

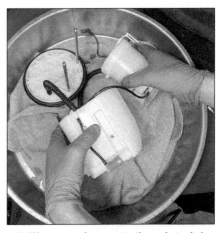

7.11b . . . and separate the sub-tank 1 from sub-tank 2

10   Remove the clips(s) from the sliders. Pull the top of the module away from the sub-tanks and remove the springs from the sliders (see illustration).

11   Pry away the two tabs at the top of sub-tank 1 and 2 and remove the sub-tank 1 from sub-tank 2 (see illustrations).

12   Pry back the retainer and remove the jet pump module at the bottom of sub-tank 2 (see illustration).

13   Use a tape-wrapped tip screwdriver to remove the jet pump from the top of sub-tank 2.

Note: *Replace the O-ring on the jet pump.*

14   Carefully pry the five tabs from the bottom of sub-tank 1 and pull the fuel pump out of sub-tank 1 (see illustration).

Note: *Sub-tank 1 is also the fuel filter.*

15   Disconnect the fuel pump electrical connector and remove the fuel pump.

Note: *Replace the O-ring on the fuel pump.*

Fuel level sending unit

16   Remove the five screws and remove fuel level sending unit assembly from the fuel tank (see illustration).

17   Carefully angle the sending unit out of

the opening without damaging the fuel level float located at the bottom of the assembly.

Installation

18   Reassemble the fuel pump module in the reverse order of disassembly.

19   Install the fuel pump/sending unit assemblies in the fuel tank.

20   The remainder of installation is the reverse of removal.

---

8   Fuel pump control module - replacement

Note: *The Fuel Pump Control Module (FPCM) is located in the rear of the vehicle, on the driver's side behind the quarter trim panel. On vehicles built before 10/2015, the FPCM is located on the floor of the vehicle. On vehicles built 10/2015 and later, the FPCM is on the side of the vehicle, just rear of the seat belt retractor.*

1   Remove the driver's rear floor trim and carefully pull back the quarter trim to access the FPCM (see Chapter 11).

2   Disconnect the FPCM connector (see illustration).

3   Remove the bolt and remove the FPCM from the vehicle.

4   Installation is reverse of removal.

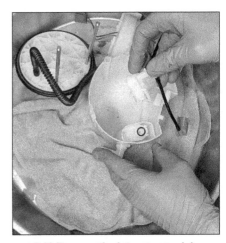

7.12 Remove the jet pump module

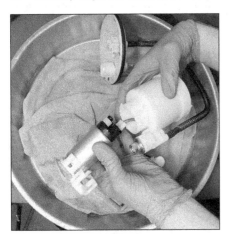

7.14 Remove the fuel pump from sub-tank 1

7.16 Fuel level sending unit

8.2 Identifying the FPCM (vehicle built prior to 10/2015 shown)

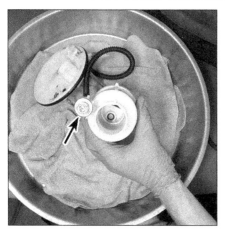

9.4 Carefully pry the fuel pressure regulator from the pump housing

9.5 Be sure to install a new O-ring(s) onto the fuel pressure regulator before installing it back into the fuel pump assembly

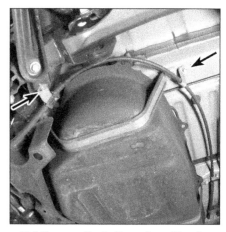

10.6 Remove the bolts and position the parking brake cable brackets out of the way

10.7 Identifying the floor undercover

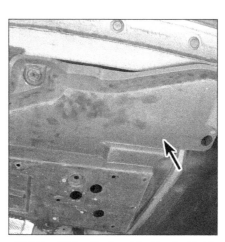

10.8 Identifying the front fuel tank shield

10.9 Identifying the rear underfloor cover

9   Fuel pressure regulator - removal and installation

Warning: *Gasoline is extremely flammable, so take extra precautions when you work on any part of the fuel system. See* Fuel system warnings *in Section 1.*

1   Relieve the fuel system pressure (see Section 3). Disconnect the cable from the negative battery terminal (see Chapter 5).
2   Remove the fuel pump from the fuel pump module (see Section 7).
3   Disconnect the fuel return jet tube from the clamp on the fuel pressure regulator.
4   Separate the fuel pressure regulator from the fuel filter (see illustration).
5   Installation is the reverse of removal. Be sure to install a new O-ring(s) (see illustration) on the fuel pressure regulator.
6   Use a scan tool to activate the fuel pump and check for leaks. If a scan tool is not available, turn the ignition to On and verify there

are no fuel leaks. If no leaks are present, start and run the engine and check for leaks again.

10   Fuel tank - removal and installation

Warning: *Gasoline is extremely flammable, so take extra precautions when you work on any part of the fuel system. See* Fuel system warnings *in Section 1.*

1   Relieve the fuel system pressure (see Section 3).
2   Remove the fuel filler cap to relieve fuel tank pressure.
3   Disconnect the cable from the negative battery terminal (see Chapter 5).
4   If the tank is full or nearly full, siphon the fuel into an approved container using a siphoning kit (available at most auto parts stores).
Warning: *Do not start the siphoning action by mouth!*

5   Raise the vehicle and support it securely on jackstands.
Warning: *To avoid personal injury, never get beneath the vehicle when it is supported only by a jack. The jack provided with your vehicle is designed solely for raising the vehicle to remove and replace the wheels. Always use jackstands to support the vehicle when it becomes necessary to place your body underneath the vehicle.*
6   Disconnect the two parking brake cable brackets from around the tank and position to the side (see illustration).
7   Remove the two nuts/bolts and multiple clips to remove the front underfloor cover (see illustration).
8   Remove the one nut and three clips and remove the front fuel tank shield (see illustration).
9   Remove the two nuts and remove the rear underfloor cover near the tank and rear trailing arm (see illustration).
10   Disconnect the fuel line, vapor return

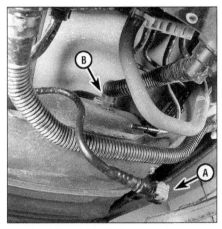

10.10 Disconnect the fuel line (A), EVAP hose (B) and vapor return line

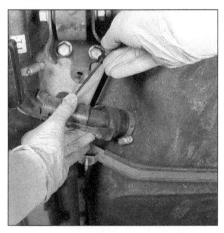

10.11 Remove the filler tube shield by disengaging the claw

10.12 Loosen both clamps but only disconnect the filler hose at the tank

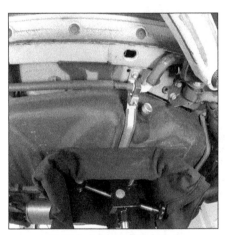

10.14 Support the fuel tank

10.15 Remove the fuel tank strap mounting bolts from the body

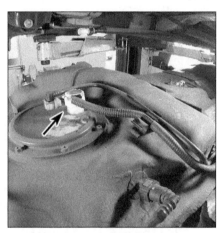

10.17a Disconnect the fuel pump module connector . . .

line, and EVAP hose (see Section 5 for fitting disconnection details) (see illustration).
Note: *Be sure to plug the hoses to prevent leakage and contamination of the fuel system.*
11   Remove the fuel filler tube shield by disengaging the claw at the top of the shield (see illustration).
12   Loosen the clamps but disconnect only the fuel tank-to-filler pipe hose from the fuel tank (see illustration).
13   Detach the wiring harness clips from the three tabs on the side of the fuel tank.
14   Support the fuel tank with a floor jack. Place a sturdy plank between the jack head and the fuel tank to protect the tank (see illustration).
15   Remove the bolts from the fuel tank retaining straps (see illustration).
16   Slowly lower the tank enough to fold back the fuel tank cushion and remove the

wiring harness from the retainer in the fuel tank.
17   Disconnect the two electrical connectors from the fuel pump module and fuel gauge sending unit (see illustrations).
Note: *It may be necessary to pull the harness through the grommet and disconnect it from inside the vehicle.*
18   Remove the tank from the vehicle.
19   Remove or replace the fuel pump module and/or sending unit as necessary (see Section 7).
20   Clean any dirt or mud from the tank before installation.
21   Installation is the reverse of removal.
22   Use a scan tool to activate the fuel pump and check for leaks. If a scan tool is not available, turn the ignition to On and verify there are no fuel leaks. If no leaks are present, start and run the engine and check for leaks again.

10.17b . . . and the fuel sending unit connector

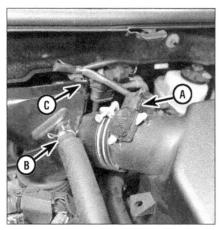

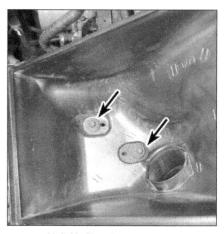

11.1 Disconnect the MAF connector (A), disconnect the air filter vent hose (B), and detach the VSV (C)

11.3 Squeeze the inlet hose spring clamp to remove from the throttle body

11.6 Air filter housing bolts

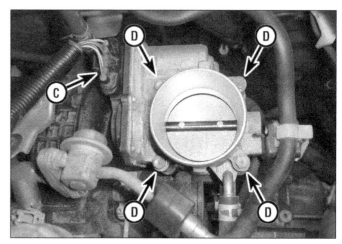

12.12 Disconnect the coolant hoses (A) and the fuel hose bracket (B)

12.14 Disconnect the throttle body electrical connector (C) and remove the throttle body bolts (D)

11   Air filter housing - removal and installation

1   Disconnect the mass air flow sensor connector, remove the screw and detach the Vacuum Switching Valve (VSV) from the intake duct, and disconnect the air filter vent hose and canister purge line hose clamp (see illustration).

2   Disconnect the PCV hose from the valve cover.

3   Disconnect the throttle body inlet hose clamp and separate the hose from the throttle body (see illustration).

4   Detach any harness clips from the air filter housing lid and base.

5   Unsnap the retainer clips and remove the air filter cover and duct as an assembly.

6   Remove the air filter element and two housing bolts (see illustration), detach the harness then remove the air filter housing from the engine compartment.

7   Installation is the reverse of removal.

12   Throttle body - check, replacement and initialization

Check

1   Turn the ignition on.

2   While listening near the throttle body, press and release the accelerator pedal. The throttle body motor should make noise, indicating operation. If excessive noise is heard, inspect the throttle body and replace as necessary.

3   Connect a scan tool and monitor the throttle sensor position.

4   Press the accelerator pedal to the floor and hold. The throttle sensor position values should be 60 percent or greater. If the value is less than 60 percent, inspect the throttle body and replace if necessary.

5   Turn the ignition off and disconnect the throttle body connector.

6   Using a digital volt-ohm meter (DVOM), measure resistance between pins 1 and 2 of

the throttle body connector. Resistance should be 0.3 ohms to 100 ohms.

Replacement

Warning: *Wait until the engine is completely cool before beginning this procedure.*

7   Disconnect the cable from the negative battery terminal (see Chapter 5).

8   Remove the engine cover.

9   Remove the intake duct and air filter cover.

10   Remove the air filter housing with the MAF sensor and bracket (see Section 11).

11   Clamp off the coolant hoses to the throttle body to minimize coolant loss. Remove the clamps and disconnect the two coolant hoses at the throttle body (see illustration). Plug the coolant hoses to prevent coolant leakage.

12   Remove the bolt for the fuel hose bracket (see illustration).

13   Disconnect the electrical connector from the throttle body (see illustration).

14   Remove the throttle body mounting bolts (see illustration) and remove the throttle body.

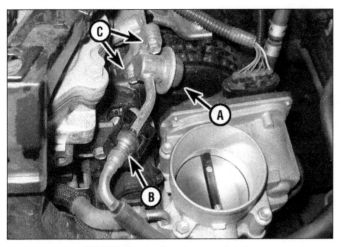

13.6 Fuel pulsation damper details

14.6 Disconnect the harness clamps from the intake manifold

A   *Fuel pulsation damper*
B   *Fuel line quick-connect fitting*
C   *Mounting bolts*

15   Remove and discard the old throttle body gasket.
16   Installation is the reverse of removal. Be sure to use a new gasket and tighten the throttle body mounting bolts to the torque listed in this Chapter's Specifications. Check the coolant level, adding as necessary (see Chapter 1).

Idle initialization

Note: *After cleaning the throttle body, perform the following procedure. After removing and installing or replacing a throttle body, perform the following procedure. After replacing the ECM, perform the following procedure.*
17   Ensure the battery is connected and the ignition is Off.
18   Remove the EFI No. 1 fuse and ETCS fuses from the under hood junction box. Wait a minimum of 60 seconds after removing the fuses.
19   Install the fuses after waiting a minimum of 60 seconds after removing the fuses.
20   Turn the ignition switch to On (DO NOT start the engine) without depressing the accelerator pedal.
21   Using a scan tool, clear any DTCs that are set.
22   Start the engine and bring to operating temperature. Monitor idle speed at operating temperature. With AC and all accessories Off, idle speed should be 720 RPM to 820 RPM.
23   Using the scan tool, monitor the throttle sensor position.
24   Press the accelerator pedal to the floor and hold. The throttle sensor position values should be 60 percent or greater.
25   Road test the vehicle and verify proper operation.

13   Fuel pulsation damper - replacement

Warning: *Gasoline is extremely flammable, so take extra precautions when you work on*

any part of the fuel system. *See* Fuel system warnings *in Section 1.*
1   Relieve the fuel system pressure (see Section 3).
2   Disconnect the cable from the negative battery terminal (see Chapter 5).
3   Remove the engine cover.
4   Remove the air filter housing (see Section 11).
5   Remove the fuel tube clamp from the fuel line and disconnect the line using a special tool (see Section 5).
6   Remove the mounting bolts from the fuel pulsation damper (see illustration).
7   Separate the fuel pulsation damper from the fuel rail.
8   Installation is the reverse of removal. Be sure to install a new O-ring onto the fuel pulsation damper. Tighten the bolts to the torque listed in this Chapter's Specifications.
9   Use a scan tool to activate the fuel pump and check for leaks. If a scan tool is not available, turn the ignition to On and verify there are no fuel leaks. If no leaks are present, start and run the engine and check for leaks again.

14   Fuel rail and injectors - removal and installation

Warning: *Gasoline is extremely flammable, so take extra precautions when you work on any part of the fuel system. See* Fuel system warnings *in Section 1.*

Removal

1   Relieve the fuel pressure (see Section 3).
2   Disconnect the cable from the negative battery terminal (see Chapter 5).
3   Remove the engine cover.
4   Remove the air filter housing (see Section 11).
5   Detach the fuel hose bracket near the throttle body.
6   Free the wiring harness from its clamps, then disconnect the electrical connectors from the fuel rail and injectors (see illustration).
7   Disconnect the fuel line from the fuel rail (see Section 5).
8   Remove the fuel rail mounting bolts (see illustration).

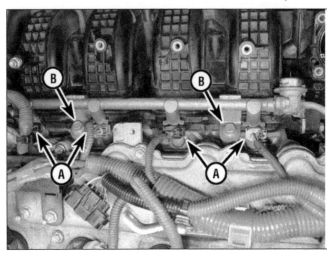

14.8 Disconnect the fuel injector connectors (A) and remove the fuel rail mounting bolts (B)

9   Remove the fuel rail with the fuel injectors attached.

Note: *There are two spacers that mount between the fuel rail and the cylinder head.*

10   Remove the fuel injector(s) from the fuel rail by firmly pulling out of the fuel rail.

11   If you intend to re-use the same injectors, replace the grommets and O-rings (see illustration).

Installation

12   Installation of the fuel injectors is the reverse of removal.

13   Apply a light coating of fuel to the injector O-rings before installation into the fuel rail AND into the cylinder head.

14   Ensure that the fuel injectors rotate freely once installed into the cylinder head with the fuel rail.

15   Tighten the fuel rail mounting bolts to the torque listed in this Chapter's Specifications.

16   Use a scan tool to activate the fuel pump and check for leaks. If a scan tool is not available, turn the ignition to On and verify there are no fuel leaks. If no leaks are present, start and run the engine and check for leaks again.

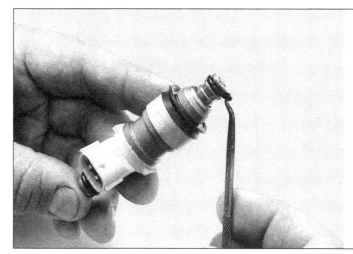

14.11 Be sure to install new O-rings on the injectors and the fuel pulsation damper

Chapter 5
Engine electrical systems

Contents

Specifications

Charging system

Battery voltage

Range	11 to 14 volts
At-rest average at 68F (20C)*	12.6 to 12.9 volts
* If battery voltage is below 9.6 volts, recharge and retest battery	
Charging voltage (at idle)	13.5 to 14.5 volts

Standard amperage

No load	10 amps or more
With load	30 amps or more

Torque Specifications

Ft-lbs (unless otherwise indicated)   **Nm**

Note: *One foot-pound (ft-lb) of torque is equivalent to 12 inch-pounds (in-lbs) of torque. Torque values below approximately 15 foot-pounds are expressed in inch-pounds, because most foot-pound torque wrenches are not accurate at these smaller values.*

	Ft-lbs	Nm
Alternator bolts	38	52
Starter bolts	27	37

1   General information and precautions

General information

Ignition system

1   The electronic ignition system consists of the Crankshaft Position (CKP) sensor, the Camshaft Position (CMP) sensor, the Knock Sensor (KS), the Powertrain Control Module (PCM), the ignition switch, the battery, the individual ignition coils or a coil pack, and the spark plugs. For more information on the CKP, CMP and KS sensors, as well as the PCM, refer to Chapter 6.

Charging system

2   The charging system includes the alternator (with an integral voltage regulator), the Powertrain Control Module (PCM), the Body Control Module (BCM), the battery current sensor, a charge indicator light on the dash, the battery, a fuse or fusible link and the wiring connecting all of these components. The charging system supplies electrical power for the ignition system, the lights, the radio, etc. The alternator is driven by a drivebelt.

Starting system

3   The starting system consists of the battery, the ignition switch, the starter relay, the Powertrain Control Module (PCM), the Body Control Module (BCM), the Transmission Range (TR) switch, the starter motor and solenoid assembly, and the wiring connecting all of the components.

Precautions

4   Always observe the following precautions when working on the electrical system:

    a)   *Be extremely careful when servicing engine electrical components. They are easily damaged if checked, connected or handled improperly.*

    b)   *Never leave the ignition switched on for long periods of time when the engine is not running.*

    c)   *Never disconnect the battery cables while the engine is running.*

    d)   *Maintain correct polarity when connecting battery cables from another vehicle during jump starting - see the Booster battery (jump) starting Section at the front of this manual.*

    e)   *Always disconnect the cable from the negative battery terminal before working on the electrical system, but read the battery disconnection procedure first (see Section 3).*

5   It's also a good idea to review the safety-related information regarding the engine electrical systems located in the *Safety first!* Section at the front of this manual before beginning any operation included in this Chapter.

Engine electrical system components

1	*Battery*	*3*	*Secondary relay box*	*5*	*Ignition coils*
2	*Main relay and fuse box*	*4*	*Alternator*	*6*	*Starter motor (below battery)*

2   Troubleshooting

Ignition system

1   If a malfunction occurs in the ignition system, do not immediately assume that any particular part is causing the problem. First, check the following items:

a) *Make sure that the cable clamps at the battery terminals are clean and tight.*

b) *Test the condition of the battery. If it doesn't pass all the tests, replace it.*

c) *Check the ignition coil or coil pack connections.*

d) *Check any relevant fuses in the engine compartment fuse and relay box (see Chapter 12). If they're burned, determine the cause and repair the circuit.*

Check

Warning: *Because of the high voltage generated by the ignition system, use extreme care when performing a procedure involving ignition components.*

Note: *The ignition system components on these vehicles are difficult to diagnose. In the event of ignition system failure that you can't diagnose, have the vehicle tested at a dealer service department or other qualified auto repair facility.*

Note: *You'll need a spark tester for the following test. Spark testers are available at most auto supply stores.*

2   If the engine turns over but won't start, verify that there is sufficient ignition voltage to fire the spark plugs as follows.

3   On models with a coil-over-plug type ignition system, remove a coil and install the tester between the boot at the lower end of the coil and the spark plug (see illustration).

4   Crank the engine and note whether or not the tester flashes.

Caution: *Do NOT crank the engine or allow it to run for more than five seconds; running the engine for more than five seconds may set a* Diagnostic Trouble Code (DTC) for a cylinder misfire.

5   If the tester flashes during cranking, the coil is delivering sufficient voltage to the spark plug to fire it. Repeat this test for each cylinder to verify that the other coils are OK.

6   If the tester doesn't flash, remove a coil from another cylinder and swap it for the one being tested. If the tester now flashes, you know that the original coil is bad. If the tester still doesn't flash, the PCM or wiring harness is probably defective. Have the PCM checked out by a dealer service department or other qualified repair shop (testing the PCM is beyond the scope of the do-it-yourselfer because it requires expensive special tools).

7   If the tester flashes during cranking but a misfire code (related to the cylinder being tested) has been stored, the spark plug could be fouled or defective.

Charging system

8   If a malfunction occurs in the charging system, do not automatically assume the alternator is causing the problem. First check the following items:

a) *Check the drivebelt tension and condition, as described in Chapter 1. Replace it if it's worn or deteriorated.*

b) *Make sure the alternator mounting bolts are tight.*

c) *Inspect the alternator wiring harness and the connectors at the alternator and voltage regulator. They must be in good condition, tight and have no corrosion.*

d) *Check the fusible link (if equipped) or main fuse in the underhood fuse/relay box. If it is burned, determine the cause, repair the circuit and replace the link or fuse (the vehicle will not start and/or the accessories will not work if the fusible link or main fuse is blown).*

e) *Start the engine and check the alternator for abnormal noises (a shrieking or squealing sound indicates a bad bearing).*

f) *Check the battery. Make sure it's fully charged and in good condition (one bad cell in a battery can cause overcharging by the alternator).*

g) *Disconnect the battery cables (negative first, then positive). Inspect the battery posts and the cable clamps for corrosion. Clean them thoroughly if necessary (see Chapter 1). Reconnect the cables (positive first, negative last).*

Alternator - check

9   Use a voltmeter to check the battery voltage with the engine off. It should be at least 12.6 volts (see illustration 2.15).

10   Start the engine and check the battery voltage again. It should now be approximately 13.5 to 15 volts.

11   If the voltage reading is more or less than the specified charging voltage, the voltage regulator is probably defective, which will require replacement of the alternator (the voltage regulator is not replaceable separately). Remove the alternator and have it bench tested (most auto parts stores will do this for you).

12   The charging system (battery) light on the instrument cluster lights up when the ignition key is turned to On, but it should go out when the engine starts.

13   If the charging system light stays on after the engine has been started, there is a problem with the charging system. Before replacing the alternator, check the battery condition, alternator belt tension and electrical cable connections.

14   If replacing the alternator doesn't restore voltage to the specified range, have the charging system tested by a dealer service department or other qualified repair shop.

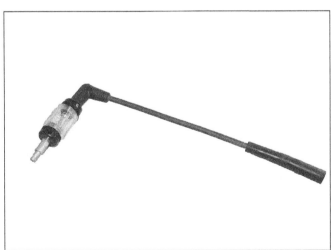

2.3 Spark tester (shown mounted with a spark plug wire, coil mounting arrangement is identical)

2.15 To test the open circuit voltage of the battery, touch the black probe of the voltmeter to the negative terminal and the red probe to the positive terminal of the battery; a fully charged battery should be at least 12.6 volts

Battery - check

Note: *The battery's surface charge must be removed before accurate voltage measurements can be made. Turn on the high beams for ten seconds, then turn them off and let the vehicle stand for two minutes.*

15   Check the battery state of charge. Visually inspect the indicator eye on the top of the battery (if equipped with one); if the indicator eye is black in color, charge the battery as described in Chapter 1. Next perform an open circuit voltage test using a digital voltmeter. With the engine and all accessories Off, touch the negative probe of the voltmeter to the negative terminal of the battery and the positive probe to the positive terminal of the battery (see illustration). The battery voltage should be 12.6 volts or slightly above. If the battery is less than the specified voltage, charge the battery before proceeding to the next test. Do not proceed with the battery load test unless the battery charge is correct.

16   Disconnect the negative battery cable, then the positive cable from the battery.

17   Perform a battery load test. An accurate check of the battery condition can only be performed with a load tester (see illustration). This test evaluates the ability of the battery to operate the starter and other accessories during periods of high current draw. Connect the load tester to the battery terminals. Load test the battery according to the tool manufacturer's instructions. This tool increases the load demand (current draw) on the battery.

18   Maintain the load on the battery for 15 seconds and observe that the battery voltage does not drop below 9.6 volts. If the battery condition is weak or defective, the tool will indicate this condition immediately.

Note: *Cold temperatures will cause the minimum voltage reading to drop slightly. Follow the chart given in the manufacturer's instructions to compensate for cold climates. Minimum load voltage for freezing temperatures (32 degrees F) should be approximately 9.1 volts.*

Starting system

The starter rotates, but the engine doesn't

19   Remove the starter (see Section 9). Check the overrunning clutch and bench test the starter to make sure the drive mechanism extends fully for proper engagement with the flywheel ring gear. If it doesn't, replace the starter.

20   Check the flywheel ring gear for missing teeth and other damage. With the ignition turned off, rotate the flywheel so you can check the entire ring gear.

The starter is noisy

21   If the solenoid is making a chattering noise, first check the battery (see Steps 15 through 18). If the battery is okay, check the cables and connections.

22   If you hear a grinding, crashing metallic sound when you turn the key to Start, check for loose starter mounting bolts. If they're tight, remove the starter and inspect the teeth on the starter pinion gear and flywheel ring gear. Look for missing or damaged teeth.

23   If the starter sounds fine when you first turn the key to Start, but then stops rotating the engine and emits a zinging sound, the problem is probably a defective starter drive that's not staying engaged with the ring gear. Replace the starter.

The starter rotates slowly

24   Check the battery (see Steps 15 through 18).

25   If the battery is okay, verify all connections (at the battery, the starter solenoid and motor) are clean, corrosion-free and tight. Make sure the cables aren't frayed or damaged.

26   Check that the starter mounting bolts are tight so it grounds properly. Also check the pinion gear and flywheel ring gear for evidence of a mechanical bind (galling, deformed gear teeth or other damage).

The starter does not rotate at all

27   Check the battery (see Steps 15 through 18).

28   If the battery is okay, verify all connections (at the battery, the starter solenoid and motor) are clean, corrosion-free and tight. Make sure the cables aren't frayed or damaged.

29   Check all of the fuses in the underhood fuse/relay box.

30   Check that the starter mounting bolts are tight so it grounds properly.

31   Check for voltage at the starter solenoid "S" terminal when the ignition key is turned to the start position. If voltage is present, replace the starter/solenoid assembly. If no voltage is present, the problem could be the starter relay or the Park/neutral Position switch (see Chapter 7A), or with an electrical connector somewhere in the circuit (see the wiring diagrams at the end of this manual). Also, on many modern vehicles, the Powertrain Control Module (PCM) and the Body Control Module (BCM) control the voltage signal to the starter solenoid; on such vehicles a special scan tool is required for diagnosis.

3   Battery - disconnection and reconnection

Disconnection

Warning: *Hydrogen gas is produced by the battery, so keep open flames and lighted cigarettes away from it at all times. Always wear eye protection when working around the battery. Rinse off spilled electrolyte immediately with large amounts of water.*

Warning: *Always disconnect the cable from the negative battery terminal FIRST and hook it up LAST or the battery may be shorted by the tool being used to loosen the cable clamps.*

1   Some systems on the vehicle require battery power to be available at all times, either to maintain continuous operation (alarm system, power door locks, etc.), or to maintain control unit memory (radio station presets, Powertrain Control Module and other control units). When the battery is disconnected, the power that maintains these systems is cut. So, before you disconnect the battery, please note that on a vehicle with power door locks, it's a wise precaution to remove the key from the ignition and to keep it with you, so that it does not get locked inside if the power door locks should engage accidentally when the battery is reconnected!

2   Devices known as "memory-savers" can be used to avoid some of these problems. Precise details vary according to the device used. The typical memory saver is plugged into the cigarette lighter and is connected to a spare battery. Then the vehicle battery can be disconnected from the electrical system. The memory saver will provide sufficient current to maintain audio unit security codes, PCM memory, etc., and will provide power to always hot circuits such as the clock and radio memory circuits.

Warning: *Some memory savers deliver a considerable amount of current in order to keep vehicle systems operational after the main battery is disconnected. If you're using a memory saver, make sure that the circuit concerned is actually open before servicing it.*

Warning: *If you're going to work near any of the airbag system components, the battery*

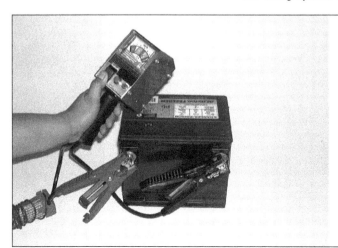

2.17 Connect a battery load tester to the battery and check the battery condition under load following the tool manufacturer's instructions

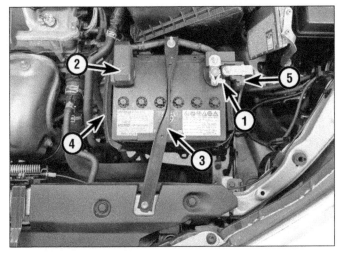

4.2 Battery details

1   *Negative battery cable*   4   *Battery insulator*
2   *Positive battery cable*   5   *Battery current sensor*
3   *Battery hold-down clamp*

4.6a Remove the four battery tray bolts and the upper battery tray . . .

MUST be disconnected and a memory saver must NOT be used. If a memory saver is used, power will be supplied to the airbag, which means that it could accidentally deploy and cause serious personal injury.

3   To disconnect the battery, turn the ignition to Off and remove the key. Ensure the headlights and any other electrical loads are off.

4   Loosen the cable end bolt and disconnect the cable from the negative battery terminal. Isolate the cable end to prevent it from coming into accidental contact with the battery terminal.

5   After disconnecting the negative battery terminal, if necessary, the positive battery terminal can now be disconnected.

Reconnection

6   Reconnect the positive battery terminal first (if disconnected).

7   Reconnect the negative battery cable last.

8   The clock, audio settings and radio station presets will need to be reset after reconnecting the battery.

9   If equipped, reset the moon roof. See Reset moon roof position procedure.

10   If the back door was open when the battery was disconnected and reconnected, perform the Reset back door close position procedure.

Reset moon roof position

11   Turn the ignition On (vehicles without smart key system) or Ignition On mode (vehicles with smart key system).

12   Press and hold the moon roof Close or Up switch until the moon roof tilts up, pauses for about one second, then tilts down and fully opens and closes. After the movement, release the switch.

13   Verify proper operation of the automatic

operation and repeat the procedure as necessary.

Reset back door close position

Note: *This procedure is not required if the back door was closed while the battery was disconnected and reconnected.*

14   With the back door in the open position, reconnect the battery.

15   Close the back door and verify proper operation.

---

4   Battery - removal and installation

---

1   If equipped, remove the fasteners securing the battery top cover, then remove the cover and set it aside.

2   Loosen the cable clamp nut and disconnect the cable from the negative battery terminal first, then disconnect the cable from the

positive battery terminal (see illustration).

3   Remove the battery hold-down clamp front fastener, then unhook the rod from the battery tray at the rear. Position the hold-down clamp and wiring harness aside.

4   Lift out the battery. Be careful - it's heavy. Remove the battery insulator, if equipped.

Note: *Battery straps and handlers are available at most auto parts stores for reasonable prices. They make it easier to remove and carry the battery.*

5   If you are replacing the battery, make sure you get one that's identical, with the same dimensions, amperage rating, cold cranking rating, terminal polarity arrangement, etc.

6   To remove the battery tray, remove the bolts attaching it to the vehicle's body (see illustrations).

7   Installation is the reverse of removal. Be sure to connect the positive cable first and the negative cable last.

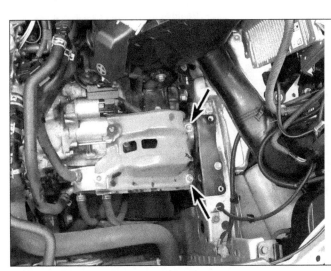

4.6b . . . and the two bolts retaining the lower battery tray

7.3 Disconnect the electrical connectors (A), remove the mounting bolts (B) then pull the coils off of the spark plugs and out of the valve cover, using a twisting motion

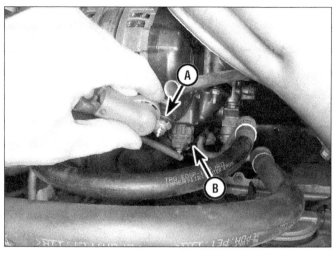

8.3 Disconnect the positive terminal (A) and the alternator connector (B)

5   Battery cables - replacement

1   When removing the cables, always disconnect the cable from the negative battery terminal first and hook it up last, or you might accidentally short out the battery with the tool you're using to loosen the cable clamps. Even if you're only replacing the cable for the positive terminal, be sure to disconnect the negative cable from the battery first.

2   Disconnect the old cables from the battery, then trace each of them to their opposite ends and disconnect them. Be sure to note the routing of each cable before disconnecting it to ensure correct installation.

3   In most cases, the battery terminals are not part of the cables and are serviced seperately.

4   If you are replacing any of the old cables, take them with you when buying new cables. It is vitally important that you replace the cables with identical parts.

5   Clean the threads of the solenoid or

8.4 Remove the bolts retaining the wiring harness brackets

ground connections with a wire brush to remove rust and corrosion. Apply a light coat of battery terminal corrosion inhibitor or petroleum jelly to the threads to prevent future corrosion.

6   Attach the cable to the solenoid or ground connection and tighten the mounting nut/bolt securely.

7   Before connecting a new cable to the battery, make sure that it reaches the battery post without having to be stretched.

8   Connect the cable to the positive battery terminal first, then connect the ground cable to the negative battery terminal.

6   Battery current sensor - replacement

Note: *The battery current sensor is attached to the negative battery cable. The battery current sensor may also be referred to as the generator control ECU assembly.*

1   Ensure the ignition switch is in the off position.

2   Remove the nut and disconnect the negative battery cable from the negative battery terminal.

3   Disconnect the electrical connector from the battery current sensor.

4   Remove the nut(s) and remove the battery current sensor from the negative battery cable (see illustration 4.2).

5   Installation is reverse of removal.

7   Ignition coil(s) - replacement

Warning: *Wear eye protection when using compressed air.*

1   Disconnect the cable from the negative battery terminal (see Section 4).

2   Remove the engine cover by pulling up to detach the retainers.

3   Each ignition coil is secured by a bolt (see illustration). Unscrew the bolt, disconnect the electrical connector and pull the coil straight up, using a twisting motion.

4   With the coils removed, use compressed air to blow out debris from inside and the area around the spark plug holes.

5   Installation is the reverse of removal. Apply a dab of dielectric compound to the inside of the coil boots to ensure a clean connection to the spark plugs.

6   For spark plug replacement, see Chapter 1.

8   Alternator - removal and installation

Removal

1   Disconnect the cable from the negative battery terminal (see Section 3).

2   Remove the drivebelt(s) (see Chapter 1).

3   Remove the protective cap, then remove the nut to disconnect the B+ terminal. Disconnect the alternator electrical connector (see illustration).

4   Remove the bolts securing the harness brackets to the alternator and engine, then position aside the wiring harness and brackets (see illustration).

5   Remove the upper and lower alternator bolts and remove the alternator from the vehicle (see illustration). A prybar may be used to help free the alternator from the upper mount.

Installation

6   If you are replacing the alternator, take the old alternator with you when purchasing a replacement unit. Make sure that the new/rebuilt unit is identical to the old alternator. Look at the terminals - they should be the same in number, size and locations as the terminals on the old alternator. Finally, look at the

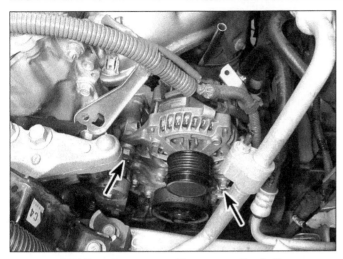

8.5 Alternator upper and lower mounting bolts

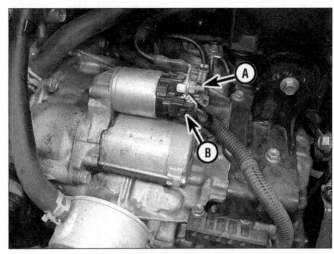

9.3 Disconnect the positive (B+) terminal (A), then disconnect the starter electrical connector (B)

identification markings - they will be stamped in the housing or printed on a tag or plaque affixed to the housing. Make sure that these numbers are the same on both alternators.

7   Installation is the reverse of removal. Tighten the bolts to the torque listed in this Chapter's Specifications.

8   Check the charging voltage to verify proper operation of the alternator (see this Chapter's Specifications).

9   Starter motor - removal and installation

1   Disconnect the cable from the negative battery terminal (see Section 3).

2   Remove the battery, battery tray and reinforcement (see Section 4).

3   Remove the terminal cover and disconnect the electrical connectors from the starter/ solenoid (see illustration). The B+ terminal cover can be easily removed by inserting a screwdriver into each slot (on the face of the

9.4 Remove the two starter mounting bolts

cover) to disengage the tabs.

4   Remove the two starter motor mounting bolts (see illustration). Remove the starter motor assembly from the engine compartment.

5   Installation is the reverse of removal. Tighten the starter motor mounting bolts to the torque listed in this Chapter's Specifications.

Notes

Chapter 6
Emissions and engine control systems

Contents

1   General information

1   To prevent pollution of the atmosphere from incompletely burned and evaporating gases, and to maintain good driveability and fuel economy, a number of emission control systems are incorporated. They include the:

Catalytic converter

2   A catalytic converter is an emission control device in the exhaust system that reduces certain pollutants in the exhaust gas stream. There are two types of converters: oxidation converters and reduction converters.

3   Oxidation converters contain a monolithic substrate (a ceramic honeycomb) coated with the semi-precious metals platinum and palladium. An oxidation catalyst reduces unburned hydrocarbons (HC) and carbon monoxide (CO) by adding oxygen to the exhaust stream as it passes through the substrate, which, in the presence of high temperature and the catalyst materials, converts the HC and CO to water vapor (H_2O) and carbon dioxide (CO_2).

4   Reduction converters contain a monolithic substrate coated with platinum and rhodium. A reduction catalyst reduces oxides of nitrogen (NOx) by removing oxygen, which in the presence of high temperature and the catalyst material produces nitrogen (N) and carbon dioxide (CO_2).

5   Catalytic converters that combine both types of catalysts in one assembly are known as "three-way catalysts" or TWCs. A TWC can reduce all three pollutants.

Evaporative Emissions Control (EVAP) system

6   The Evaporative Emissions Control (EVAP) system prevents fuel system vapors (which contain unburned hydrocarbons) from

escaping into the atmosphere. On warm days, vapors trapped inside the fuel tank expand until the pressure reaches a certain threshold. Then the fuel vapors are routed from the fuel tank through the fuel vapor vent valve and the fuel vapor control valve to the EVAP canister, where they're stored temporarily until the next time the vehicle is operated. When the conditions are right (engine warmed up, vehicle up to speed, moderate or heavy load on the engine, etc.) the PCM opens the canister purge valve, which allows fuel vapors to be drawn from the canister into the intake manifold. Once in the intake manifold, the fuel vapors mix with incoming air before being drawn through the intake ports into the combustion chambers where they're burned up with the rest of the air/fuel mixture. The EVAP system is complex and virtually impossible to troubleshoot without the right tools and training.

Powertrain Control Module (PCM)

7   The Powertrain Control Module (PCM) is the brain of the engine management system. It also controls a wide variety of other vehicle systems. In order to program the new PCM, the dealer needs the vehicle as well as the new PCM. If you're planning to replace the PCM with a new one, there is no point in trying to do so at home because you won't be able to program it yourself.

Positive Crankcase Ventilation (PCV) system

8   The Positive Crankcase Ventilation (PCV) system reduces hydrocarbon emissions by scavenging crankcase vapors, which are rich in unburned hydrocarbons. A PCV valve or orifice regulates the flow of gases into the intake manifold in proportion to the amount of intake vacuum available.

9   The PCV system generally consists of the fresh air inlet hose, the PCV valve or orifice and the crankcase ventilation hose (or PCV hose). The fresh air inlet hose connects the air intake duct to a pipe on the valve cover.

The crankcase ventilation hose (or PCV hose) connects the PCV valve or orifice in the valve cover to the intake manifold.

2   On-Board Diagnostic (OBD) system and Diagnostic Trouble Codes (DTCs)

OBD system general description

1   All models are equipped with the second generation OBD-II system. This system consists of an on-board computer known as the Powertrain Control Module (PCM), and information sensors, which monitor various functions of the engine and send data to the PCM. This system incorporates a series of diagnostic monitors that detect and identify fuel injection and emissions control systems faults and store the information in the computer memory. This updated system also tests sensors and output actuators, diagnoses drive cycles, freezes data and clears codes. This powerful diagnostic computer must be accessed using the new OBD-II scan tool and 16 pin Data Link Connector (DLC) located under the driver's dash area.

2   The PCM is the "brain" of the electronically controlled fuel and emissions system. It receives data from a number of sensors and other electronic components (switches, relays, etc.). Based on the information it receives, the PCM generates output signals to control various relays, solenoids (i.e., fuel injectors) and other actuators. The PCM is specifically calibrated to optimize the emissions, fuel economy and driveability of the vehicle.

3   It isn't a good idea to attempt diagnosis or replacement of the PCM or emission control components at home while the vehicle is under warranty. Because of a Federally mandated warranty which covers the emissions system components and because any owner-induced damage to the PCM, the sensors and/or the control devices may void this warranty, take the vehicle to a dealer service department if the PCM or a system component malfunctions.

Scan tool information

4   Because extracting the Diagnostic Trouble Codes (DTCs) from an engine management system is now the first step in troubleshooting many computer-controlled systems and components, a code reader, at the very least, will be required (see illustration). More powerful scan tools can also perform many of the diagnostics once associated with expensive factory scan tools (see illustration). If you're planning to obtain a generic scan tool for your vehicle, make sure that it's compatible with OBD-II systems. If you don't plan to purchase a code reader or scan tool and don't have access to one, you can have the codes extracted by a dealer service department or an independent repair shop.

Note: *Some auto parts stores even provide this service free of charge*

Accessing the DTCs

5   The Diagnostic Trouble Codes (DTCs) can only be accessed with a code reader or scan tool. Professional scan tools are expensive, but relatively inexpensive generic code readers or scan tools (see illustrations 2.4a and 2.4b) are available at most auto parts stores. Simply plug the connector of the scan tool into the diagnostic connector (see illustration). Then follow the instructions included with the scan tool to extract the DTCs.

6   Once you have outputted all of the stored DTCs, look them up on the accompanying DTC chart.

7   After troubleshooting the source of each DTC, make any necessary repairs or replace the defective component(s).

Clearing the DTCs

8   Clear the DTCs with the code reader or scan tool in accordance with the instructions provided by the tool's manufacturer.

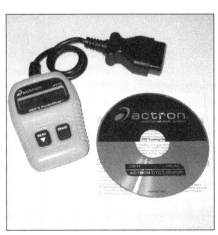

2.4a Simple code readers are an economical way to extract trouble codes when the CHECK ENGINE light comes on

2.4b Hand-held scan tools like these can extract computer codes and also perform diagnostics

2.5 Connect the scan tool to the Data Link Connector (DLC) located under the left side of the dash

Information Sensors

Accelerator Pedal Position (APP) sensor - as you press the accelerator pedal, the APP sensor alters its voltage signal to the PCM in proportion to the angle of the pedal, and the PCM commands a motor inside the throttle body to open or close the throttle plate accordingly

Camshaft Position (CMP) sensor - produces a signal that the PCM uses to identify the number 1 cylinder and to time the firing sequence of the fuel injectors

Crankshaft Position (CKP) sensor - produces a signal that the PCM uses to calculate engine speed and crankshaft position, which enables it to synchronize ignition timing with fuel injector timing, and to detect misfires

Engine Coolant Temperature (ECT) sensor - a thermistor (temperature-sensitive variable resistor) that sends a voltage signal to the PCM, which uses this data to determine the temperature of the engine coolant

Fuel tank pressure sensor - measures the fuel tank pressure and controls fuel tank pressure by signaling the EVAP system to purge the fuel tank vapors when the pressure becomes excessive

Intake Air Temperature (IAT) sensor - monitors the temperature of the air entering the engine and sends a signal to the PCM to determine injector pulse-width (the duration of each injector's on-time) and to adjust spark timing (to prevent spark knock)

Knock sensor - a piezoelectric crystal that oscillates in proportion to engine vibration which produces a voltage output that is monitored by the PCM. This retards the ignition timing when the oscillation exceeds a certain threshold

Manifold Absolute Pressure (MAP) sensor - monitors the pressure or vacuum inside the intake manifold. The PCM uses this data to determine engine load so that it can alter the ignition advance and fuel enrichment

Mass Air Flow (MAF) sensor - measures the amount of intake air drawn into the engine. It uses a hot-wire sensing element to measure the amount of air entering the engine

Oxygen sensors - generates a small variable voltage signal in proportion to the difference between the oxygen content in the exhaust stream and the oxygen content in the ambient air. The PCM uses this information to maintain the proper air/fuel ratio. A second oxygen sensor monitors the efficiency of the catalytic converter

Throttle Position (TP) sensor - a potentiometer that generates a voltage signal that varies in relation to the opening angle of the throttle plate inside the throttle body. Works with the PCM and other sensors to calculate injector pulse width (the duration of each injector's on-time)

Photos courtesy of Wells Manufacturing, except APP and MAF sensors.

Diagnostic Trouble Codes

OBD-II trouble codes

Note: *Not all trouble codes apply to all models.*

Code	Probable cause
P0010	Camshaft position "A" actuator circuit (bank 1)
P0011	Camshaft position "A" - timing over-advanced or system performance (bank 1)
P0012	Camshaft position "A" - timing over-retarded (bank 1)
P0013	Camshaft position "B" actuator circuit open (bank 1)
P0014	Camshaft position "B" - timing over-advanced or system performance (bank 1)
P0015	Camshaft position "B" - timing over-retarded (bank 1)
P0016	Crankshaft position - camshaft position correlation (bank 1 sensor A)
P0017	Crankshaft position - camshaft position correlation (bank 1 sensor B)
P0031	Oxygen (A/F) sensor heater control circuit low (bank 1 sensor 1)
P0032	Oxygen (A/F) sensor heater control circuit high (bank 1 sensor 1)
P0037	Oxygen sensor heater control circuit low (bank 1 sensor 2)
P0038	Oxygen sensor heater control circuit high (bank 1 sensor 2)
P0100	Mass airflow sensor or circuit fault
P0101	Mass airflow sensor range or performance problem
P0102	Mass or volume air flow circuit low input
P0103	Mass or volume air flow circuit high input
P0105	Manifold absolute pressure sensor or circuit fault
P0106	Manifold absolute pressure range or performance problem
P0110	Intake air temperature sensor or circuit fault
P0111	Intake air temperature sensor gradient too high
P0112	Intake air temperature circuit low input
P0113	Intake air temperature circuit high input
P0115	Engine coolant temperature sensor or circuit fault
P0116	Engine coolant temperature sensor range or performance problem
P0117	Engine coolant temperature (ECT) circuit low input
P0118	Engine coolant temperature (ECT) circuit high input
P011B	Engine coolant temperature (ECT), intake air temperature (IAT) correlation
P0120	Throttle position sensor or circuit fault

Code	Probable cause
P0121	Throttle Position sensor range or performance problem
P0122	Throttle Position sensor or switch "a" circuit low input
P0123	Throttle Position sensor or switch "a" circuit high input
P0125	Insufficient coolant temperature for closed loop; O2 sensor heater malfunction
P0128	Thermostat malfunction
P0130	Pre-converter O2 sensor or circuit fault
P0133	Pre-converter O2 sensor circuit slow response fault
P0135	Pre-converter O2 sensor heater fault
P0136	Post-converter O2 sensor or circuit fault
P0137	Oxygen sensor circuit low voltage (bank 1 sensor 2)
P0138	Oxygen sensor circuit high voltage (bank 1 sensor 2)
P0139	Oxygen sensor circuit slow response (bank 1 sensor 2)
P0141	Post-converter oxygen sensor heater or circuit fault
P0158	Post-converter oxygen sensor heater or circuit fault
P0161	Post-converter oxygen sensor heater or circuit fault
P0171	Fuel injection system lean
P0172	Fuel injection system rich
P0174	Fuel injection system lean
P0175	Fuel injection system rich
P0220	Throttle pedal position sensor or switch "b" circuit fault
P0222	Throttle pedal position sensor or switch "b" circuit low input
P0223	Throttle pedal position sensor or switch "b" circuit high input
P0230	Fuel pump primary circuit
P0300	Multiple cylinder misfire detected
P0301	Cylinder no. 1 misfire detected
P0302	Cylinder no. 2 misfire detected
P0303	Cylinder no. 3 misfire detected
P0304	Cylinder no. 4 misfire detected
P0325	Knock sensor or circuit fault
P0327	Knock sensor 1 circuit low input (bank 1 or single sensor)

OBD-II trouble codes (continued)

Note: *Not all trouble codes apply to all models.*

Code	Probable cause
P0328	Knock sensor 1 circuit high input (bank 1 or single sensor)
P0335	Crankshaft position sensor or circuit fault
P0339	Crankshaft position sensor "a" circuit intermittent
P0340	Camshaft position sensor or circuit fault
P0342	Camshaft position sensor "a" circuit low input (bank 1 or single sensor)
P0343	Camshaft position sensor "a" circuit high input (bank 1 or single sensor)
P0351	Ignition coil "a" primary or secondary circuit fault
P0352	Ignition coil "b" primary or secondary circuit fault
P0353	Ignition coil "c" primary or secondary circuit fault
P0354	Ignition coil "d" primary or secondary circuit fault
P0365	Camshaft position sensor "b" circuit (bank 1)
P0367	Camshaft position sensor "b" circuit low input (bank 1)
P0368	Camshaft position sensor "b" circuit high input (bank 1)
P0420	Catalytic converter system fault
P043E	EVAP system reference orifice clog up
P043F	EVAP system reference orifice high flow
P0430	Catalytic converter system fault
P0440	EVAP system malfunction
P0441	EVAP system incorrect purge flow detected
P0442	EVAP system leak detected
P0443	EVAP system purge control valve circuit fault
P0446	EVAP canister vent control valve circuit fault
P0450	EVAP system pressure sensor or circuit fault
P0451	EVAP system pressure sensor range or performance problem
P0452	EVAP pressure sensor or switch low input
P0453	EVAP system pressure sensor or switch high input
P0455	EVAP system leak detected (gross leak)
P0456	EVAP system leak detected (very small leak)
P0500	Vehicle speed sensor or circuit fault

Code	Probable cause
P0503	Vehicle speed sensor "A" intermittent/erratic/high
P0504	Brake switch "a" or "b" correlation
P0505	Idle air control valve or circuit fault
P050A	Cold start idle air control system performance
P050B	Cold start ignition timing performance
P0516	Battery temperature sensor circuit low
P0517	Battery temperature sensor circuit high
P0560	System voltage problem
P0571	Brake switch "A" circuit
P0575	Cruise control input circuit
P0604	Random access memory (ram)
P0606	ECM/PCM processor fault
P0607	Control module performance
P060A	ICM monitoring processor performance
P060B	ICM a/d processing performance
P060D	ICM accelerator pedal position performance
P060E	ICM throttle position performance
P0617	Starter relay circuit high
P062F	E-prom error - permanent DTC
P0630	VIN not programmed or mismatch - ECM/PCM
P0657	Actuator supply voltage circuit fault or open
P0705	Transmission range sensor circuit malfunction (PRNDL input)
P0710	Automatic transaxle fluid temperature sensor or circuit fault
P0711	Automatic transaxle fluid temperature sensor range performance or circuit fault
P0712	Transmission fluid temperature sensor "A" circuit low input
P0713	Transmission fluid temperature sensor "A" circuit high input
P0715	Input/turbine speed sensor circuit malfunction
P0717	Input speed sensor circuit no signal
P0724	Brake switch "B" circuit high
P0741	Torque converter clutch solenoid performance (shift solenoid valve SL)

OBD-II trouble codes (continued)

Note: *Not all trouble codes apply to all models.*

Code	Probable cause
P0746	Pressure control solenoid "A" performance (shift solenoid valve SL1)
P0748	Pressure control solenoid "A" electrical (shift solenoid valve SL1)
P0750	Automatic transaxle shift solenoid A stuck open or closed
P0753	Automatic transaxle shift solenoid A circuit fault
P0755	Automatic transaxle shift solenoid B stuck open or closed
P0758	Automatic transaxle shift solenoid B circuit fault
P0765	Automatic transaxle shift solenoid D stuck open or closed
P0768	Automatic transaxle shift solenoid D circuit fault
P0770	Automatic transaxle shift solenoid E stuck open or closed
P0773	Automatic transaxle shift solenoid E circuit fault
P0776	Pressure control solenoid "B" performance (shift solenoid valve SL2)
P0778	Pressure control solenoid "B" electrical (shift solenoid valve SL2)
P0791	Intermediate shaft speed sensor "A" circuit
P0793	Intermediate shaft speed sensor "A"
P0796	Pressure control solenoid "C" performance (shift solenoid valve SL3)
P0798	Pressure control solenoid "C" electrical (shift solenoid valve SL3)
P1550	Battery current sensor circuit
P1551	Battery current sensor circuit low
P1552	Battery current sensor circuit high
P1602	Deterioration of battery
P1603	Engine stall history
P1604	Startability malfunction
P1605	Rough idling
P1607	Cruise control input processor
P2004	Intake manifold runner control stuck open (bank 1)
P2006	Intake manifold runner control stuck closed (bank 1)
P2009	Intake manifold runner control circuit low (bank 1)
P2010	Intake manifold runner control circuit high (bank 1)
P2014	Intake manifold runner position sensor or switch circuit (bank 1)

Code	Probable cause
P2016	Intake manifold runner position sensor or switch circuit low (bank 1)
P2017	Intake manifold runner position sensor or switch circuit high (bank 1)
P2102	Throttle actuator control motor circuit low
P2103	Throttle actuator control motor circuit high
P2109	Intake manifold runner control circuit low (bank 1)
P2010	Intake manifold runner control circuit high (bank 1)
P2111	Throttle actuator control system - stuck open
P2112	Throttle actuator control system - stuck closed
P2118	Throttle actuator control motor current range or performance
P2119	Throttle actuator control throttle body range or performance
P2120	Throttle or pedal position sensor or switch "d" circuit fault
P2121	Throttle or pedal position sensor or switch "d" circuit range or performance
P2122	Throttle or pedal position sensor or switch "d" circuit low input
P2123	Throttle or pedal position sensor or switch "d" circuit high input
P2125	Throttle or pedal position sensor or switch "e" circuit fault
P2127	Throttle or pedal position sensor or switch "e" circuit low input
P2128	Throttle or pedal position sensor or switch "e" circuit high input
P2135	Throttle or pedal position sensor or switch "a" or "b" voltage correlation
P2138	Throttle or pedal position sensor or switch "d" or "e" voltage correlation
P2195	Oxygen (A/F) sensor signal stuck lean (bank 1 sensor 1)
P2196	Oxygen (A/F) sensor signal stuck rich (bank 1 sensor 1)
P2237	Oxygen (A/F) sensor pumping current circuit or open (bank 1 sensor 1)
P2238	Oxygen (A/F) sensor pumping current circuit low (bank 1 sensor 1)
P2239	Oxygen (A/F) sensor pumping current circuit high (bank 1 sensor 1)
P2252	Oxygen (A/F) sensor reference ground circuit low (bank 1 sensor 1)
P2253	Oxygen (A/F) sensor reference ground circuit high (bank 1 sensor 1)
P2401	EVAP leak detection pump stuck off
P2402	EVAP leak detection pump stuck on
P2419	EVAP switching valve control circuit low
P2420	EVAP switching valve control circuit high
P2610	ECM/PCM internal engine off timer performance

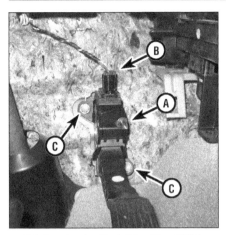

3.2 Accelerator Pedal Position (APP) sensor details

A   Accelerator Pedal Position (APP) sensor
B   Electrical connector
C   Mounting bolts

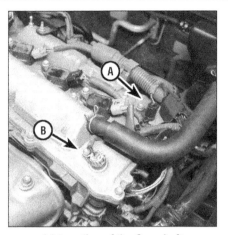

4.3 Location of the Camshaft Position Sensors

A   Intake Camshaft Position Sensor
B   Exhaust Camshaft Position Sensor

6.3 Location of the Crankshaft Position Sensor

3   Accelerator Pedal Position (APP) sensor - replacement

Caution: *Do not apply any oil or spray lubricant to the APP sensor assembly, as this could damage the sensor.*
Note: *The APP sensor and the accelerator pedal are replaced as a single assembly.*
1   Make sure the ignition key is in the Off position.
2   Disconnect the APP sensor electrical connector, then remove the mounting bolts (see illustration) and remove the sensor assembly.
3   Installation is the reverse of removal. Be sure to tighten the accelerator pedal assembly mounting bolts securely.

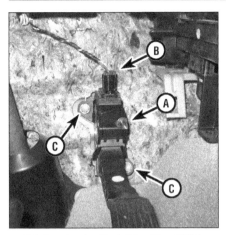

7.3 Location of the ECT sensor

4   Camshaft Position (CMP) sensors - replacement

Note: *The CMP sensors are located on the top rear (driver's side) of the valve cover. See Section 15 for information regarding the camshaft VVT (variable valve timing) system and VVT oil control solenoids.*
1   Make sure the ignition key is in the Off position.
2   Remove the engine cover.
3   The sensors are mounted at the left end (driver's side) of the valve cover (see illustration).
4   Disconnect the sensor electrical connector, remove the mounting bolt and remove the camshaft position sensor from the valve cover, pulling it straight out.
5   Installation is the reverse of removal.

5   Throttle Position Sensor (TPS) - replacement

1   All models are equipped with electronic throttle bodies; the TPS is an integral component of the throttle body. If the TPS is defective, the throttle body will have to be replaced (see Chapter 4).

6   Crankshaft Position (CKP) sensor - replacement and testing

Note: *The sensor is mounted on the timing chain cover next to the crankshaft pulley.*

Replacement
1   Make sure the ignition key is in the Off position.

2   Working under the vehicle, remove the inner fender shield (see Chapter 11).
3   Disconnect the CKP sensor electrical connector, then remove the bolt and detach the sensor (see illustration).
4   Installation is the reverse of removal.
5   If reusing the old sensor, check the O-ring for damage and replace if necessary.

Testing
6   Disconnect the CKP sensor connector.
7   Using a digital volt / ohm meter preset to measure resistance, measure between the two terminals of the CKP sensor. With the engine cold (14 to 122 degrees-F), resistance should be 1630 to 2740 ohms. With the engine hot (122 to 212 degrees-F), resistance should be 2065 to 3225 ohms. If the resistance value does not fall within the ranges specified for the proper temperature, replace the sensor.

7   Engine Coolant Temperature (ECT) sensor - replacement and testing

Replacement
Warning: *Wait until the engine has cooled completely before beginning this procedure.*
Caution: *Handle the coolant sensor with care. Damage to this sensor will affect the operation of the entire fuel injection system.*
1   Make sure the ignition key is in the Off position.
2   Drain approximately 1/4 of the engine coolant (see Chapter 1), to bring it below the level of the ECT sensor.
3   Disconnect the electrical connector and carefully unscrew the sensor from the rear (driver's end) of the cylinder head (see illustration). Also be sure to withdraw and discard

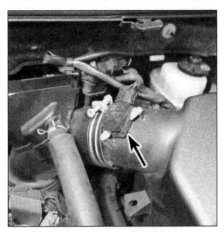

8.2 Location of the MAF sensor

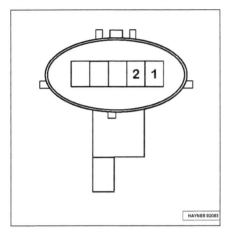

8.7 Measure terminals 1 and 2 of the MAF sensor with an ohmmeter

9.5 Knock sensor location

the old sealing ring.

Note: *On these models, it is NOT necessary to wrap the threads of the new sensor with Teflon sealing tape to prevent leakage - the new O-ring should be able to accomplish this sufficiently*

4   Ensure a new sealing ring is used upon installation.

5   Installation is the reverse of removal. Tighten the sensor securely.

Testing

Note: *A food thermometer or laser gauge can be used to check the water temperature*

6   Remove the ECT sensor (see previous Steps).

7   Partially submerge the sensor in a suitable pan/pot of water (only the end of the sensor submerged - up to the hex fitting), making sure the electrical terminals are dry. The sensor's voltage will be measured with the water's varying temperature. This step is performed easiest with sensor suspended by a length of wire.

8   Using a digital volt or ohm meter set to measure resistance, measure between the two terminals of the ECT sensor. With the water in the pot/pan at room temperature (68 degrees-F), resistance should be 2.32k to 2.59k ohms. With the water heated to 176 degrees-F, resistance should be 0.310k to 0.326k ohms.

9   If the resistance of the sensor in water does not fall within the specified range at room temperature or when heated, replace the sensor.

8   Mass Airflow (MAF) sensor - replacement and testing

Replacement

1   Make sure the ignition key is in the Off position.

2   The Mass Airflow (MAF) sensor is

located on the air filter housing (see illustration).

3   Disconnect the electrical connector from the MAF sensor.

4   Remove the two sensor retaining screws and remove the MAF sensor and O-ring.

5   Installation is the reverse of removal. Be sure to inspect the MAF sensor O-ring and replace it if damaged.

Testing

Note: *Before performing the test (with the MAF sensor removed), visually inspect the bottom end of the sensor. If there is a buildup of dirt or debris on the sensor element (it resembles the look of a resistor(s) recessed into the sensor), clean it off with aerosol MAF sensor cleaner or electronics-safe cleaner.*

6   Remove the MAF sensor (see previous Steps).

7   Using a Digital Volt Ohm Meter (DVOM) set to measure resistance, measure between terminals 1 and 2 of the MAF sensor (see illustration).

8   In below freezing temperatures (-4 degrees-F), resistance should be 13.6k to 18.4k ohms. At room temperature (68 degrees-F), resistance should be 2.21k to 2.69k ohms. With the engine warmed up and sensor installed (140 degrees-F), resistance should be 0.49k to 0.67k ohms.

9   Knock sensor - replacement and testing

Warning: *Wait for the engine to cool completely before performing this procedure.*

Warning: *Gasoline is extremely flammable, so take extra precautions when you work on any part of the fuel system. The fuel system is under pressure and MUST be relieved before beginning the procedure*

Note: *The knock sensor is located on the rear of the engine block, directly below the cylinder head.*

Replacement

1   Relieve the fuel system pressure (see Chapter 4).

2   Disconnect the cable from the negative battery terminal (see Chapter 5).

3   Drain the cooling system (see Chapter 1).

4   Remove the intake manifold to access the sensor (see Chapter 4).

5   Disconnect the electrical connector. Unscrew the mounting bolt and remove the sensor (see illustration).

6   Install the knock sensor noting the following specifications:

a)   *Install the sensor in a parallel line to the engine block (connector of sensor should be no more than a 7-degree angle (upward or downward) from a horizontal plane).*

b)   *Tighten the sensor mounting bolt to 15 ft-lbs (20 Nm).*

7   Don't overtighten the sensor or damage may occur. The remainder of installation is the reverse of removal.

Testing

8   Disconnect the knock sensor connector.

9   Using a Digital Volt Ohm Meter (DVOM) set to measure resistance, measure between the two terminals of the knock sensor. With the engine cold (68 degrees-F), resistance should be 120k to 280k ohms.

10   Oxygen sensors - replacement

Note: *Because it is installed in the exhaust manifold or pipe, which contracts when cool, the oxygen sensor may be very difficult to loosen when the engine is cold. Rather than risk damage to the sensor (assuming you are planning to reuse it in another manifold or pipe), start and run the engine for a minute or two, then shut it off. Be careful not to burn yourself during the following procedure.*

10.2 Upstream oxygen (air/fuel ratio) sensor details (radiator hose disconnected for clarity)

A   *Electrical connector*   B   *Sensor*

10.3 Downstream oxygen sensor connector

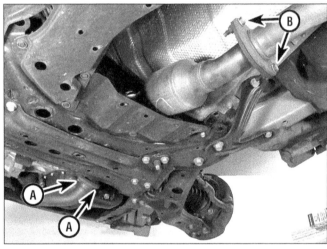

10.4 Remove the front (A) and rear (B) exhaust pipe section fasteners

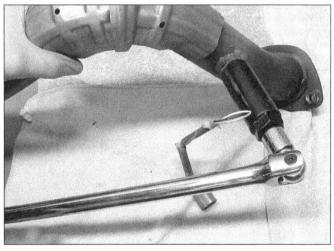

10.5 Use a slotted socket to unscrew the oxygen sensor

Note: *"Upstream" refers to the sensor before the catalytic converter and is also referred to as the Air/Fuel Ratio Sensor (Bank 1, Sensor 1). "Downstream" refers to the sensor after the catalytic converter and is also referred to as the Heated Oxygen Sensor (Bank 1, Sensor 2).*

1   Disconnect the cable from the negative battery terminal (see Chapter 5).

2   The upstream sensor (see illustration) can be replaced without raising the vehicle. Locate the sensor in the engine compartment and unplug the sensor electrical connector. Also unclip (but don't disconnect) any interfering coolant hoses from their retainers, as necessary. Unscrew the sensor using a special oxygen sensor socket (tool shown in illustration 10.5).

3   If you're replacing the downstream sensor, raise the vehicle and support it on jackstands. Access the oxygen sensor harness and unplug the electrical connector (see illustration).

4   At first glance, it may look possible to remove the downstream sensor at this stage. However, there might be a lack of 'wrenching room' to allow for unscrewing. If this happens to be the case, this section of the exhaust pipe will have to be removed (see illustration). A penetrating lubricant sprayed onto the bolts/threads will help with removing the bolts.

Note: *The best tool for removing an oxygen sensor is a special slotted socket, especially if you're planning to reuse a sensor. If you don't have this tool, and you plan to reuse the sensor, be extremely careful not to round-off the hex fitting when unscrewing the sensor.*

5   Unscrew the sensor from the exhaust manifold or exhaust pipe (see illustration).

6   Apply anti-seize compound to the threads of the sensor to facilitate future removal. The threads of new sensors should already be coated with this compound, but if you're planning to reuse an old sensor, re-coat the threads. Install the sensor and tighten

it securely. For the downstream sensor, install the exhaust pipe section and tighten the fasteners securely.

7   Reconnect the electrical connector of the pigtail lead to the main wiring harness (and attach any connector retaining clips, if necessary).

8   Lower the vehicle (if it was raised), then test drive it and verify that no trouble codes have been set.

11   Powertrain Control Module (PCM) - removal and installation

Warning: *The models covered by this manual are equipped with a Supplemental Restraint System (SRS), more commonly known as airbags. Always disable the airbag system before working in the vicinity of any airbag system components to avoid the possibility of accidental deployment of the airbag, which could*

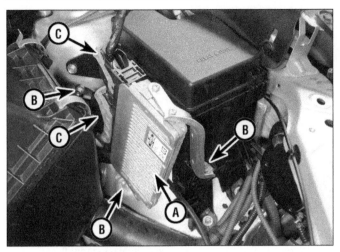

11.4 Typical PCM mounting details

A   PCM
B   PCM mounting bracket fasteners
C   PCM harness connectors

12.1 EVAP canister purge valve location (air filter housing removed for clarity)

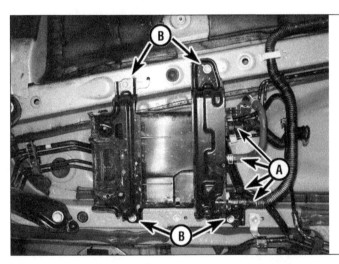

12.7 Disconnect the EVAP canister hoses and electrical connector (A) and remove the bracket bolts (B)

cause personal injury (see Chapter 12).

Caution: *To avoid electrostatic discharge damage to the PCM, handle the PCM only by its case. Do not touch the electrical terminals during removal and installation. If available, ground yourself to the vehicle with an anti-static ground strap, available at computer supply stores.*

Note: *The Powertrain Control Module (PCM) may also be referred to as the Engine Control Module (ECM) or Engine Control Unit (ECU).*

1   The Powertrain Control Module (PCM) is located in the engine compartment on the driver's side, next to the relay and fuse box.

2   Disconnect the cable from the negative battery terminal (see Chapter 5).

3   Remove the air filter housing, if necessary, to provide working clearance (see Chapter 4).

4   Disconnect the electrical connectors from the PCM (see illustration). Release the lever safety tabs, and at the same time, pull the levers outward. Remove the PCM mounting bolts.

5   Carefully remove the PCM. Transfer the

mounting brackets onto the new PCM, if necessary.

Caution: *Avoid any static electricity damage to the computer by grounding yourself to the body before touching the PCM and using a special anti-static pad to store the PCM on / in once it is removed.*

6   Installation is the reverse of removal.

7   The TRANSAXLE COMPENSATION CODE and RESET MEMORY procedures (A/T initialization) must be performed by a qualified technician or repair facility.

8   Perform the idle initialization procedure (see Chapter 4, Section 12).

---

12 Evaporative Emissions Control (EVAP) system - component replacement

---

Warning: *Gasoline and gasoline vapors are extremely flammable, so take extra precautions when you work on any part of the fuel system. Don't smoke or allow open flames or*

bare light bulbs near the work area, and don't work in a garage where a gas-type appliance (such as a water heater or clothes dryer) is present. Since gasoline is carcinogenic, wear fuel-resistant gloves when there's a possibility of being exposed to fuel, and, if you spill any fuel on your skin, rinse it off immediately with soap and water. Mop up any spills immediately and do not store fuel-soaked rags where they could ignite. When you perform any kind of work on the fuel system, wear safety glasses and have a Class B type fire extinguisher on hand.

EVAP canister purge valve

Note: *The EVAP canister purge valve is located on the side of the air intake tube. The canister purge valve may also be referred to as the purge Vacuum Switching Valve (VSV) - not to be confused with the ACIS VSV.*

1   Release the purge valve from the air intake tube by detaching the claw at the base of the valve (see illustration).

2   Disconnect the electrical connector and harness clip, then disconnect the EVAP hoses from the purge valve.

3   Installation is the reverse of removal.

EVAP canister

4   Disconnect the cable from the negative battery terminal.

5   Raise the vehicle and support it securely on jackstands.

6   Remove the inner and outer under-vehicle splash shields located in front of the fuel tank (see Chapter 4, Section 7).

Note: *See Chapter 4, Section 5 for general examples on fuel/EVAP line connection types.*

7   Disconnect the three EVAP hoses and leak detection pump electrical connector from the canister. Remove the four bolts and separate the canister from the underside of the vehicle (see illustration).

8   If a new canister is to be installed,

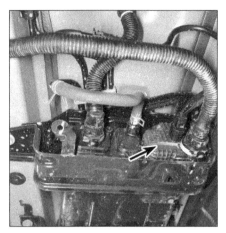

12.11 Location of the leak detection pump on the EVAP canister

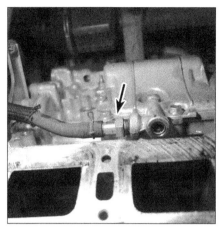

13.2 Location of the PCV valve (viewed from above - intake manifold removed)

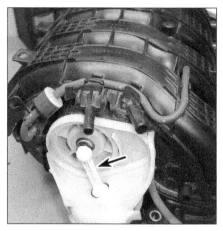

15.2a ACIS actuator bell-crank (integral with intake manifold)

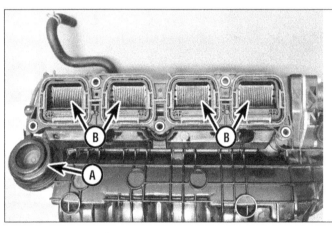

15.2b Identifying the ACIS vacuum tank (A) and intake air control valve (B) inside of the intake manifold

remove the brackets (held together by plastic claws to be detached) and leak detection pump, transferring to the new canister.

9   Installation is the reverse of removal.

Leak Detection Pump (LDP)

Note: *The LDP is also referred to as the canister pump module. The LDP is attached to the EVAP canister.*

10   Remove the EVAP canister (see previous Steps).

11   Clean the canister of dirt and debris before removing the LDP (see illustration).

12   Depress the locking tabs on each side securing the LDP to the canister. Then, at the same time, pull the LDP off of the canister, using a screwdriver engaged into the prying slot for leverage.

13   Installation is the reverse of removal.

13   Positive Crankcase Ventilation (PCV) system - general information

1   The Positive Crankcase Ventilation (PCV) system reduces hydrocarbon emissions by scavenging crankcase vapors. It does this by circulating fresh air from the air cleaner through the crankcase, where it mixes

with blow-by gases and is then rerouted through a PCV valve to the intake manifold. The PCV system uses intake manifold vacuum to draw crankcase vapors from the crankcase into the intake manifold. The PCV valve varies the amount of blow-by gases that can be returned to the intake manifold in proportion to the amount of intake manifold vacuum available, but in the event of a backfire it prevents the entry of combustion gases into the manifold. To maintain idle quality, the PCV valve restricts the flow when intake manifold vacuum is high.

2   The PCV system consists of a fresh-air inlet hose between the air intake duct and the crankcase, the PCV valve itself and the PCV hose (also referred to as the crankcase ventilation hose), which connects the PCV valve to the intake manifold. The PCV valve is attached to a crankcase vent oil separator, located underneath the intake manifold (see illustration). The crankcase vent oil separator acts as a baffle to prevent oil in the crankcase from entering the PCV valve.

3   Inspect the fresh-air inlet hose and the PCV valve hose (crankcase ventilation hose) for cracks, leaks and other damage. Disconnect the hose from the crankcase vent oil separator and the intake manifold and check the inside for obstructions. If a hose is clogged, replace it. If the PCV valve is clogged or oth-

erwise not working properly, replace it - do not try to clean it.

4   PCV valve replacement is covered in Chapter 1.

14   Catalytic converter - replacement

Note: *Because of a Federally mandated extended warranty which covers emissions-related components such as the catalytic converter, check with a dealer service department before replacing the converter at your own expense.*

The catalytic converter is incorporated into the exhaust manifold and is replaced as one unit. Refer to Chapter 2A, Section 9.

15   Acoustic Control Induction System (ACIS) - description and component replacement

Description

1   The PCM-controlled Acoustic Control Induction System (ACIS) varies the effective length of the intake manifold runners in response to engine speed and the angle of the throttle plate inside the throttle body. This capability increases efficiency and power at low and high speeds.

2   The ACIS consists of a PCM-controlled Vacuum Switching Valve (VSV), an actuator, a vacuum tank and an intake air control valve. The VSV controls the intake vacuum applied to the actuator. The actuator is a vacuum diaphragm that uses a pushrod and bellcrank to open and close the intake air control valve (see illustration). The vacuum tank, bellcrank and pushrod are an integral part of the intake manifold. The vacuum tank is used to maintain a constant source of vacuum to keep the intake air control valve closed during deceleration (low vacuum). The intake air control valve, which is also an integral part of the intake manifold (see illustration), opens

and closes to alter the effective length of the intake manifold runners in two stages. There is one large intake air control valve for all four intake runners.

3   At low-to-medium speeds, the PCM activates the VSV, sending vacuum to the actuator diaphragm. The actuator closes the intake air control valve, increasing the length of the intake manifold and improving intake efficiency.

4   At higher speeds, the PCM deactivates the VSV, cutting vacuum to the actuator diaphragm. The actuator opens the intake air control valve, decreasing the length of the intake manifold and improving engine power.

VSV replacement

Note: *The vacuum tank/bell crank and the intake air control valve are integral components of the intake manifold. If these particular components fail, replace the intake manifold (see Chapter 2A). For the replacement of the IAC valve actuator, see Section 16.*

Note: *The VSV is located at the top end of the intake manifold (passenger's side)*

5   Remove the engine cover.

6   Locate the VSV (see illustration).

7   Disconnect the electrical connector from the VSV.

8   Clearly label the vacuum hoses, then disconnect both hoses from the VSV.

9   Remove the VSV mounting bolt and remove the VSV.

10   Installation is the reverse of removal.

16 Intake air control valve actuator - replacement

Warning: *Gasoline is extremely flammable, so take extra precautions when you work on any part of the fuel system. The fuel system is under pressure and MUST be relieved before beginning the procedure.*

Note: *The intake air control valve actuator is*

located on the intake manifold, near the throttle body.

1   Remove the intake manifold (see Chapter 2A).

2   Remove the two bolts and the actuator from the intake manifold (see illustration).

Caution: *DO NOT remove the two nuts or bolt on the intake manifold - only remove the two bolts as shown.*

3   If reusing the same actuator, inspect the O-ring and replace if damaged.

4   Installation is the reverse of removal.

17 Variable Valve Timing-intelligent (VVT-i) system - description and component replacement

Description

1   Variable Valve Timing-intelligent (VVT-i) system is used on all models. The VVT-i system varies intake camshaft timing to produce valve timing that is optimized for the driving conditions. The VVT-i system achieves this by using engine oil pressure to advance or retard the controller on the front end of each intake camshaft.

2   The VVT-i system consists of two CMP sensors, the Powertrain Control Module (PCM), two VVT-i oil control valves and two controllers (intake camshaft sprocket/actuator assemblies).

3   The controller consists of a timing rotor (for the VVT-i sensor), a housing with an impeller-type vane inside it, a lock pin and the actual timing chain sprocket for the intake camshaft. The vane is fixed on the end of the camshaft.

4   The camshaft timing oil control valve is a PCM-controlled device that controls and directs the flow of oil to the advance or retard passages leading to the controller. There is one oil control valve for each camshaft. A spring-loaded spool valve inside the oil control

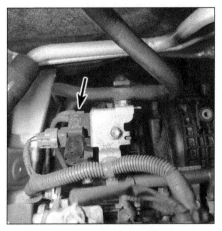

15.6 Identifying the Acoustic Control Induction System Vacuum Switching Valve (ACIS VSV)

valve directs oil pumped into the valve toward either the advance outlet or the retard outlet port, depending on the engine condition.

Oil Control Valve (OCV)

Note: *Make sure the ignition key is in the OFF position before before beginning*

Replacement

Note: *Two OCVs are located on top of the valve cover - one for each camshaft. The intake side OCV is mounted at the rear, the exhaust side OCV is mounted toward the front.*

5   Remove the engine cover.

6   Disconnect the electrical connector from the camshaft timing oil control valve (see illustration). Remove the engine harness mounting fasteners and move the harness to the side.

7   Remove the camshaft timing oil control valve mounting bolt and remove the oil control valve from the valve cover.

8   Installation is the reverse of removal.

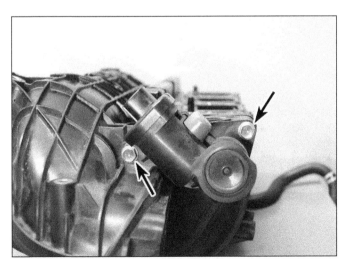

16.2 Remove the bolts and the actuator from the manifold

17.6 Identifying camshaft timing oil control valves

A   Intake side camshaft timing oil control valve
B   Exhaust side camshaft timing oil control valve

Testing

9   Remove the OCV (see previous Steps).

10   Using a Digital Volt Ohm Meter (DVOM) set to measure resistance, measure between the two terminals of the OCV. Resistance of the OCV should be 6.9 to 7.9 Ohms, at room temperature (68 degrees-F).

11   Next, apply battery voltage to the OCV to check solenoid functionality. Viewing the OCV from the connector terminals with the valve held upright (connector at the top; bracket facing downward), positive (+) voltage should be applied to the right side terminal, and negative (-) voltage at the left. Do not apply the battery voltage for more than a couple seconds.

12   If the OCV does not measure within the specified value or operate properly, replace it.

Chapter 7 Part A
Automatic transaxle

Contents

Specifications

Torque specifications   Ft-lbs (unless otherwise indicated)   Nm

Note: *One foot-pound (ft-lb) of torque is equivalent to 12 inch-pounds (in-lbs) of torque. Torque values below approximately 15 foot-pounds are expressed in inch-pounds, because most foot-pound torque wrenches are not accurate at these smaller values.*

	Ft-lbs (unless otherwise indicated)	Nm
Torque converter-to-driveplate bolts	32	44
Transaxle-to-engine bolts		
Three upper bolts	47	64
Two lower side bolts	34	46
Four bottom bolts	32	44
Transaxle-to-transfer assembly fasteners	51	68.6
Transaxle oil cooler bolts	10	13.5
Transaxle shift lever nut	9	12
Shift cable-to-shift lever nut	9	12
Engine mounts		
Rear engine mount through-bolt	70	95
Front mount-to-subframe bolts	70	95
Front subframe bolts	See Chapter 10	

1   General information

1   All information on the automatic transaxle is included in this Chapter.

2   Because of the complexity of the automatic transaxles and the specialized equipment necessary to perform most service operations, this Chapter contains only those procedures related to general diagnosis, routine maintenance, adjustment and removal and installation.

3   If the transaxle requires major repair work, it should be left to a dealer service department or an automotive or transmission repair shop. Once properly diagnosed you can, however, remove and install the transaxle yourself and save the expense, even if the repair work is done by a transmission shop.

2   Diagnosis - general

Note: *Automatic transaxle malfunctions may be caused by five general conditions: poor engine performance, improper adjustments, hydraulic malfunctions, mechanical malfunctions or malfunctions in the computer or its signal network. Diagnosis of these problems should always begin with a check of the easily repaired items: fluid level and condition (see Chapter 1) and shift linkage adjustment. Next, perform a road test to determine if the problem has been corrected or if more diagnosis is necessary. Because the transaxle relies on many sensors in the engine control system, and since the transaxle shift points are controlled by the Powertrain Control Module, you'll also want to check to see if any trouble codes have been stored on the PCM (see Chapter 6 for a list of trouble codes and how to extract them). If the problem persists after the preliminary tests and corrections are completed, additional diagnosis should be done by a dealer service department or transmission repair shop. Refer to the* Troubleshooting *Section at the front of this manual for information on symptoms of transaxle problems.*

Note: *If the engine is malfunctioning, do not proceed with the preliminary checks until it has been repaired and runs normally.*

Preliminary checks

1   Drive the vehicle to warm the transaxle to normal operating temperature.

2   Check the fluid level as described in Chapter 1:

a) *If the fluid level is unusually low, add fluid to bring it to the appropriate level, then check for external leaks (see below).*

b) *If the fluid level is abnormally high, drain off the excess, then check the drained fluid for contamination by coolant. The presence of engine coolant in the automatic transmission fluid indicates that a failure has occurred in the internal radiator walls that separate the coolant from the transmission fluid (see Chapter 3).*

c) *If the fluid is foaming, drain it and refill the transaxle, then check for coolant in the fluid, or a high fluid level.*

3   Inspect the shift cable (see Section 4). Make sure that it's properly adjusted and that the cable operates smoothly.

Fluid leak diagnosis

4   Most fluid leaks are easy to locate visually. Repair usually consists of replacing a seal or gasket. If a leak is difficult to find, the following procedure may help.

5   Identify the fluid. Make sure it's transmission fluid and not engine oil or brake fluid (automatic transmission fluid is a deep red color).

6   Try to pinpoint the source of the leak. Drive the vehicle several miles, then park it over a large sheet of cardboard. After a minute or two, you should be able to locate the leak by determining the source of the fluid dripping onto the cardboard.

7   Make a careful visual inspection of the suspected component and the area immediately around it. Pay particular attention to gasket mating surfaces. A mirror is often helpful for finding leaks in areas that are hard to see.

8   If the leak still cannot be found, clean the suspected area thoroughly with a degreaser or solvent, then dry it.

9   Drive the vehicle for several miles at normal operating temperature and varying speeds. After driving the vehicle, visually inspect the suspected component again.

10   Once the leak has been located, the cause must be determined before it can be properly repaired. If a gasket is replaced but the sealing flange is bent, the new gasket will not stop the leak. The bent flange must be straightened.

11   Before attempting to repair a leak, check to make sure that the following conditions are corrected or they may cause another leak.

Note: *Some of the following conditions cannot be fixed without highly specialized tools and expertise. Such problems must be referred to a transmission shop or a dealer service department.*

Gasket leaks

12   Check the pan periodically. Make sure the bolts are tight, no bolts are missing, the

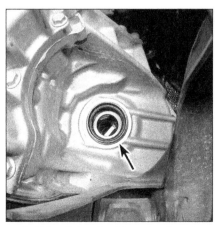

3.5 Identifying the driveaxle oil seal

gasket is in good condition and the pan is flat (dents in the pan may indicate damage to the valve body inside).

13   If the pan gasket is leaking, the fluid level or the fluid pressure may be too high, the vent may be plugged, the pan bolts may be too tight, the pan sealing flange may be warped, the sealing surface of the transaxle housing may be damaged, the gasket may be damaged or the transaxle casting may be cracked or porous. If sealant instead of gasket material has been used to form a seal between the pan and the transaxle housing, it may be the wrong sealant.

Seal leaks

14   If a transaxle seal is leaking, the fluid level or pressure may be too high, the vent may be plugged, the seal bore may be damaged, the seal itself may be damaged or improperly installed, the surface of the shaft protruding through the seal may be damaged or a loose bearing may be causing excessive shaft movement.

15   Make sure the dipstick tube seal is in good condition and the tube is properly seated. Periodically check the area around the speedometer gear or sensor for leakage. If transmission fluid is evident, check the O-ring for damage.

Case leaks

16   If the case itself appears to be leaking, the casting is porous and will have to be repaired or replaced.

17   Make sure the oil cooler hose fittings are tight and in good condition.

Fluid comes out vent pipe

18   If this condition occurs, the transaxle is overfilled, there is coolant in the fluid, the vent is plugged or the drain-back holes are plugged.

3   Driveaxle oil seal replacement

1   Oil leaks frequently occur due to wear of the driveaxle oil seals and/or the speedometer drive gear oil seal and O-rings. Replacement of these seals is relatively easy, since the repairs can be performed without removing the transaxle from the vehicle.

2   The driveaxle oil seals are located on the sides of the transaxle, where the inner ends of the driveaxles are splined into the differential side gears. If you suspect that a driveaxle oil seal is leaking, raise the vehicle and support it securely on jackstands. If the seal is leaking, you'll see lubricant on the side of the transaxle, below the seal.

3   Remove the driveaxle (see Chapter 8).

4   On 4WD models, if servicing the passenger's seal, remove the transfer unit (see Chapter 7B, Section 5).

5   On all models, using a screwdriver or prybar, carefully pry the oil seal out of the transaxle bore (see illustration).

6   If the oil seal cannot be removed with a screwdriver or prybar, a special oil seal

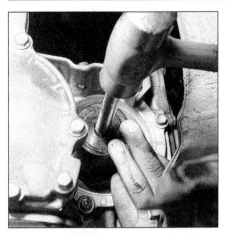

3.7 Use a seal installer, a large socket or a piece of pipe to install the new seal

4.6 Remove the nut (A) and the clip (B) and disconnect the shift cable from the transaxle

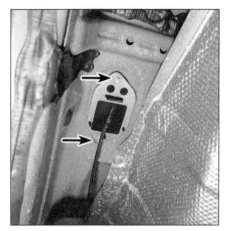

4.7 Remove the two bolts and the shift cable grommet

removal tool (available at auto parts stores) will be required.

7   Carefully install a new seal using a seal installer or large socket to a depth of -0.0197 to 0.0197 in (-0.5 to 0.5 mm) (see illustration).

8   Lubricate the lip of the new seal with multi-purpose grease, then install the drive-axle (see Chapter 8). Be careful not to damage the lip of the new seal.

---

4   Shift cable - removal, installation and adjustment

Warning: *These models are equipped with airbags. Always disable the airbag system before working in the vicinity of any airbag system component to avoid the possibility of accidental deployment of the airbag(s), which could cause personal injury (see Chapter 12).*

Removal and installation

1   If helpful for access, raise the front of the vehicle and support it securely on jackstands.

2   Remove the center console (see Chapter 11).

3   Remove the cable from the shift lever assembly and bracket as described in Section 6.

4   Remove the air filter housing (see Chapter 4, Section 11).

5   Remove the nut and disconnect the shift cable from the transaxle shift lever (see illustration).

6   Remove the clip and detach the shift cable housing from the bracket on the transaxle (see illustration).

7   Remove the two bolts attaching the shift cable grommet to the vehicle body (see illustration).

8   Remove the cable through the engine compartment.

9   Installation is the reverse of removal.

10   Adjust the shift cable.

Adjustment

11   Shift the shift lever to the Neutral position.

12   Locate the cable adjuster in the center console. Slide the white cover toward the front of the vehicle, and pull out the green lock piece.

13   Gently pull the shift cable rod toward the rear of the vehicle by hand to pull the cable tight (see illustration).

14   Press the lock piece into the adjuster case and lock it, then slide the cover back into place (see illustration).

Note: *Slide the cover past the protrusion of the lock piece.*

15   Check the operation of the transaxle in each shift lever position (try to start the engine in each gear - the starter should operate in the Park and Neutral positions only).

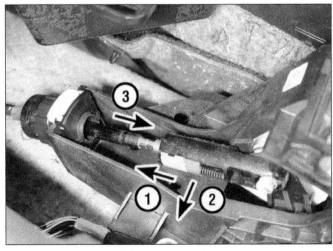

4.13 Adjust the cable

1   *Slide the cover toward the front*
2   *Pull out the lock*
3   *Pull the shift cable rod toward the rear of the vehicle*

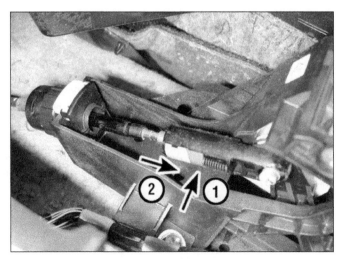

4.14 Lock the cable after adjustment

1   *Press the lock in*
2   *Slide the cover toward the rear*

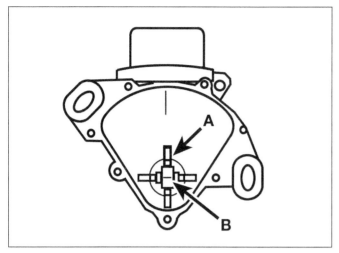

5.9 To adjust the PNP, align the line (A) with the groove (B)

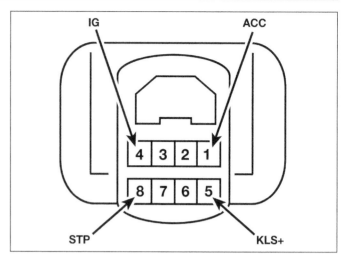

6.4 Shift lever assembly connector terminals

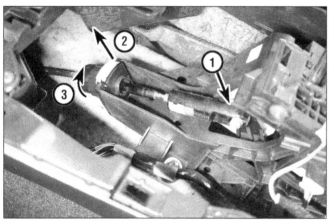

6.9 Disconnect the shift cable
1   Pry the cable off
2   Pull up the white stopper
3   Hold stopper up while rotating the knob 180 degrees

Terminal No.	Condition	DVOM reading
1 (ACC)	Ignition ON	11 to 14 volts
	Ignition in ACC	11 to 14 volts
	Ignition OFF	Less than 1 volt
8 (STP)	Brake pedal pressed	11 to 14 volts
	Brake pedal released	Less than 1 volt
5 (KLS+)	Ignition in ACC and shift lever in Park	Less than 1 volt
	Ignition in ACC and shift lever NOT in Park	11 to 14 volts
4 (IG)	Ignition ON	11 to 14 volts
	Ignition OFF	Less than 1 volt

5   Park/Neutral Position (PNP) switch - replacement and adjustment

1   If the engine will start with the shift lever in any position other than Park or Neutral, adjust the Park/Neutral Position switch.
2   Remove the air filter housing (see Chapter 4, Section 11).
3   Apply the parking brake and block the rear wheels and shift the transaxle into Neutral.
4   Remove the nut and disconnect the shift cable from the lever.
5   Remove the nut and lift off the manual shift lever.
6   If the switch is to be adjusted, loosen the switch retaining bolts.
7   If the switch is to be replaced, remove the bolts, disconnect the electrical connector, and remove the switch.
8   Install the new switch as necessary and loosely install the bolts.

9   To adjust the switch, align the line on the PNP switch body with the groove in the PNP switch pivot (see illustration).
10   Tighten the switch retaining bolts securely.
11   Install the components previously removed.
12   Check the operation of the switch. If the vehicle still starts in any position other than Park or Neutral, replace the switch.

6   Shift lever assembly - check, removal and installation

Check

1   Set the parking brake and block the wheels to prevent the vehicle from rolling.
2   Remove the center console (see Chapter 11, Section 22).
3   Access the shift lever assembly 8-pin electrical connector. Do not disconnect the connector.

4   Using a digital volt/ohm meter (DVOM) set to measure voltage, connect the Black lead to the vehicle body (ground) and use the Red lead to measure at the following terminals by back-probing from the rear of the connector (see illustration). If the results are not as shown, replace the shift lever assembly.

Removal

5   Set the parking brake and block the wheels to prevent the vehicle from rolling.
6   Unscrew the shift knob to remove it.
7   Remove the center console (see Chapter 11, Section 22).
8   Place the shift lever in Neutral and carefully pry the shift cable end from the shift lever assembly.
9   Pull up on the cable stopper and hold while rotating the knob 180 degrees counterclockwise (see illustration) to remove the cable from the shift lever assembly bracket.
10   Disconnect both connectors and the harness clips and secure the harness to the side of the shift lever housing.

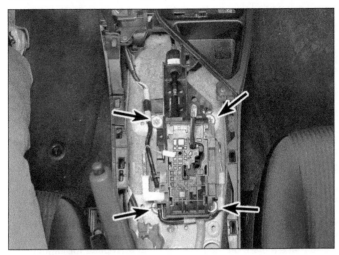

6.11 Remove the bolts and the shift lever assembly

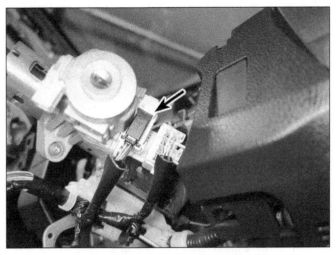

7.15 Locating the key interlock solenoid

11   Remove the four bolts securing the shift lever assembly to the floor, and remove the shift lever assembly from the vehicle (see illustration).

Installation

12   Installation is reverse of removal. Ensure the transaxle and the shift lever are still in Neutral.

13   To attach the cable to the shift lever assembly bracket, insert the cable, rotate the knob 180 degrees clockwise and press the lock down. The knob may need to be rotated more or less to get the lock to slide down.

14   Press the shift cable end over the shift lever stud until it clicks into place.

15   Adjust the shift cable (see Section 4).

16   Verify the transaxle shift into all positions properly and the shift lock/key interlock operate correctly (see Section 7).

7   Shift lock system - description, check and replacement

Warning: *These models are equipped with airbags. Always disable the airbag system before working in the vicinity of any airbag system component to avoid the possibility of accidental deployment of the airbag(s), which could cause personal injury (see Chapter 12).*

Description

1   The shift lock system prevents the shift lever from being shifted out of Park or Neutral until the brake pedal is applied. It also prevents the key from being removed from the ignition lock cylinder until the shift lever is placed in the Park position.

2   The shift lock control solenoid and ECU are part of the shift lever assembly and not serviced separately. The key interlock solenoid is attached to the ignition lock cylinder housing.

Functional check

3   If the components are not functioning properly, check the shift lever assembly electrical operation (see Section 6).

Shift lock operation

4   Move the shift lever to the Park position.

5   Turn the ignition switch to the Off position.

6   Check that the shift lever cannot be moved to any position other than the Park position.

7   Turn the ignition switch On, depress the brake pedal and check that the shift lever can be moved to other positions.

Shift lock release button operation

8   Move the shift lever with the shift lock release button pressed, check that the lever can be moved to any position other than Park. If this cannot be done, check the shift lever assembly.

Note: *Do not disconnect the shift lock control ECU connector.*

Key interlock operation

9   Turn the ignition switch to the On position.

10   Depress the brake pedal and move the shift lever to any position other than Park.

11   Check that the ignition switch cannot be turned to the Off position.

12   Move the shift lever to the Park position, turn the ignition switch to the Off position, and check that the key can be removed.

Replacement

Shift lock solenoid and ECU

13   The shift lock control solenoid and ECU are part of the shift lever assembly, and not serviceable separately. If either of these components are diagnosed as faulty, the shift lever assembly must be replaced (see Section 6).

Key interlock solenoid

14   Remove the steering column covers (see Chapter 11, Section 24).

15   Locate the key interlock solenoid attached to the steering lock actuator and disconnect the electrical connector (see illustration).

16   Remove the single screw and remove the solenoid from the lock actuator (see illustration).

17   Installation is the reverse of removal.

18   Verify proper operation.

8   Automatic transaxle - removal and installation

Warning: *Gasoline is extremely flammable so take extra precautions when you work on any part of the fuel system. See* Fuel system warnings *in Chapter 4, Section 1.*

Note: *The TRANSAXLE COMPENSATION CODE and RESET MEMORY procedures (A/T initialization) must be performed when replacing the automatic transaxle assembly, engine assembly or ECM. The TRANSAXLE COMPENSATION CODE and RESET MEMORY can be performed only with the intelligent tester and should be done by a dealer service department or transmission repair shop.*

Removal

1   Disconnect the negative battery cable (see Chapter 5, Section 3).

2   Remove the wipers and cowl (see Chapter 11, Section 13).

3   Remove the engine cover.

4   Remove the air filter housing (see Chapter 4, Section 11).

5   Loosen the front wheel lug nuts, then raise and support the vehicle on a lift or jackstands. Remove the front wheels.

6   Remove the center, left and right vehicle undercovers.

8.9 Remove the bolts at both collectors for the exhaust pipe and remove the pipe

8.10 Place a mark on the converter and driveplate for alignment during installation

8.12 Remove the clamps and disconnect the transaxle oil cooler hoses from the transaxle

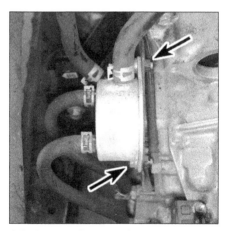

8.13 Remove the two bolts and detach the transaxle oil cooler (with the hoses connected)

8.19a Remove the two top transaxle-to-engine bolts near the starter opening . . .

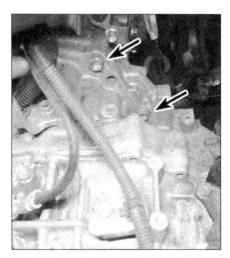

8.19b . . . then remove the two top transaxle-to-engine bolts on the opposite side from the starter . . .

7   Remove the starter (see Chapter 5, Section 9).

8   Drain the automatic transaxle fluid (see Chapter 1), then remove the driveaxles (see Chapter 8).

9   Remove the four bolts and the exhaust pipe that runs under the engine (see illustration). Remove the exhaust gaskets.

10   Remove the cover from the bellhousing and mark the relationship of the torque converter to the driveplate (see illustration).

11   Remove the driveplate-to-torque converter bolts. Turn the crankshaft for access to all six bolts. Turn the crankshaft in a clockwise direction only (as viewed from the front).

12   Slide back the clamps and disconnect the transaxle oil cooler hoses from the transaxle (see illustration).

Note: *Place a drain pan under the transaxle to catch any fluid while removing the transaxle oil cooler and hoses.*

13   Remove the two bolts and remove the transaxle oil cooler assembly from the transaxle (leave the hoses connected) (see illustration).

14   Disconnect the transaxle connectors,

harness routing clips/clamps, and disconnect the PCM connectors.

15   Disconnect the shift cable at the transaxle (see Section 4).

16   Attach an engine support fixture to the lifting hook at the transaxle end of the engine. If no hook is provided, use a suitable bolt of the proper size and thread pitch to attach the support fixture chain to a hole at the end of the cylinder head.

Note: *Engine support fixtures can be obtained at most equipment rental yards and some auto parts stores.*

17   Remove the front subframe (see Chapter 10, Section 19).

18   Support the transaxle with a jack - preferably a jack made for this purpose (available at most tool rental yards). Safety chains will help steady the transaxle on the jack.

19   Remove the nine engine-to-transaxle mounting bolts (see illustrations).

Note: *Different length bolts are used - be sure to note the location of each bolt so they can be returned to their original positions when the transaxle is installed.*

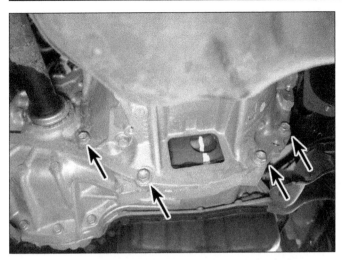

8.19c . . . then the four lower transaxle-to-engine bolts . . .

8.19d . . . and last, the engine-to-transaxle bolt on the back side of the engine block

20   Remove the two bolts attaching the front engine mount to the front crossmember.

21   Remove the four bolts attaching the front crossmember to the vehicle, and remove the crossmember.

22   On AWD models, remove the driveshaft (see Chapter 8).

23   Remove the transaxle mount through-bolt (see Chapter 2A, Section 17).

24   Move the transaxle away from the engine to disengage it from the engine block dowel pins. Make sure the torque converter is detached from the driveplate. Clamp a pair of locking pliers on the bellhousing case. The pliers will prevent the torque converter from falling out while you're removing the transaxle.

25   Remove any remaining brackets or components from the transaxle that are to be reused on the new transaxle.

Installation

26   Installation is the reverse of removal noting the following:

a)   *Tighten the transaxle fasteners to the torque listed in this Chapter's Specifications.*

b)   *Fill the transaxle fluid to the proper level (see Chapter 1, Section 30).*

c)   *On AWD models, fill the transfer case with the proper type and amount of lubricant (see Chapter 1).*

d)   *Adjust the shift cable (see Section 4).*

e)   *The TRANSAXLE COMPENSATION CODE and RESET MEMORY procedures (A/T initialization) must be performed by a qualified technician or repair facility.*

f)   *Road test the vehicle and verify proper operation.*

---

9   Automatic transaxle overhaul - general information

---

1   In the event of a problem occurring, it will be necessary to establish whether the fault is electrical, mechanical or hydraulic in nature, before repair work can be contemplated. Diagnosis requires detailed knowledge of the transaxle's operation and construction, as well as access to specialized test equipment, and so is deemed to be beyond the scope of this manual. It is therefore essential that problems with the automatic transaxle are referred to a dealer service department or other qualified repair facility for assessment.

2   Note that a faulty transaxle should not be removed before the vehicle has been diagnosed by a knowledgeable technician equipped with the proper tools, as troubleshooting must be performed with the transaxle installed in the vehicle.

Notes

Chapter 7 Part B
Transfer case

Contents

Specifications

Torque specifications

Ft-lbs (unless otherwise indicated)   **Nm**

Note: *One foot-pound (ft-lb) of torque is equivalent to 12 inch-pounds (in-lbs) of torque. Torque values below approximately 15 ft-lbs are expressed in inch-pounds, because most foot-pound torque wrenches are not accurate at these smaller values.*

	Ft-lbs	Nm
Transfer unit-to-transaxle fasteners	51	68.6
Driveaxle bracket-to-engine fasteners	47	63.7
Rear engine mount bracket-to-transaxle fasteners	33	45

1   General information

Note: *The 4WD system may also be referred to as the dynamic torque control AWD system.*

1   All-wheel drive (AWD) models are equipped with a transfer unit mounted to the transaxle. Drive is transmitted from the engine, through the transaxle to the front wheels by driveaxles, and through a transfer unit and a driveshaft to the rear differential, then to the rear wheels by driveaxles.

2   The AWD system detects driving conditions based on input signals from each control module, switch, steering angle sensor, wheel speed sensors, and the acceleration sensor. The system controls the current passing through the solenoid of the electromagnetic coupling, and controls the optimum torque to the rear wheels.

3   We don't recommend trying to rebuild a transfer unit at home. It's difficult to overhaul without the special tools, and units are available for less than it would cost to rebuild your own.

Between terminals	Switch position	Resistance (ohms)
6 and 9	Pressed (On)	Less than 1 ohm (continuity)
6 and 9	Released (Off)	10 ohm or greater (open)

2.4 4WD control switch test details

---

2   4WD control switch - replacement

1   Remove the switch panel trim on the left of the steering column (see Chapter 11, Section 23).
2   Depress the two tabs using a small, flat-bladed screwdriver and pull the switch out from the instrument panel enough to disconnect the electrical connector and remove the switch.
3   Installation is reverse of removal.

4WD control switch test

4   Using a digital ohmmeter (DVOM) set to measure resistance, test the switch as described in the table (see illustration). If the readings do not match, replace the switch.

---

3   4WD control unit - replacement

Note: *The 4WD control unit is located at the top of the instrument panel.*
1   Disconnect the negative battery cable (see Chapter 5, Section 3).
2   Remove the upper instrument panel (see Chapter 11, Section 25).
3   Remove the harness clips from the defroster nozzle assembly, detach the three retainers and remove the defroster nozzle assembly from the vehicle.
4   Below the defroster nozzle assembly, remove the two retainers (down at the bottom) and pull the heater-to-register duct upwards to detach the six retainers and remove the heater-to-register duct from the vehicle.
5   Locate the 4WD control unit and disconnect the electrical connector.
6   Remove the mounting bolt and remove the 4WD control unit from the vehicle.
7   Installation is reverse of removal.
8   If the back door was open during this procedure, after reconnecting the negative battery terminal and connections, close the back door to relearn the closed position.

---

4   Transfer unit oil seal - replacement

1   Raise and support the vehicle on hoist or jackstands.

Left seal

Note: *The left seal is between the transaxle and the transfer unit.*
2   Remove the transfer unit (see Section 5).
3   Secure the transfer unit and use a seal puller to remove the left seal between the transfer unit and the transaxle.
4   Carefully install a new seal using a seal installer or large socket to a depth of 0.375 to 0.413 in (9.5 to 10.5 mm).
5   Installation is the reverse of removal. Tighten the fasteners to the specifications in this Chapter's Specifications.
6   Fill the transfer unit with the correct lubricant to the proper level (see Chapter 1, Section 27).

Right seal

7   Drain the transfer unit oil (see Chapter 1).
8   Remove the passenger's front driveaxle (see Chapter 8, Section 2).
9   Remove the driveaxle bearing bracket fasteners and the bracket.
10   Use a seal puller to remove the left seal between the transfer unit and the transaxle.
11   Carefully install a new seal using a seal installer or large socket to a depth of 0.296 to 0.334 in (7.5 to 8.5 mm).
12   Installation is the reverse of removal. Tighten the fasteners to the specifications in this Chapter's Specifications.
13   Fill the transfer unit with the correct oil to the proper level.

Output shaft

14   The output shaft seal is part of the pinion assembly and should be serviced by a qualified technician as special tools and procedures are necessary to ensure the pinion is properly set up after removal.

---

5   Transfer unit - removal and installation

Note: *The transfer unit is attached to the transaxle. To remove the transfer unit, the transaxle must first be removed.*

Removal

1   Disconnect the cable from the negative battery terminal (see Chapter 5).
2   Drain the transfer unit oil (see Chapter 1, Section 27).
3   Remove the transaxle assembly (see Chapter 7A, Section 8).
4   Remove the rear engine mounting bracket fasteners and bracket.
5   Remove the transfer unit mounting fasteners and separate the transfer unit from the transaxle.

Note: *It is recommended to replace the studs for the transfer unit-to-transaxle mounting when removed.*

Installation

6   Install the transfer unit to the transaxle.
7   Install the mounting fasteners and tighten the fasteners to the torque listed in this Chapter's Specifications.
8   Install the rear engine mounting bracket fasteners and tighten them to the torque listed in this Chapter's Specifications.
9   Install the transaxle assembly (see Chapter 7A, Section 8).
10   The remainder of installation is the reverse of removal.
11   Fill the transfer unit with the correct oil to the proper level.

---

6   Transfer unit overhaul - general information

1   In the event of a problem occurring, it will be necessary to establish whether the fault is electrical, mechanical or hydraulic in nature before repair work can be contemplated. Diagnosis requires detailed knowledge of the transaxle's operation and construction, as well as access to specialized test equipment, and so is deemed to be beyond the scope of this manual. It is therefore essential that problems with the transfer unit are referred to a dealer service department or other qualified repair facility for assessment.
2   Note that a faulty transfer unit should not be removed before the vehicle has been diagnosed by a knowledgeable technician equipped with the proper tools, as troubleshooting must be performed with the transaxle and transfer unit installed in the vehicle.

Chapter 8
Driveline

Contents

Specifications

Torque specifications

Ft-lbs (unless otherwise indicated)   **Nm**

Note: *One foot-pound (ft-lb) of torque is equivalent to 12 inch-pounds (in-lbs) of torque. Torque values below approximately 15 foot-pounds are expressed in inch-pounds, because most foot-pound torque wrenches are not accurate at these smaller values.*

	Ft-lbs	Nm
Differential mounting bracket-to-differential bolts (4WD models)	72	98
Differential mounting bracket-to-body bolt(s) (4WD models)	41	55
Differential pinion flange nut (4WD models)	181	245
Driveaxle/hub nut		
Front (all models)	215	292
Rear (4WD models)	159	216
Driveaxle (right) bearing support bracket bolt (see illustration 2.12)	24	32
Driveshaft flange-to-rear differential nuts (4WD models)	26	35
Driveshaft flange-to-transfer unit nuts (4WD models)	26	35
Driveshaft center support bearing bolts (4WD models)	27	37
Electromagnetic coupling-to-differential bolts (4WD models)	15	20

1   General information

1   The information in this Chapter deals with the components from the transaxle to the front wheels (and rear wheels on AWD models).

2   Since some of the procedures covered in this Chapter involve working under the vehicle, make sure it's securely supported on sturdy jackstands or on a hoist where the vehicle can be easily raised and lowered.

2   Driveaxles - removal and installation

Front

Removal

1   Remove the wheel cover or hub cap. Unstake the nut with a punch or chisel (see illustration).

2   Break the driveaxle/hub nut loose with a socket and large breaker bar (see illustration).

Note: *If the design of the wheel prevents installing a socket on the nut, skip to Step 4.*

3   Loosen the wheel lug nuts, raise the vehicle and support it securely on jackstands. Remove the wheel. Drain the transaxle lubricant (see Chapter 1).

4   If, due to wheel design, you could not loosen the driveaxle hub nut with the wheel installed, insert a punch into the cooling vanes of the brake disc and allow it to rest against the brake caliper mounting bracket, then loosen the nut (see illustration).

5   Remove the engine undercovers to allow access to the driveaxle being removed.

6   Disconnect the ABS speed sensor connector.

7   Remove the driveaxle/hub nut. Mark the relationship of the axle to the hub for installation.

8   Disconnect the stabilizer bar link from the stabilizer bar.

9   Remove the nuts and bolt securing the balljoint to the control arm, then pry the control arm down and separate the lower control arm from the balljoint (see Chapter 10).

10   If reinstalling the same axle and hub, mark the relationship of the driveaxle to the hub for installation (see illustration).

11   Swing the knuckle/hub assembly out (away from the vehicle) until the end of the driveaxle is free of the hub. Support the outer end of the driveaxle with a piece of wire to avoid unnecessary strain on the inner CV joint.

Note: *If the driveaxle splines stick in the hub, tap on the end of the driveaxle with a plastic hammer.*

12   To remove the passenger's side driveaxle, remove the snap-ring and single bolt from the bearing bracket (see illustration) and pull the driveaxle out of the transaxle (2WD) or transfer unit (AWD). Support the CV joints and carefully remove the driveaxle from the vehicle.

13   To remove the driver's side driveaxle,

2.1 If the driveaxle nut is staked, use a center punch to unstake it (wheel removed for clarity)

2.2 Loosen the driveaxle/hub nut with a long breaker bar

2.4 Prevent the hub from turning by inserting a punch into the brake disc cooling vanes

2.10 Mark the driveaxle and hub for installation

2.12 Squeeze the snap ring (A) to remove, then remove the bolt from the bearing bracket (B)

use a slide hammer with an angled claw attachment engaged with one of the slots in the inner CV joint housing to pull the joint out of the transaxle (see illustration). Support the CV joints and carefully remove the driveaxle from the vehicle.

Installation

14   Pry the old spring clip from the inner end of the driveaxle and install a new one (see illustrations). Lubricate the differential or intermediate shaft seal with multi-purpose grease and raise the driveaxle into position while supporting the CV joints.

15   If you're installing the driver's side driveaxle, insert the inner CV joint splines into the differential side gear and make sure the spring clip locks in its groove.

16   If you're installing the passenger's side driveaxle, install a new bearing support snapring onto the driveaxle, insert the splined end of the inner CV joint's shaft into the differential side gear, then install the center support bearing bolt and snap-ring. Tighten the bolt to the torque listed in this Chapter's Specifications.

17   Apply a light coat of multi-purpose grease to the outer CV joint splines, pull out on the strut/steering knuckle assembly and install the stub axle into the hub. Be sure to align the reference marks (if installing the original driveaxle).

18   Reconnect the balljoint to the lower control arm and tighten the nuts and bolt (see the torque specifications in Chapter 10).

19   Install a new driveaxle/hub nut. Tighten the hub nut securely, but don't try to tighten it to the actual torque specification until you've lowered the vehicle to the ground.

20   Grasp the inner CV joint housing (not the driveaxle) and pull out to make sure the driveaxle has seated securely in the transaxle.

21   Install the wheel and lug nuts, then lower the vehicle.

Note: *If the wheels prevented you from loosening the nut without removing them, prevent the hub from turning by using the method described in Step 4, then install the wheel.*

22   Tighten the lug nuts to the torque listed in the Chapter 1 Specifications. Tighten the hub nut to the torque listed in this Chapter's Specifications. Stake the nut to the groove in the driveaxle, using a hammer and punch.

23   Refill the transaxle (see Chapter 1).

Rear (AWD models)

Removal

24   Remove the wheel cover or hub cap. Remove the cotter pin and nut lock from the driveaxle/hub nut. Break the hub nut loose with a socket and large breaker bar.

Note: *If the design of the wheel prevents you from getting the socket on the nut, loosen it after the wheel has been removed. To prevent the hub from turning, brace a large prybar between two of the wheel lugs and allow it to contact the ground.*

25   Block the front wheels to prevent the vehicle from rolling. Loosen the wheel lug nuts, raise the rear of the vehicle and support it securely on jackstands. Remove the wheel.

26   Disconnect the ABS speed sensor connector.

27   Remove the driveaxle/hub nut. Mark the relationship of the axle to the hub for installation.

28   Remove the rear differential support braces to gain access to the inner CV joints.

29   Disconnect the stabilizer bar links from the bar, then pivot the bar up.

30   Remove the rear knuckle (see Chapter 10).

31   Carefully pry the inner end of the driveaxle from the differential, using a large screwdriver or prybar positioned between the differential and CV joint housing.

32   Pull the outer driveaxle CV joint splines from the hub. Support the CV joints and carefully remove the driveaxle from the vehicle.

Note: *If the driveaxle splines stick in the hub, tap on the end of the driveaxle with a plastic hammer.*

Installation

33   Apply a light coat of multi-purpose grease to the outer CV joint splines and install the stub axle into the hub. Be sure to align reference marks.

34   Insert the splined end of the inner CV joint into the differential side gear and make sure the spring clip locks in its groove.

35   Install the rear knuckle and reconnect the stabilizer bar links (see Chapter 10).

36   Install a new driveaxle/hub nut. Tighten the hub nut securely, but don't try to tighten it to the actual torque specification until you've lowered the vehicle.

Note: *Alternatively, prevent the hub from turning by using the method described in Step 24.*

37   Install the wheel and lug nuts, then lower the vehicle. Tighten the lug nuts to the torque listed in the Chapter 1 Specifications.

38   Tighten the driveaxle/hub nut to the torque listed in this Chapter's Specifications, then stake the nut.

---

3   Driveshaft (AWD models) - check, removal and installation

---

Check

1   Raise the rear of the vehicle and support it securely on jackstands. Block the front wheels to keep the vehicle from rolling off the stands. Release the parking brake and place the transmission in Neutral.

2   Crawl under the vehicle and visually inspect the driveshaft. Look for any dents or cracks in the tubing. If any are found, the driveshaft must be replaced.

3   Check for oil leakage at the front and rear of the driveshaft. Leakage where the driveshaft connects to the transfer case indicates a defective transfer case seal. Leakage where the driveshaft connects to the differential indicates a defective pinion seal.

4   While under the vehicle, have an assistant rotate a rear wheel so the driveshaft will

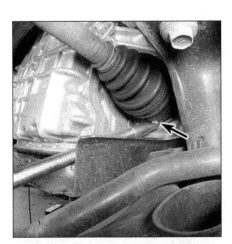

2.13 To separate the inner end of the driveaxle from the transaxle, use a slide hammer with a claw attachment to pull the inner joint from the transaxle

2.14a Pry the old spring clip from the inner end of the driveaxle with a small screwdriver or awl

2.14b To install the new spring clip, start one end in the groove and work the clip over the shaft end, into the groove

rotate. As it does, make sure the universal joints are operating properly without binding, noise or looseness. Listen for any noise from the center bearing, indicating it's worn or damaged. Also check the rubber portion of the center bearing for cracking or separation.

5   The universal joints can also be checked with the driveshaft motionless, by gripping your hands on either side of the joint and attempting to twist the joint. Any movement at all in the joint is a sign of considerable wear. Lifting up on the shaft will also indicate movement in the universal joints. If the joints are worn, front or rear portion of the driveshaft must be replaced as an assembly.

6   Finally, check the driveshaft mounting bolts at the ends to make sure they're tight.

Removal and installation

Note: *If the center support bearing requires service, it is recommended to take to a driveshaft/driveline specialist for repair.*

7   Raise the rear of the vehicle and support it securely on jackstands. Block the front wheels to prevent the vehicle from rolling.

8   Place the transmission in Neutral with

the parking brake off.

9   Make reference marks on the driveshaft flange and the differential pinion flange in line with each other (see illustration). This is to make sure the driveshaft is reinstalled in the same position to preserve the balance (see illustration).

10   Remove the bolts attaching the center support bearing to the vehicle and carefully lower the center support bearing (see illustration).
Caution: *DO NOT allow the angle of the center support bearing and driveshaft to exceed 20 degrees to prevent damage.*

11   Remove the rear universal joint flange nuts and washers. Turn the driveshaft (or wheels) as necessary to bring the nuts into the most accessible position. Remove the four nuts and washers (see illustration).

12   Remove the transfer unit output shaft flange nuts and washers. Turn the driveshaft (or wheels) as necessary to bring the nuts into the most accessible position. Remove the four nuts and washers.

13   Remove the driveshaft assembly.

14   Installation is the reverse of removal.

15   Reconnect the driveshaft first to the

transfer unit flange, then the differential flange. Be sure to align the marks and tighten the fasteners to the torque listed in this Chapter's Specifications.

16   Position the center support bearing bracket on the floorpan and install the bolts, then tighten the bolts to the torque listed in this Chapter's Specifications.

---

4   Differential oil seals (4WD models) - replacement

Pinion oil seal

Removal

1   Raise the rear of the vehicle and support it securely on jackstands. Block the front wheels to prevent the vehicle from rolling. Place the transmission in Neutral with the parking brake off.

2   Mark the relationship of the driveshaft to the pinion flange or coupling, then unbolt the driveshaft from the flange (see Section 3). Suspend the driveshaft with a piece of wire (don't let it hang by the center support bearing).

3   Drain the differential oil (see Chapter 1).

4   Remove the electromagnetic coupling mounting fasteners. Using a plastic hammer, tap the coupling cover and remove the coupling from the differential.

5   Remove the linear solenoid retaining snap-ring from the differential case, using a pair of expanding snap ring pliers.

6   Remove the 4WD linear solenoid from the yoke and discard the O-ring.

7   Remove the yoke mounting fasteners and remove the yoke from the differential.

8   Remove the coupling conical spring washer, noting the direction the washer is facing and remove the coupling shim.

9   Carefully pry out the diaphragm oil seal with a seal removal tool or a large screwdriver; make sure you don't scratch the seal bore.

3.9 Mark the relationship of the driveshaft to the differential pinion flange

3.10 The driveshaft center support bearing is retained by two bolts

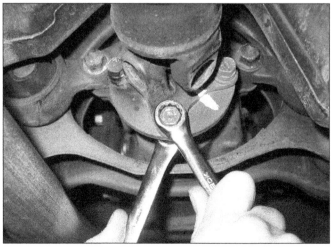

3.11 Using a back-up wrench to hold each bolt, break loose all four bolts

5.5 Remove the rear differential support bolts

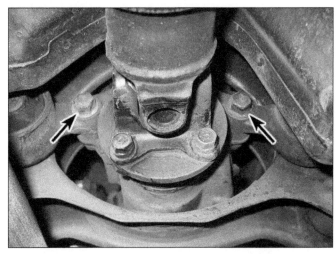

5.8 Remove the front differential support bolts

Installation

10   Lubricate the lips of the new seal and fill the space between the seal lips with wheel bearing grease.

11   Carefully install a new seal using a seal installer or large socket to a depth of 0.263 to 0.303 in (6.7 to 7.7 mm).

12   Install the coupling shim, and conical spring washer.

13   Install the yoke mounting fasteners and tighten the fasteners to the torque listed in this Chapter's Specifications.

14   Install a new O-ring onto the solenoid and install the solenoid and snap-ring.

15   The remainder of installation is the reverse of removal. Check the differential lubricant level and add some, if necessary, to bring it to the appropriate level (see Chapter 1).

Driveaxle oil seals

16   Raise the rear of the vehicle and support it securely on jackstands. Block the front wheels to prevent the vehicle from rolling. Place the transmission in Neutral with the parking brake off.

17   Remove the rear driveaxle(s) (see Section 2).

18   Carefully pry out the side gear shaft oil seal with a seal removal tool or a large screwdriver; make sure you don't scratch the seal bore.

19   Carefully install a new seal using a seal installer or large socket to a depth of 0.263 to 0.303 in (6.7 to 7.7 mm).

20   Lubricate the lip of the new seal with multi-purpose grease, then install the driveaxle(s). Be careful not to damage the lip of the new seal.

21   Check the differential lubricant level and add some, if necessary, to bring it to the appropriate level (see Chapter 1).

5   Differential (4WD models) - removal and installation

Removal

1   Raise the rear of the vehicle and support it securely on jackstands. Block the front wheels to prevent the vehicle from rolling. Place the transaxle in Neutral with the parking brake off.

2   Drain the differential lubricant (see Chapter 1).

3   Remove the fuel tank and rear exhaust pipe (see Chapter 4).

4   Remove the driveshaft (see Section 3).

5   Remove the rear differential support brace bolts (see illustration).

6   Disconnect the electromagnetic harness and breather clamp, then remove the breather tube.

7   Remove the rear driveaxles from the differential and secure out of the way (see Section 3).

8   Support the differential with a floor jack and remove the front and rear differential-to-body mounting fasteners (see illustration).

9   Slowly lower the jack and remove the differential from the vehicle.

Installation

10   Installation is the reverse of removal. Tighten all fasteners to the torque values listed in this Chapter's Specifications. Fill the differential with the proper lubricant (see Chapter 1).

Notes

Chapter 9
Brakes

Contents

Specifications

General

Brake fluid type	See Chapter 1
Brake pedal height	7.437 to 7.830 inches (188.9 to 198.9 mm)
Brake pedal freeplay	0.039 to 0.236 inch (1.0 to 6.0 mm)
Brake pedal reserve distance (minimum)	3.74 inches (95 mm)
Brake light switch-to-pedal clearance	0.0197 to 0.102 inch (0.5 to 2.6 mm)
Power brake booster pushrod-to-master cylinder piston clearance	0.0 inch (0.0 mm)

Disc brakes

Minimum brake pad thickness	See Chapter 1
Disc minimum thickness	Cast or stamped into disc
Disc runout limit	0.002 inch (0.05 mm)

Parking brake

Parking brake lever travel (with 45 lbs of force on the lever)	6 to 8 clicks

Torque specifications

Note: *One foot-pound (ft-lb) of torque is equivalent to 12 inch-pounds (in-lbs) of torque. Torque values below approximately 15 ft-lbs are expressed in inch-pounds, because most foot-pound torque wrenches are not accurate at these smaller values.*

	Ft-lbs (unless otherwise indicated)	Nm
Brake hose-to-caliper banjo bolts	22	30
Caliper mounting bolts		
Front	25	34
Rear	20	27
Caliper mounting bracket bolts		
Front	79	107
Rear	65	88
Deceleration sensor mounting nuts	41 in-lbs	4.5
Master cylinder-to-brake booster nuts	108 in-lbs	12
Power brake booster mounting nuts	108 in-lbs	12
Wheel cylinder mounting bolts	84 in-lbs	9.5
Wheel speed sensor bolt (front and rear)	75 in-lbs	8.5
Wheel lug nuts	See Chapter 1	

1 General information

1 The vehicles covered by this manual are equipped with hydraulically operated front and rear brake systems. The brakes are disc type front and rear. Both the front and rear brakes are self adjusting.

Hydraulic system

2 The hydraulic system consists of two separate circuits. The master cylinder has separate reservoirs for the two circuits, and, in the event of a leak or failure in one hydraulic circuit, the other circuit will remain operative.

Power brake booster

3 The power brake booster, utilizing engine manifold vacuum and atmospheric pressure to provide assistance to the hydraulically operated brakes, is mounted on the firewall in the engine compartment.

Parking brake

4 The parking brake operates the rear brakes only, through cable actuation. It's activated by a lever mounted in the center console.

Service

5 After completing any operation involving disassembly of any part of the brake system, always test drive the vehicle to check for proper braking performance before resuming normal driving. When testing the brakes, perform the tests on a clean, dry, flat surface. Conditions other than these can lead to inaccurate test results.

6 Test the brakes at various speeds with both light and heavy pedal pressure. The vehicle should stop evenly without pulling to one side or the other. Avoid locking the brakes, because this slides the tires and diminishes braking efficiency and control of the vehicle.

7 Tires, vehicle load and wheel alignment are factors which also affect braking performance.

Precautions

8 There are some general cautions and warnings involving the brake system on this vehicle:

a) *Use only brake fluid conforming to DOT 3 specifications.*

b) *The brake pads and linings contain fibers that are hazardous to your health if inhaled. Whenever you work on brake system components, clean all parts with brake system cleaner. Do not allow the fine dust to become airborne. Also, wear an approved filtering mask.*

c) *Safety should be paramount whenever any servicing of the brake components is performed. Do not use parts or fasteners that are not in perfect condition, and be sure that all clearances and torque specifications are adhered to. If you are at all unsure about a certain procedure, seek professional advice. Upon completion of any brake system work, test the brakes carefully in a controlled area before putting the vehicle into normal service. If a problem is suspected in the brake system, don't drive the vehicle until it's fixed.*

d) *Used brake fluid is considered a hazardous waste and it must be disposed of in accordance with federal, state and local laws. DO NOT pour it down the sink, into septic tanks or storm drains, or on the ground. Clean up any spilled brake fluid immediately and then wash the area with large amounts of water. This is especially true for any finished or painted surfaces.*

2 Troubleshooting

PROBABLE CAUSE	CORRECTIVE ACTION

No brakes - pedal travels to floor

1 Low fluid level 2 Air in system	1 and 2 Low fluid level and air in the system are symptoms of another problem a leak somewhere in the hydraulic system. Locate and repair the leak
3 Defective seals in master cylinder	3 Replace master cylinder
4 Fluid overheated and vaporized due to heavy braking	4 Bleed hydraulic system (temporary fix). Replace brake fluid (proper fix)

Brake pedal slowly travels to floor under braking or at a stop

1 Defective seals in master cylinder	1 Replace master cylinder
2 Leak in a hose, line, caliper or wheel cylinder	2 Locate and repair leak
3 Air in hydraulic system	3 Bleed the system, inspect system for a leak

Brake pedal feels spongy when depressed

1 Air in hydraulic system	1 Bleed the system, inspect system for a leak
2 Master cylinder or power booster loose	2 Tighten fasteners
3 Brake fluid overheated (beginning to boil)	3 Bleed the system (temporary fix). Replace the brake fluid (proper fix)
4 Deteriorated brake hoses (ballooning under pressure)	4 Inspect hoses, replace as necessary (it's a good idea to replace all of them if one hose shows signs of deterioration)

PROBABLE CAUSE | **CORRECTIVE ACTION**

Brake pedal feels hard when depressed and/or excessive effort required to stop vehicle

1 Power booster faulty	1 Replace booster
2 Engine not producing sufficient vacuum, or hose to booster clogged, collapsed or cracked	2 Check vacuum to booster with a vacuum gauge. Replace hose if cracked or clogged, repair engine if vacuum is extremely low
3 Brake linings contaminated by grease or brake fluid	3 Locate and repair source of contamination, replace brake pads or shoes
4 Brake linings glazed	4 Replace brake pads or shoes, check discs and drums for glazing, service as necessary
5 Caliper piston(s) or wheel cylinder(s) binding or frozen	5 Replace calipers or wheel cylinders
6 Brakes wet	6 Apply pedal to boil-off water (this should only be a momentary problem)
7 Kinked, clogged or internally split brake hose or line	7 Inspect lines and hoses, replace as necessary

Excessive brake pedal travel (but will pump up)

1 Drum brakes out of adjustment	1 Adjust brakes
2 Air in hydraulic system	2 Bleed system, inspect system for a leak

Excessive brake pedal travel (but will not pump up)

1 Master cylinder pushrod misadjusted	1 Adjust pushrod
2 Master cylinder seals defective	2 Replace master cylinder
3 Brake linings worn out	3 Inspect brakes, replace pads and/or shoes
4 Hydraulic system leak	4 Locate and repair leak

Brake pedal doesn't return

1 Brake pedal binding	1 Inspect pivot bushing and pushrod, repair or lubricate
2 Defective master cylinder	2 Replace master cylinder

Brake pedal pulsates during brake application

1 Brake drums out-of-round	1 Have drums machined by an automotive machine shop
2 Excessive brake disc runout or disc surfaces out-of-parallel	2 Have discs machined by an automotive machine shop
3 Loose or worn wheel bearings	3 Adjust or replace wheel bearings
4 Loose lug nuts	4 Tighten lug nuts

Brakes slow to release

1 Malfunctioning power booster	1 Replace booster
2 Pedal linkage binding	2 Inspect pedal pivot bushing and pushrod, repair/lubricate
3 Malfunctioning proportioning valve	3 Replace proportioning valve
4 Sticking caliper or wheel cylinder	4 Repair or replace calipers or wheel cylinders
5 Kinked or internally split brake hose	5 Locate and replace faulty brake hose

Brakes grab (one or more wheels)

1 Grease or brake fluid on brake lining	1 Locate and repair cause of contamination, replace lining
2 Brake lining glazed	2 Replace lining, deglaze disc or drum

Troubleshooting (continued)

PROBABLE CAUSE	CORRECTIVE ACTION

Vehicle pulls to one side during braking

1 Grease or brake fluid on brake lining	1 Locate and repair cause of contamination, replace lining
2 Brake lining glazed	2 Deglaze or replace lining, deglaze disc or drum
3 Restricted brake line or hose	3 Repair line or replace hose
4 Tire pressures incorrect	4 Adjust tire pressures
5 Caliper or wheel cylinder sticking	5 Repair or replace calipers or wheel cylinders
6 Wheels out of alignment	6 Have wheels aligned
7 Weak suspension spring	7 Replace springs
8 Weak or broken shock absorber	8 Replace shock absorbers

Brakes drag (indicated by sluggish engine performance or wheels being very hot after driving)

1 Brake pedal pushrod incorrectly adjusted	1 Adjust pushrod
2 Master cylinder pushrod (between booster and master cylinder)	2 Adjust pushrod incorrectly adjusted
3 Obstructed compensating port in master cylinder	3 Replace master cylinder
4 Master cylinder piston seized in bore	4 Replace master cylinder
5 Contaminated fluid causing swollen seals throughout system	5 Flush system, replace all hydraulic components
6 Clogged brake lines or internally split brake hose(s)	6 Flush hydraulic system, replace defective hose(s)
7 Sticking caliper(s) or wheel cylinder(s)	7 Replace calipers or wheel cylinders
8 Parking brake not releasing	8 Inspect parking brake linkage and parking brake mechanism, repair as required
9 Improper shoe-to-drum clearance	9 Adjust brake shoes
10 Faulty proportioning valve	10 Replace proportioning valve

Brakes fade (due to excessive heat)

1 Brake linings excessively worn or glazed	1 Deglaze or replace brake pads and/or shoes
2 Excessive use of brakes	2 Downshift into a lower gear, maintain a constant slower speed (going down hills)
3 Vehicle overloaded	3 Reduce load
4 Brake drums or discs worn too thin	4 Measure drum diameter and disc thickness, replace drums or discs as required
5 Contaminated brake fluid	5 Flush system, replace fluid
6 Brakes drag	6 Repair cause of dragging brakes
7 Driver resting left foot on brake pedal	7 Don't ride the brakes

Brakes noisy (high-pitched squeal)

1 Glazed lining	1 Deglaze or replace lining
2 Contaminated lining (brake fluid, grease, etc.)	2 Repair source of contamination, replace linings
3 Weak or broken brake shoe hold-down or return spring	3 Replace springs
4 Rivets securing lining to shoe or backing plate loose	4 Replace shoes or pads
5 Excessive dust buildup on brake linings	5 Wash brakes off with brake system cleaner
6 Brake drums worn too thin	6 Measure diameter of drums, replace if necessary
7 Wear indicator on disc brake pads contacting disc	7 Replace brake pads
8 Anti-squeal shims missing or installed improperly	8 Install shims correctly

PROBABLE CAUSE   **CORRECTIVE ACTION**

Brakes noisy (scraping sound)

1 Brake pads or shoes worn out; rivets, backing plate or brake metal contacting disc or drum

1 Replace linings, have discs and/or drums machined (or replace) shoe metal contacting disc or drum

Brakes chatter

Probable Cause	Corrective Action
1 Worn brake lining	1 Inspect brakes, replace shoes or pads as necessary
2 Glazed or scored discs or drums	2 Deglaze discs or drums with sandpaper (if glazing is severe, machining will be required)
3 Drums or discs heat checked	3 Check discs and/or drums for hard spots, heat checking, etc. Have discs/drums machined or replace them
4 Disc runout or drum out-of-round excessive	4 Measure disc runout and/or drum out-of-round, have discs or drums machined or replace them
5 Loose or worn wheel bearings	5 Adjust or replace wheel bearings
6 Loose or bent brake backing plate (drum brakes)	6 Tighten or replace backing plate
7 Grooves worn in discs or drums	7 Have discs or drums machined, if within limits (if not, replace them)
8 Brake linings contaminated (brake fluid, grease, etc.)	8 Locate and repair source of contamination, replace pads or shoes
9 Excessive dust buildup on linings	9 Wash brakes with brake system cleaner
10 Surface finish on discs or drums too rough after machining	10 Have discs or drums properly machined (especially on vehicles with sliding calipers)
11 Brake pads or shoes glazed	11 Deglaze or replace brake pads or shoes

Brake pads or shoes click

Probable Cause	Corrective Action
1 Shoe support pads on brake backing plate grooved or	1 Replace brake backing plate excessively worn
2 Brake pads loose in caliper	2 Loose pad retainers or anti-rattle clips
3 Also see items listed under Brakes chatter	

Brakes make groaning noise at end of stop

Probable Cause	Corrective Action
1 Brake pads and/or shoes worn out	1 Replace pads and/or shoes
2 Brake linings contaminated (brake fluid, grease, etc.)	2 Locate and repair cause of contamination, replace brake pads or shoes
3 Brake linings glazed	3 Deglaze or replace brake pads or shoes
4 Excessive dust buildup on linings	4 Wash brakes with brake system cleaner
5 Scored or heat-checked discs or drums	5 Inspect discs/drums, have machined if within limits (if not, replace discs or drums)
6 Broken or missing brake shoe attaching hardware	6 Inspect drum brakes, replace missing hardware

Rear brakes lock up under light brake application

Probable Cause	Corrective Action
1 Tire pressures too high	1 Adjust tire pressures
2 Tires excessively worn	2 Replace tires
3 Defective proportioning valve	3 Replace proportioning valve

Brake warning light on instrument panel comes on (or stays on)

Probable Cause	Corrective Action
1 Low fluid level in master cylinder reservoir (reservoirs with fluid level sensor)	1 Add fluid, inspect system for leak, check the thickness of the brake pads and shoes
2 Failure in one half of the hydraulic system	2 Inspect hydraulic system for a leak
3 Piston in pressure differential warning valve not centered	3 Center piston by bleeding one circuit or the other (close bleeder valve as soon as the light goes out)

Troubleshooting (continued)

PROBABLE CAUSE	CORRECTIVE ACTION

Brake warning light on instrument panel comes on (or stays on) (continued)

4 Defective pressure differential valve or warning switch	4 Replace valve or switch
5 Air in the hydraulic system	5 Bleed the system, check for leaks
6 Brake pads worn out (vehicles with electric wear sensors - small probes that fit into the brake pads and ground out on the disc when the pads get thin)	6 Replace brake pads (and sensors)

Brakes do not self adjust

Disc brakes

1 Defective caliper piston seals	1 Replace calipers. Also, possible contaminated fluid causing soft or swollen seals (flush system and fill with new fluid if in doubt)
2 Corroded caliper piston(s)	2 Same as above

Drum brakes

1 Adjuster screw frozen	1 Remove adjuster, disassemble, clean and lubricate with high-temperature grease
2 Adjuster lever does not contact star wheel or is binding	2 Inspect drum brakes, assemble correctly or clean or replace parts as required
3 Adjusters mixed up (installed on wrong wheels after brake job)	3 Reassemble correctly
4 Adjuster cable broken or installed incorrectly (cable-type adjusters)	4 Install new cable or assemble correctly

Rapid brake lining wear

1 Driver resting left foot on brake pedal	1 Don't ride the brakes
2 Surface finish on discs or drums too rough	2 Have discs or drums properly machined
3 Also see Brakes drag	

---

3   Anti-lock Brake System (ABS) and Vehicle Stability Control (VSC) - general information, trouble codes and component removal and installation

1   The Anti-lock Brake System (ABS) and Vehicle Stability Control (VSC) are designed to maintain vehicle steerability, directional stability and optimum deceleration under severe braking conditions and on most road surfaces. The ABS system is primarily designed to prevent wheel lockup during heavy or panic braking situations. It works by monitoring the rotational speed of each wheel and controlling the brake line pressure to each wheel when engaged. Data provided by the ABS wheel sensors is shared with the VSC electronic control unit (ECU). This very sophisticated system helps with traction control, over/understeering, down hill assist and acceleration control and handling. Another system added to vehicles with VSC is the Electronic Brake Force Distribution (EBD) system. This system

works with ABS system applications in accordance with the driving conditions and vehicle load. In addition, when the brakes are applied while cornering, it also controls the braking forces of the right and left wheels, helping to maintain vehicle stability.

General information
Actuator assembly
2   The actuator assembly consists of an electric hydraulic pump and four solenoid valves. The electric pump provides hydraulic pressure to charge the reservoirs in the actuator, which supplies pressure to the braking system. The pump and reservoirs are housed in the actuator assembly. The solenoid valves modulate brake line pressure during ABS operation.

Speed sensors
3   The speed sensors, which are located at each wheel, generate small electrical pulsations when the toothed sensor rotors are turning, sending a variable voltage signal to the ABS electronic control unit (ECU) indicating

wheel rotational speed.
4   The front speed sensors (see illustration) are mounted on the steering knuckles in close relationship to the toothed sensor rotors, which are integral with the outer constant velocity (CV) joints.

3.4 ABS front wheel speed sensor location

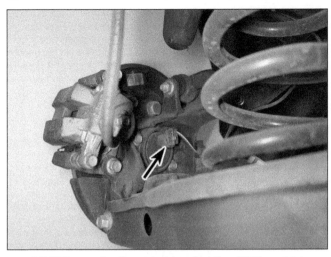

3.5 ABS rear wheel speed sensor location (2WD models)

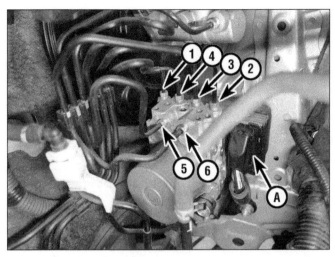

**3.12 12/2012 to 10/2015 ABS actuator brake line positions shown.
NOTE: 10/2015 and later models lines 5 and 6 have been reversed.**

A	Electrical connector	5	From rear connection of the master cylinder
1	To Right-front wheel		
2	To left-front wheel	6	From front connection of the master cylinder
3	To right-rear wheel		
4	To left-rear wheel		

5   On AWD models, the rear speed sensors are mounted to the rear knuckles. On 2WD models, the sensors are integral with the rear hub and bearing assemblies (see illustration).

Yaw rate and acceleration sensor

6   This sensor, which is part of the airbag sensor, relays negative acceleration data to the ABS computer, which uses this information to determine the forward pitch (yaw rate) of the vehicle during panic stops.

ABS and VSC computer

7   The function of the ECU is to accept and process information received from the wheel speed sensors to control the hydraulic line pressure, avoiding wheel lock up and slipping. The ECU also constantly monitors the system, even under normal driving conditions, to find faults within the system. If a problem develops within the system, an "ABS" light will glow on the dashboard. A diagnostic code will also be stored in the ECU, which will indicate the problem area or component. A scan tool is necessary to retrieve the codes.

Diagnosis and repair

8   If the dashboard warning light comes on and stays on while the vehicle is in operation, the ABS system requires attention and the system should be checked for stored trouble codes. Before checking for trouble codes, however, you should perform a few simple checks.

a) Check the brake fluid level in the reservoir.
b) Check that all electrical connectors are securely connected.
c) Check the fuses.

9   If the above preliminary checks do not rectify the problem, or if any stored trouble codes don't lead you to the problem, the vehicle should be diagnosed and repaired by a dealer service department or other repair shop.

Components - removal and installation

ABS actuator

10   Disconnect the cable from the negative terminal of the battery (see Chapter 5).
Note: *Wait at least 90 seconds for the SRS system to deactivate.*
11   Remove the air filter housing (see Chapter 2B).
12   Disconnect the ABS electrical connector by releasing the locking lever, then place some rags under the ABS actuator to catch any brake fluid that spills. Using a flare-nut wrench, unscrew the brake line fittings from the ABS actuator (see illustration).
Note: *It's a good idea to mark the individual brake lines so you can be sure to put them back into the same position. Use the list provided for reference when reinstalling the brake lines.*
Note: *10/2015 and later models the front and rear brake lines to the master cylinder are reversed from the illustration shown (see illustration 3.12)*
13   Remove any brackets securing the brake lines to the ABS bracket, then unbolt the ABS actuator bracket from the body and remove the actuator and bracket as one unit from the engine compartment.
14   Installation is the reverse of removal. Bleed the brake system (see Section 9).

Front wheel speed sensor

15   Loosen the wheel lug nuts, raise the front of the vehicle and support it securely on jackstands. Block the wheels at the opposite end.
16   Remove the inner fender liner (see Chapter 11).
17   Follow the sensor wiring harness up to the electrical connector, then unplug the connector.
18   Unbolt the wiring harness securing brackets.
19   Remove the sensor mounting bolt and detach the sensor and bushing from the steering knuckle (see illustration 3.4).
20   Installation is the reverse of removal. Be sure to tighten the sensor mounting bolt to the torque listed in this Chapter's Specifications. Tighten the wheel lug nuts to the torque listed in the Chapter 1 Specifications.

Rear wheel speed sensor

21   Loosen the wheel lug nuts, raise the rear of the vehicle and support it securely on jackstands. Block the wheels at the opposite end.
22   Remove the rear wheel.
23   Remove the rear deck trim side panel (see Chapter 11).
24   Disconnect the speed sensor connection and push the grommet through the wheel house along with the speed sensor harness.
25   Follow the speed sensor harness to the harness brackets, then remove the bolts securing the brackets to the vehicle.

2WD models

26   On these models the sensor is integral with the rear hub and bearing assembly. See

Chapter 10 for the hub and bearing replacement procedure.

AWD models

27   Remove the bolt securing the speed sensor to the rear knuckle.

28   Remove the sensor

29   Installation is the reverse of removal. Be sure to tighten the sensor mounting bolt to the torque listed in this Chapter's Specifications. Tighten the wheel lug nuts to the torque listed in the Chapter 1 Specifications.

4   Disc brake pads - replacement

Warning: *Disc brake pads must be replaced on both front or rear wheels at the same time - never replace the pads on only one wheel. Also, the dust created by the brake system is harmful to your health. Never blow it out with compressed air and don't inhale any of it. An approved filtering mask should be worn when working on the brakes. Do not, under any circumstances, use petroleum-based solvents to clean brake parts. Use brake system cleaner only!*

Front

1   Remove the cap from the brake fluid reservoir.

2   Loosen the wheel lug nuts, raise the vehicle and support it securely on jackstands. Block the wheels at the opposite end.

3   Remove the wheels. Work on one brake assembly at a time, using the assembled brake for reference if necessary.

4   Inspect the brake disc carefully as outlined in Section 6. If machining or replacement is necessary, follow the information in that Section to remove the disc, at which time the pads can be removed as well.

5   Push the piston back into its bore to provide room for the new brake pads. A C-clamp can be used to accomplish this (see illustration). As the piston is depressed to the bottom of the caliper bore, the fluid in the master cylinder will rise. Make sure that it doesn't overflow. If necessary, siphon off some of the fluid.

6   Follow the accompanying photos (illustrations 4.6a through 4.6k) for the actual pad replacement procedure. Be sure to stay in order and read the caption under each illustration.

7   When reinstalling the caliper, be sure to tighten the mounting bolts to the torque listed in this Chapter's Specifications. After the job has been completed, firmly depress the brake pedal a few times to bring the pads into contact with the disc. Check the level of the brake fluid, adding some if necessary. Check the operation of the brakes carefully before placing the vehicle into normal service.

4.5 Before removing the caliper, be sure to depress the piston into the bottom of its bore in the caliper with a large C-clamp to make room for the new pads

4.6a Always wash the brakes with brake cleaner before disassembling anything

4.6b To remove the caliper, remove the lower bolt (A) while holding the sliding pin with an open-end wrench; also remove the upper bolt (B)

4.6c Slide the caliper off of the brake pads and secure it so that the flexible brake hose is not being stretched or pinched

4.6d Remove the outer brake pad . . .

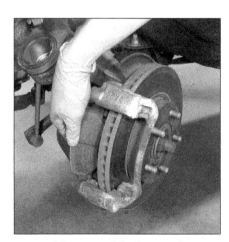

4.6e . . . and the inner pad

Rear disc brakes

8   Wash the brakes with brake cleaner before disassembling anything (see illustration).

9   Push the piston back into its bore to provide room for the new brake pads. A C-clamp can be used to accomplish this (see illustration). As the piston is depressed to the bot-

tom of the caliper bore, the fluid in the master cylinder will rise. Make sure that it doesn't overflow. If necessary, siphon off some of the fluid.

10   Follow the accompanying photos (illustrations 4.10a through 4.10h) for the actual pad replacement procedure. Be sure to stay

in order and read the caption under each illustration.

11   When reinstalling the caliper, be sure to tighten the mounting bolts to the torque listed in this Chapter's Specifications. After the job has been completed, firmly depress the brake pedal a few times to bring the pads into con-

4.6f Remove the brake pad support plates

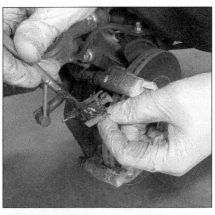

4.6g Lubricate the support plates where the pad will contact them

4.6h Install the support plates

4.6i Lubricate the caliper guide pins

4.6j Install the replacement brake pads in the caliper mounting bracket

4.6k Tighten the caliper mounting bolts to the specified torque

4.8 Rinse off any brake dust and debris with brake cleaner. Do NOT use compressed air

4.9 Press the caliper piston back in with a C clamp

4.10a Remove the upper caliper bolt by holding the pin with an open end wrench while using a socket on the retaining bolt

tact with the disc. Check the level of the brake fluid, adding some if necessary. Check the operation of the brakes carefully before placing the vehicle into normal service.

---

5   Disc brake caliper - removal and installation

Warning: *Dust created by the brake system is harmful to your health. Never blow it out with*

compressed air and don't inhale any of it. An approved filtering mask should be worn when working on the brakes. Do not, under any circumstances, use petroleum-based solvents to clean brake parts. Use brake system cleaner only.

Note: *If replacement is indicated (usually because of fluid leakage), it is recommended that the calipers be replaced, not overhauled. New and factory rebuilt units are available on an exchange basis, which makes this job*

quite easy. Always replace the calipers in pairs - never replace just one of them.

Removal

1   Loosen the front wheel lug nuts, raise the front of the vehicle and place it securely on jackstands. Remove the wheel.

Note: *If the caliper is being removed just to gain access to another component, don't disconnect the hose.*

2   Remove the bolt and disconnect the

4.10b You can gain access to the brake pads by rotating the caliper on the lower pin and hang the caliper in this position with a wire

4.10c Remove the brake pads, then remove the brake pad retaining clips by prying them out with a screwdriver

4.10d Before installing the replacement retaining clips, lubricate where the brake pads will come in contact with the clips.

4.10e Don't forget to lubricate the upper guide pin

4.10f Slide the lower guide pin out of the housing and lubricate the lower pin as well

4.10g With the lower pin back in place, install the replacement brake pads

4.10h Tighten the caliper mounting bolts to the specified torque

brake hose from the caliper. Plug the brake hose to keep contaminants out of the brake system and to prevent losing any more brake fluid than is necessary (see illustration). Discard the sealing washers - new ones should be used during installation.

3   Remove the caliper mounting bolts while holding the slide pin with an open-end wrench and a socket or boxed end wrench on the bolt (see illustration).

4   With the caliper mounting bolts removed, and the line disconnected, remove the caliper.

Installation

5   Install the caliper by reversing the removal procedure. Remember to replace the sealing washers (gaskets) at the brake hose-to-caliper connection. Tighten the cali-

per mounting bolts and the brake hose banjo fitting bolt to the torque listed in this Chapter's Specifications.

6   Bleed the brake circuit (see Section 9). Make sure there are no leaks from the hose connections. Test the brakes carefully before returning the vehicle to normal service.

---

6   Brake disc - inspection, removal and installation

Inspection

1   Loosen the wheel lug nuts, raise the vehicle and support it securely on jackstands. Remove the wheel and install the lug nuts to hold the disc in place against the hub flange.

Note: *If the lug nuts don't contact the disc*

when screwed on all the way, install washers under them.

2   Remove the brake caliper as outlined in Section 5. It isn't necessary to disconnect the brake hose. After removing the caliper bolts, suspend the caliper out of the way with a piece of wire (see illustration 4.6c). Remove the two mounting bracket-to-steering knuckle bolts (see illustration) and detach the mounting bracket.

3   Visually inspect the disc surface for score marks and other damage. Light scratches and shallow grooves are normal after use and may not always be detrimental to brake operation, but deep scoring requires disc removal and refinishing by an automotive machine shop. Be sure to check both sides of the disc (see illustration). If pulsating has been noticed during application of the brakes, suspect disc runout.

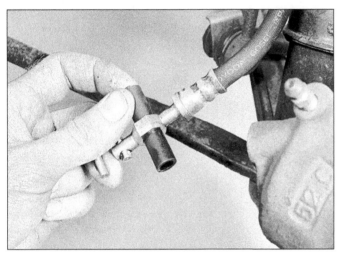

5.2 Using a piece of rubber hose of the appropriate size, plug the brake line; this will prevent brake fluid from leaking out and dirt and moisture from contaminating the system

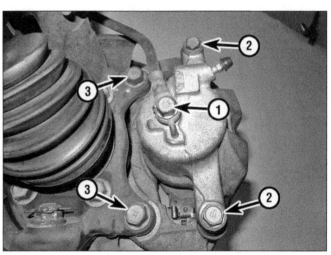

5.3 Caliper mounting details

1   Brake hose banjo bolt
2   Caliper mounting bolts
3   Caliper mounting bracket bolts

6.2 To remove the mounting bracket, remove these two bolts; be careful not to lose the pad support plates

6.3 The brake pads on this vehicle were obviously neglected, as they wore down completely and cut deep grooves into the disc - wear this severe means the disc must be replaced

4   To check disc runout, place a dial indicator at a point about 1/2-inch from the outer edge of the disc (see illustration). Set the indicator to zero and turn the disc. The indicator reading should not exceed the specified allowable run-out limit. If it does, the disc should be refinished by an automotive machine shop. **Note:** *The discs should be resurfaced regardless of the dial indicator reading, as this will impart a smooth finish and ensure a perfectly flat surface, eliminating any brake pedal pulsation or other undesirable symptoms related to questionable discs. At the very least, if you elect not to have the discs resurfaced, remove the glaze from the surface with emery cloth or sandpaper, using a swirling motion (see illustration).*

5   It's absolutely critical that the disc not be machined to a thickness under the specified minimum thickness. The minimum (or discard) thickness is cast or stamped into the inside of the disc (see illustration). The disc thickness can be checked with a micrometer (see illustration).

Removal

6   Remove the lug nuts that were put on to hold the disc in place and slide the disc off the hub. If you're removing a rear disc and it won't come off, it may be interfering with the parking brake shoes; remove the plug (see illustration) and rotate the adjuster to back the parking brake shoes away from the drum surface within the disc (see illustration).

Installation

7   Place the disc in position over the threaded studs.
8   Install the mounting bracket and caliper, tightening the bolts to the torque values listed in this Chapter's Specifications.
9   Install the wheel, then lower the vehicle to the ground. Tighten the lug nuts to the torque listed in the Chapter 1 Specifications. Depress the brake pedal a few times to bring the brake pads into contact with the disc. Bleeding won't be necessary unless the brake hose was disconnected from the caliper. Check the operation of the brakes carefully before driving the vehicle.

6.4a To check disc runout, mount a dial indicator as shown and rotate the disc

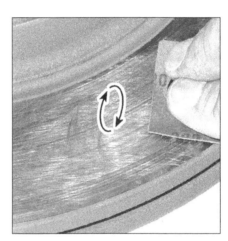

6.4b Using a swirling motion, remove the glaze from the disc surface with sandpaper or emery cloth

6.5a The minimum thickness dimension is cast into the back side of the disc (typical)

6.5b Use a micrometer to measure disc thickness

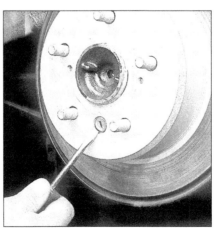

6.6a If the rear disc is difficult to remove, remove this plug . . .

6.6b . . . insert a screwdriver through the hole (the hole must be at the 6 o'clock position, because that's where the adjuster is located) and rotate the adjuster to back the parking brake shoe away from the drum surface in the disc

7 Master cylinder - removal, installation and reservoir/ grommet replacement

Removal

Note: *With the engine off, pump the brake pedal several times to deplete the vacuum in the power brake booster.*

1   Remove the air filter housing cover (see Chapter 4).

2   Unplug the electrical connector for the fluid level warning switch (see illustration).

3   Remove as much fluid as possible from the reservoir with a suction gun, large syringe or a poultry baster.

Warning: *If a poultry baster is used, never again use it for the preparation of food.*

4   Place rags under the fittings and prepare caps or plastic bags to cover the ends of the lines once they're disconnected.

Warning: *Brake fluid will damage paint. Cover all body parts and be careful not to spill fluid during this procedure. Loosen the fittings at the ends of the brake lines where they enter the master cylinder. To prevent rounding off the flats, use a flare-nut wrench, which wraps around the fitting hex.*

5   Pull the brake lines away from the master cylinder and plug the ends to prevent contamination.

6   Remove the two nuts attaching the master cylinder to the power booster. Pull the master cylinder off the studs to remove it. Again, be careful not to spill the fluid as this is done. Remove and discard the old gasket between the master cylinder and the power brake booster. Also check the O-ring on the end of the master cylinder, replacing it if it is cracked or hardened.

Installation

Note: *Before installing a new or rebuilt master cylinder, check the clearance between the*

booster pushrod and the pocket in the master cylinder piston. If necessary, adjust the length of the power brake booster pushrod (see Section 10).

7   Bench bleed the master cylinder before installing it. Because it will be necessary to apply pressure to the master cylinder piston and, at the same time, control flow from the brake line outlets, it is recommended that the master cylinder be mounted in a vise, with the jaws of the vise clamping on the mounting flange.

8   Attach a pair of bleeder tubes (available at most auto parts stores) to the outlet ports of the master cylinder (see illustration).

9   Fill the reservoir with brake fluid of the recommended type (see Chapter 1).

10   Slowly push the pistons into the master cylinder (a large Phillips screwdriver can be used for this) - air will be expelled from the pressure chambers and into the reservoir. Because the tubes are submerged in fluid, air can't be drawn back into the master cylinder when you release the pistons.

11   Repeat the procedure until no more air bubbles are present.

12   Remove the bleed tubes, one at a time, and install plugs in the open ports to prevent fluid leakage and air from entering. Install the reservoir cap.

13   Install the master cylinder over the studs on the power brake booster and tighten the nuts only finger-tight at this time. Don't forget to use a new gasket.

14   Thread the brake line fittings into the master cylinder. Since the master cylinder is still a bit loose, it can be moved slightly so the fittings thread in easily. Don't strip the threads as the fittings are tightened.

15   Tighten the mounting nuts to the torque listed in this Chapter's Specifications. Tighten the brake line fittings securely.

16   Plug in the electrical connector to the fluid level warning switch.

17   Fill the master cylinder reservoir with fluid, then bleed the master cylinder and the brake system (see Section 9). To bleed the master cylinder on the vehicle, have an assistant depress the brake pedal and hold it down. Loosen the fitting to allow air and fluid to escape (see illustration). Tighten the fitting, then allow your assistant to return the pedal

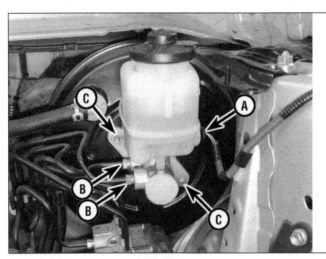

7.2 Master cylinder mounting details

A   *Electrical connector*
B   *Brake line fittings*
C   *Mounting nuts*

7.8 The best way to bleed air from the master cylinder before installing it on the vehicle is with a pair of bleeder tubes that direct brake fluid into the reservoir during bleeding

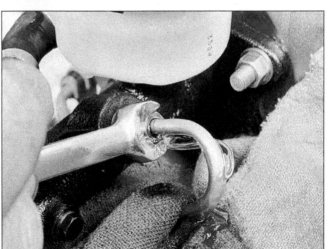

7.17 Have an assistant depress the brake pedal and hold it down, then loosen the fitting nut, allowing the air and fluid to escape; repeat this procedure on both fittings until the fluid is clear of air bubbles

to its rest position. Repeat this procedure on both fittings until the fluid is free of air bubbles. Check the operation of the brake system carefully before driving the vehicle.

Reservoir/grommet replacement

Note: *The brake fluid reservoir can be replaced separately from the master cylinder body if it becomes damaged. If there is leakage between the reservoir and the master cylinder body, the grommets on the reservoir can be replaced.*

18   Remove as much fluid as possible from the reservoir with a suction gun, large syringe or a poultry baster.

Warning: *If a poultry baster is used, never again use it for the preparation of food.*

19   Place rags under the master cylinder to absorb any fluid that may spill out once the reservoir is detached from the master cylinder.

Caution: *Brake fluid will damage paint. Cover all body parts and be careful not to spill fluid during this procedure.*

20   Remove the screw that retains the reservoir to the master cylinder.

Note: *A roll pin is used to retain the reservoir to the master cylinder that must be driven out with a hammer and a punch.*

21   Pull the reservoir out of the master cylinder body.

22   Pull the grommets out of the master cylinder (see illustration).

23   Lubricate the new grommets with clean brake fluid, then press them into place.

24   Push the reservoir into the grommets and secure it with the screw.

25   Refill the reservoir with the recommended brake fluid (see Chapter 1) and check for leaks.

26   Bleed the master cylinder (see illustration 8.17), followed by the remainder of the system (see Section 9).

8   Brake hoses and lines - inspection and replacement

Inspection

1   About every six months, with the vehicle raised and supported securely on jackstands, the rubber hoses which connect the steel brake lines with the front and rear brake assemblies should be inspected for cracks, chafing of the outer cover, leaks, blisters and other damage. These are important and vulnerable parts of the brake system and inspection should be complete. A light and mirror will be helpful for a thorough check. If a hose exhibits any of the above conditions, replace it with a new one.

Replacement

Disc brake hose (front or rear)

2   Loosen the wheel lug nuts, raise the vehicle and support it securely on jackstands. Remove the wheel.

3   At the frame bracket, unscrew the brake line fitting from the hose (see illustration). Use a flare-nut wrench to prevent rounding off the corners. If the bracket begins to bend, hold the hose fitting with an open-end wrench.

4   Remove the U-clip from the female fitting at the bracket with a pair of pliers (see illustration), then pass the hose through the bracket.

5   At the caliper end of the hose, remove the banjo fitting bolt, then separate the hose from the caliper. Note that there are two copper sealing washers on either side of the fitting - they should be replaced with new ones during installation.

6   Unbolt the hose from the bracket on the strut, where applicable.

7   Installation is the reverse of removal. Make sure the hose isn't twisted. Tighten the banjo bolt to the torque listed in this Chapter's Specifications, and tighten the brake hose-to-

brake line fitting securely.

8   Bleed the caliper (see Section 9).

9   Install the wheel and lug nuts, lower the vehicle and tighten the lug nuts to the torque listed in the Chapter 1 Specifications.

Metal brake lines

10   When replacing brake lines, be sure to use the correct parts. Don't use copper tubing for any brake system components. Purchase steel brake lines from a dealer or auto parts store.

11   Prefabricated brake line, with the tube ends already flared and fittings installed, is available at auto parts stores and dealer parts departments. These lines can be bent to the proper shape with a tubing bender.

12   When installing the new line, make sure it's securely supported in the brackets and has plenty of clearance between moving or hot components.

13   After installation, check the master cylinder fluid level and add fluid as necessary. Bleed the brake system (see Section 9) and test the brakes carefully before driving the vehicle in traffic.

9   Brake hydraulic system - bleeding

Warning: *Wear eye protection when bleeding the brake system. If the fluid comes in contact with your eyes, immediately rinse them with water and seek medical attention.*

Note: *Bleeding the hydraulic system is necessary to remove any air that manages to find its way into the system when it's been opened during removal and installation of a hose, line, caliper or master cylinder.*

1   You'll probably have to bleed the system at all four brakes if air has entered it due to low fluid level, or if the brake lines have been disconnected at the master cylinder or ABS

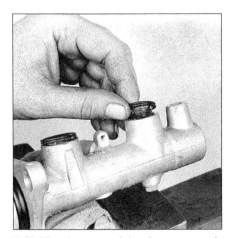

7.22 After the reservoir has been removed, pull the grommets from the master cylinder body; if they're hard, cracked or damaged, or have been leaking, replace them

8.3 Unscrew the brake line threaded fitting with a flare-nut wrench to protect the fitting corners from being rounded off

8.4 Pull off the U-clip with a pair of pliers

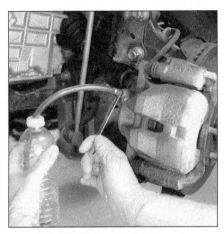

9.8 When bleeding the brakes, a hose is connected to the bleed screw at the caliper or wheel cylinder and then submerged in brake fluid - air will be seen as bubbles in the tube and container (all air must be expelled before moving to the next wheel)

10.13a Left side booster retaining nuts

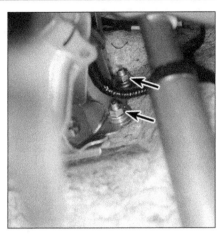

10.13b Right side booster retaining nuts

hydraulic actuator.

2   If a brake line was disconnected only at a wheel, then only that caliper or wheel cylinder must be bled.

3   If a brake line is disconnected at a fitting located between the master cylinder and any of the brakes, the entire system must be bled. And, if the master cylinder has run dry or has been replaced, bleed the master cylinder as described in Section 7, Step 17, followed by the remainder of the system.

4   Remove any residual vacuum from the brake power booster by applying the brake several times with the engine off.

5   Remove the master cylinder reservoir cap and fill the reservoir with brake fluid. Reinstall the cap.

Note: Check the fluid level often during the bleeding operation and add fluid as necessary to prevent the fluid level from falling low enough to allow air bubbles into the master cylinder.

6   Have an assistant on hand, as well as a supply of new brake fluid, a clear plastic container partially filled with clean brake fluid, a length of clear tubing to fit over the bleeder valve and a wrench to open and close the bleeder valve.

7   Beginning at the right rear wheel, loosen the bleeder valve slightly, then tighten it to a point where it's snug but can still be loosened quickly and easily.

8   Place one end of the tubing over the bleeder valve and submerge the other end in brake fluid in the container (see illustration).

9   Have the assistant depress the brake pedal slowly and hold the pedal down firmly.

10   While the pedal is held down, open the bleeder valve just enough to allow a flow of fluid to leave the valve. Watch for air bubbles to exit the submerged end of the tube. When the fluid flow slows after a couple of seconds, close the valve and have your assistant

release the pedal.

11   Repeat Steps 9 and 10 until no more air is seen leaving the tube, then tighten the bleeder valve and proceed to the left rear wheel, the right front wheel and the left front wheel, in that order, and perform the same procedure. Be sure to check the fluid in the master cylinder reservoir frequently.

Warning: Brake fluid is "Hydroscopic" meaning: Brake fluid will mix with water and can absorb moisture from the surrounding air. Always use brake fluid from a sealed container and never one that has been on a shelf for any length of time. Brake fluid can gain (on average) about 4% water per year. Under the extreme pressures in a brake system as well as the heat developed, the moisture in the brake fluid can actually boil which can leave you with brake fade or loss of functioning brakes entirely. Most manufacturers recommend that the brake fluid should be flushed out and replaced every 3 to 5 years.

12   Refill the master cylinder to the MAX level with fluid at the end of the operation.

13   Check the operation of the brakes. The pedal should feel solid when depressed, with no sponginess. If necessary, repeat the entire process.

Warning: Do not operate the vehicle if the ABS light or BRAKE light fails to go out, if the brakes feel low or spongy, or if you have any doubts as to the effectiveness of the brake system.

---

10   Power brake booster - check, removal and installation

Operating check

1   Depress the brake pedal several times with the engine off and make sure there's no change in the pedal reserve distance.

2   The booster reserve vacuum should be entirely used at this point.

3   While still holding the pedal down firmly, start the engine. The pedal should continue to fall to the floor. Let up on the pedal, (engine still running) wait a few seconds and apply the

brake pedal again. The pedal should return to its normal position.

4   If the pedal does not return to the normal position the booster is not working correctly. Check the vacuum to the booster. If the vacuum is good, the booster is faulty. Replace the booster.

Airtightness check

5   Start the engine and turn it off after one or two minutes. Depress the brake pedal slowly several times. If the pedal depresses less each time, the booster is airtight.

6   Depress the brake pedal while the engine is running, then stop the engine with the pedal depressed. If there's no change in the pedal reserve travel after holding the pedal for 30 seconds, the booster is airtight.

Removal

Note: With the engine off, pump the brake pedal several times to deplete the vacuum in the power brake booster.

7   Power brake booster units shouldn't be disassembled. They require special tools not normally found in most automotive repair stations or shops. Because of its critical relationship to brake performance, the booster should be replaced with a new or rebuilt one.

8   Disconnect the cable from the negative terminal of the battery (see Chapter 5), then remove the windshield wiper motor (see Chapter 12) and cowl/cowl reinforcement (see Chapter 11). Remove the brake master cylinder (see Section 7).

9   Disconnect the hose leading from the engine to the booster. Be careful not to damage the hose when removing it from the booster fitting.

10   Remove the driver's knee bolster/airbag (see Chapter 11).

11   Remove the ABS actuator assembly (see Section 3).

12   Remove the pedal return spring. Locate the pushrod clevis connecting the booster to the brake pedal. Remove the clevis pin retaining clip with pliers and pull out the pin.

13   Remove the four nuts holding the brake booster to the firewall (see illustrations); you

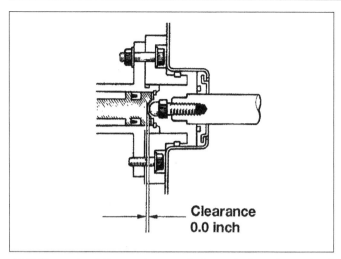

10.16 There should be no clearance between the booster pushrod and the master cylinder pushrod, but no interference either. If there is interference between the two, the brakes may drag; if there is clearance, there will be excessive brake pedal travel

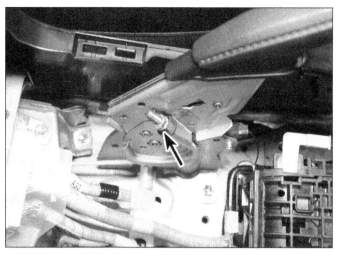

11.3 Loosen the locknut, then turn the adjusting nut until the desired handle travel is obtained

may need a light to see them.

14　Slide the booster away from the firewall until the studs clear the holes, then remove it.

Installation

15　Installation procedures are basically the reverse of removal. Tighten the clevis locknut securely and the booster mounting nuts to the torque listed in this Chapter's Specifications.

16　If a new power brake booster unit is being installed, check the pushrod clearance (see illustration) as follows:

a) *Measure the distance that the pushrod is recessed from the master cylinder mounting surface on the front of the power brake booster. Record this measurement.*

b) *This is "dimension A."*

c) *Measure the distance from the mounting flange to the end of the master cylinder.*

d) *Record this measurement. This is "dimension B."*

e) *Measure the distance from the end of the master cylinder to the bottom of the pocket in the piston.*

f) *Record this measurement. This is "dimension C."*

g) *Subtract measurement B from measurement C, then subtract measurement A from the difference between B and C. This is the pushrod clearance.*

h) *Compare your calculated pushrod clearance to the pushrod clearance listed in this Chapter's Specifications. If necessary, adjust the pushrod length to achieve the correct clearance.*

17　After the final installation of the master cylinder and brake hoses and lines, the brake pedal height and freeplay must be adjusted and the system must be bled. See the appropriate Sections of this Chapter for the procedures.

11　Parking brake - adjustment

1　The parking brake lever, when properly adjusted, should travel (on average) six to nine clicks, when a moderate pulling force is applied (45 ft-lbs). If it travels less than the specified minimum number of clicks, there's a chance the parking brake might not be releasing completely and might be dragging. If the lever can be pulled up more than the specified maximum number of clicks, the parking brake may not hold adequately on an incline, allowing the car to roll.

2　Remove the center console rear box sub-assembly.

3　Loosen the locknut (the upper nut) while holding the adjusting nut (lower nut) with a wrench (see illustration). Turn the adjusting nut until the desired travel is attained. Tighten the locknut.

4　If the lever is still not within the number of clicks as per the year of the vehicle, check the rear brakes for proper adjustment or the condition of the parking brake cables.

12　Parking brake cables - replacement

Warning: *These models are equipped with a Supplemental Restraint System (SRS), more commonly known as airbags. Always disable the airbag system before working in the vicinity of any airbag system component to avoid the possibility of accidental deployment of the airbag(s), which could cause personal injury (see Chapter 12).*

Warning: *Do not use a memory saving device to preserve the PCM or radio memory when working on or near airbag system components.*

1　Disconnect the cable from the negative terminal of the battery (see Chapter 5).

Equalizer-to-parking brake cable

2　Loosen the rear wheel lug nuts, raise the rear of the vehicle and support it securely on jackstands. Block the front wheels. Remove the wheel. Make sure the parking brake is completely released, then remove the brake drum.

3　Remove the parking brake shoes (see Section 15) and disconnect the cable from the parking brake lever.

4　Remove the cable retaining bolt from the backing plate and pull the cable through the backing plate.

5　Unbolt the cable clamps from the trailing arm and floor pan.

6　Remove the center console (see Chapter 11).

7　Unbolt the SRS control unit and remove the center console rear mounting bracket.

8　Detach the cable(s) from the equalizer, then push the cables through the floor pan.

9　Installation is the reverse of removal. Apply a light coat of grease to the portion of the cable end that engages with the equalizer.

10　Adjust the parking brake (see Section 11).

Equalizer-to-brake lever cable

11　Remove the center console (see Chapter 11).

12　With the lever in the down (off) position, remove the locknut and the adjusting nut (see Section 11).

13　Unbolt the SRS control unit and remove the center console rear mounting bracket.

14　Disconnect the cable from the equalizer.

15　Remove the bolts and detach the parking brake lever along with the cable.

16　Bend the cable retaining claw up and detach the cable from the lever.

17　Installation is the reverse of removal.

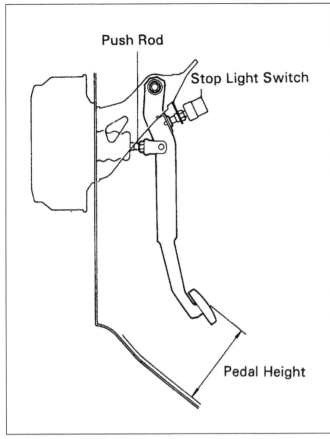

13.1 Brake pedal height is the distance between the pedal and the firewall when the pedal is released

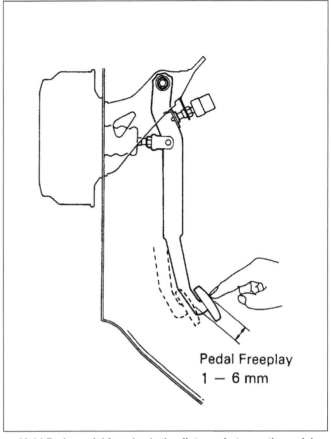

13.14 Brake pedal freeplay is the distance between the pedal when it's released and the point at which some resistance is first felt when the pedal is depressed

Apply a light coat of grease to the portion of the cable end that engages with the equalizer.
18   Adjust the parking brake (see Section 11).

13   Brake pedal - check and adjustment

Warning: *These models are equipped with a Supplemental Restraint System (SRS), more commonly known as airbags. Always disable the airbag system before working in the vicinity of any airbag system component to avoid the possibility of accidental deployment of the airbag(s), which could cause personal injury (see Chapter 12).*
Warning: *Do not use a memory saving device to preserve the PCM or radio memory when working on or near airbag system components.*

Pedal height

1   Measure the pedal height (see illustration) and compare your measurement to the pedal height listed in this Chapter's Specifications. If the pedal height is incorrect, adjust it as follows:

2   Remove the knee bolster/airbag module (see Chapter 11, Section 23).
3   Unplug the electrical connector from the brake light switch.
4   Remove the brake light switch (see Section 14).
5   Loosen the pushrod locknut.
6   Adjust the pedal height by turning the pedal pushrod.
7   Tighten the pushrod locknut.
8   Install the brake light switch until the switch body just touches the brake pedal, then turn the switch clockwise 1/4 turn to lock it in place.
9   Measure the distance between the switch body and the pedal and compare your measurement to the clearance listed in this Chapter's Specifications. If the clearance is not as specified, repeat the adjustment and try again.
10   Plug in the brake light switch electrical connector.
11   Verify that brake lights come on when the brake pedal is depressed, and go off when the brake pedal is released.
12   Check the pedal freeplay.

Pedal freeplay

13   Stop the engine, if it's running, and depress the brake pedal several times until

there's no more vacuum left in the booster.
14   Push in the pedal until you feel some resistance, then measure the distance between the release pedal and this point at which you can feel resistance (see illustration). Compare your measurement with the pedal freeplay listed in this Chapter's Specifications. If the pedal freeplay is incorrect, check the clearance between the switch body and the pedal. If the clearance is correct, the clearance between the brake booster pushrod and the master cylinder piston might be incorrect (see Section 10).

Pedal reserve distance

15   Start the engine, depress the brake pedal a few times, then press down hard and hold it.
16   Pedal reserve travel is measured from the floor to the top of the pedal while it's being depressed. Compare your measurement to the pedal reserve distance listed in this Chapter's Specifications.
17   If the pedal reserve is less than specified, check the adjustment of the power brake booster pushrod-to-master cylinder piston clearance (see Section 10). If the brake pedal feels spongy, bleed the brake system (see Section 9).

14.1 Brake switch location

14.10 Adjuster bracket location

15.4 Before disassembling it, be sure to wash the parking brake assembly with brake cleaner

14　Brake light switch - check and replacement

Check

1　The brake light switch is located on a bracket near the top of the brake pedal (see illustration). The switch activates the brake lights at the rear of the vehicle when the pedal is depressed.

2　To check the brake light switch, simply note whether the brake lights come on when the pedal is depressed and go off when the pedal is released. If they don't, check the fuse first (see Chapter 12). If the fuse is good, adjust the switch as described in Section 13 (adjusting the switch is part of brake pedal adjustment).

3　If the lights still don't come on, either the switch is not getting voltage, the switch itself is defective, or the circuit between the switch and the lights is defective. There is always the remote possibility that all of the brake light bulbs are burned out, but this is not very likely.

4　Use a voltmeter or test light to verify that there's voltage present at one side of the switch connector. If no voltage is present, troubleshoot the circuit from the switch to the fuse box. If there is voltage present, check for voltage on the other terminal when the brake pedal is depressed. If no voltage is present, replace the switch. If there is voltage present, troubleshoot the circuit from the switch to the brake lights (see the wiring diagrams at the end of Chapter 12).

Replacement

Brake light switch

5　Unplug the electrical connector for the brake light switch.

6　Rotate the switch counterclockwise about a 1/4 turn to remove the switch from the bracket.

7　Installation is the reverse of removal.

8　Brake light switch will self adjust once it is installed properly.

Brake light switch adjuster bracket (plastic)

9　Remove the brake light switch from the bracket.

10　The light switch adjuster bracket is a plastic component pressed into the metal brake light switch bracket (see illustration).

11　Press the two adjuster clips (plastic) inward, then pull the adjuster out of the metal bracket.

12　Installation is the reverse of removal.

15　Parking brake shoes - inspection and replacement

Warning: *Dust created by the brake system is hazardous to your health. Never blow it out with compressed air and don't inhale any of it. An approved filtering mask should be worn when working on the brakes. Do not, under any circumstances, use petroleum-based solvents to clean brake parts. Use brake system cleaner only!*

Warning: *Parking brake shoes must be replaced on both wheels at the same time - never replace the shoes on only one wheel.*

1　Remove the brake disc (see Section 6).

2　Inspect the thickness of the lining material on the shoes. If the lining has worn down to 1/32-inch or less, the shoes must be replaced.

3　Remove the hub and bearing assembly (see Chapter 10).

Note: *It is possible to perform the shoe replacement procedure without removing the hub and bearing assembly, although working room is limited.*

4　Wash off the brake parts with brake system cleaner (see illustration).

5　Follow the accompanying illustrations for the brake shoe replacement procedure (see

illustrations 15.5a through 15.5u). Be sure to stay in order and read the caption under each illustration.

6　Install the brake disc. Temporarily thread three of the wheel lug nuts onto the studs to hold the disc in place.

7　Remove the hole plug from the brake disc. Adjust the parking brake shoe clearance by turning the adjuster star wheel with a brake adjusting tool or screwdriver until the shoes contact the disc and the disc can't be turned (see illustrations 6.6a and 6.6b). Back off the adjuster eight notches, then install the hole plug.

8　Install the caliper mounting bracket and brake caliper (see Section 5). Be sure to tighten the bolts to the torque listed in this Chapter's Specifications.

9　Install the wheel and tighten the lug nuts to the torque listed in the Chapter 1 Specifications.

10　Set the parking brake and count the number of clicks that it travels. It should be as listed in this Chapter's Specifications - if it's not, adjust the parking brake (see Section 11).

15.5a Remove the rear parking brake shoe return spring from the anchor pin . . .

15.5b . . . and unhook it from the rear shoe

15.5c Remove the front parking brake shoe return spring from the anchor pin . . .

15.5d . . . and unhook it from the front shoe

15.5e Remove the rear shoe hold-down spring and pull out the pin

Note: *Some models are equipped with U-shaped spring steel retainers. Depress the open end of the retainer and rotate the pin 90 degrees to remove it.*

15.5f Remove the shoe strut from between the shoes

15.5g Remove the front shoe hold-down spring and pull out the pin

15.5h Remove the adjuster and the tension spring (the tension spring, which is not visible in this photo, is behind the adjuster and is attached to both shoes)

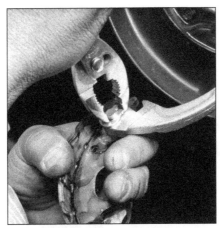

15.5i Pop the C-washer off the pivot pin on the back of the rear shoe . . .

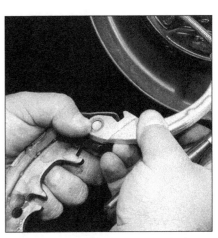

15.5j . . . and pull the parking brake lever off the pivot pin

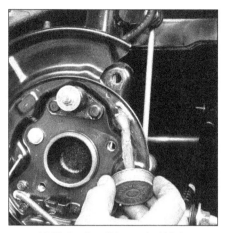

15.5k Apply a thin coat of high-temperature grease to the contact surfaces of the backing plate

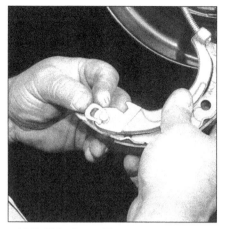

15.5l Slide the parking brake lever onto the pivot pin and install a new C-washer

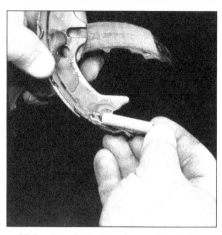

15.5m Attach the tension spring to the back side of the rear shoe . . .

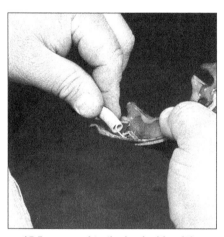

15.5n . . . and to the back side of the front shoe

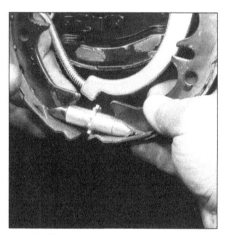

15.5o Flip the shoes around and install the adjuster; make sure both ends of the adjuster are properly engaged with the shoes as shown

15.5p Place the shoes in position and install the strut and spring as shown; make sure the ends of the strut are properly engaged with the shoes as shown

15.5q Install the front shoe return spring . . .

15.5r . . . and the rear shoe return spring

15.5s Install the rear shoe hold-down spring . . .

15.5t . . . and the front shoe hold-down spring

15.5u This is how the parking brake assembly should look when you're done

Notes

Chapter 10
Suspension and steering

Contents

Specifications

Torque specifications

Ft-lbs (unless otherwise indicated)   **Nm**

Note: *One foot-pound (ft-lb) of torque is equivalent to 12 inch-pounds (in-lbs) of torque. Torque values below approximately 15 ft-lbs are expressed in inch-pounds, because most foot-pound torque wrenches are not accurate at these smaller values.*

Front suspension

	Ft-lbs	Nm
Balljoints		
Balljoint-to-control arm bolt/nuts	68	92
Balljoint-to-steering knuckle nut	98	133
Control arm		
Front pivot bolt	172	233
Rear pivot stud nut	101	137
Rear bushing bolt	172	233
Rear bushing bracket bolts	101	137
Rear bushing bracket nut	21	28
Front suspension braces		
Rear brace to body (rear)	69	94
Rear brace to front suspension crossmember (front)	107	145
Stabilizer bar		
Brace bolts	64	87
Stabilizer end link nuts	55	75
Strut		
Strut-to-steering knuckle bolts/nuts	177	240
Strut upper mounting nuts	37	50
Damper shaft nut	35	46
Subframe		
Subframe main mounting bolts	101	137
Rear brace-to-subframe	101	137
Rear brace-to-body	69	93
Subframe front reinforcement bolts	73	99
Hub and bearing assembly-to-steering knuckle bolts	71	96
Driveaxle/hub nut	See Chapter 8	

Rear suspension

Suspension arms

Trailing arm

Trailing arm-to-knuckle bolts	148	201
Trailing arm-to-bracket bolt/nut	111	150
Trailing arm bracket-to-body bolts	59	80

Upper arm

Upper arm-to-rear knuckle fasteners	66	89
Upper arm-to-subframe fasteners	66	89

Lower arm A

Lower arm-to-subframe bolt/nut	89	120
Lower arm-to-knuckle nut	74	100
Lower arm B (all fasteners)	66	89
Hub and bearing assembly-to-knuckle bolts	66	90

Stabilizer bar

Link nuts

Upper	55	75
Lower	22	30
Bracket fasteners	44	60
Shock absorber fasteners	59	80

Steering

Power steering ECU bolts	15	20
ECU wiring sub-assembly to body	74 in-lbs	8
ECU electrical terminal bolt	28 in-lbs	3.2
Steering wheel nut	37	50
Tie-rod end-to-steering knuckle nut	65	88
Steering intermediate shaft fasteners	26	35

Steering column

Upper mounting nuts	18	25
Lower mounting bolts	27	36
Power steering motor bolts	15	20
Steering gear mounting bolts	81	110

1   General information

1   The front suspension (see illustration) is a McPherson strut design. The upper end of each strut/coil spring assembly is attached to the vehicle's body strut support. The lower end of the strut assembly is connected to the upper end of the steering knuckle. The steering knuckle is attached to a balljoint mounted on the outer end of the suspension control arm. A stabilizer bar reduces body roll.

2   The rear suspension (see illustrations) employs a trailing arm, upper suspension arm, lower suspension arms "A" and "B", a coil spring and shock absorber (per side) and a stabilizer bar.

1.1 Front suspension and steering components

1	Strut/coil spring assembly	4	Control arm	6	Stabilizer bar
2	Steering knuckle	5	Tie-rod end	7	Subframe
3	Balljoint				

1.2a Rear suspension components

1	Subframe	4	Upper suspension arm
2	Lower suspension arm "B"	5	Lower suspension arm "A"
3	Coil spring		

6	Shock absorber
7	Trailing arm

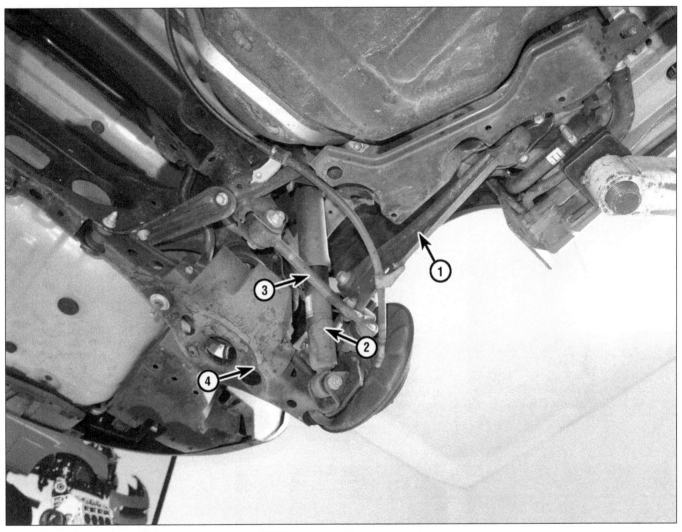

1.2b Rear suspension components

1	Trailing arm	3	Lower suspension arm "A"
2	Shock absorber	4	Lower suspension arm "B"

Electronic steering system

3   The electric power steering system consists of a serviceable power steering motor that is incorporated into the steering column as well as the ECU, torque sensor and steering angle sensor.

4   The use of an electronic power steering reduces weight as well as the need for hydraulic fluids. For the most part, the electronic power steering is virtually maintenance free. Replacement of the electronic steering system or components will require re-calibrating the steering angle sensor and the ECU to the data system of your car. See the appropriate repair facility to accomplish the reprogramming.

Precautions

5   Frequently, when working on the suspension or steering system components, you may come across fasteners which seem impossible to loosen. These fasteners on the underside of the vehicle are continually subjected to water, road grime, mud, etc., and can become

2.2 Brake hose connection attached to the strut assembly

rusted or frozen, making them extremely difficult to remove. In order to unscrew these stubborn fasteners without damaging them (or other components), be sure to use lots of penetrating oil and allow it to soak in for a while. Using a wire brush to clean exposed threads will also ease removal of the nut or bolt and prevent damage to the threads. Sometimes a sharp blow with a hammer and punch will break the bond between a nut and bolt threads, but care must be taken to prevent the punch from slipping off the fastener and ruining the threads. Heating the stuck fastener and surrounding area with a torch sometimes helps too, but isn't recommended because of the obvious dangers associated with fire. Long breaker bars and extension, or cheater, pipes will increase leverage, but never use an extension pipe on a ratchet - the ratcheting mechanism could be damaged. Sometimes tightening the nut or bolt first will help to break it loose. Fasteners that require drastic measures to remove should always be replaced with new ones.

6   Since most of the procedures dealt with in this Chapter involve jacking up the vehicle and working underneath it, a good pair of jackstands will be needed. A hydraulic floor jack is the preferred type of jack to lift the vehicle, and it can also be used to support certain components during various operations.
Warning: *Never, under any circumstances, rely on a jack to support the vehicle while working on it. Whenever any of the suspension or steering fasteners are loosened or removed they must be inspected and, if necessary, replaced with new ones of the same part number or of original equipment quality and design. Torque specifications must be followed for proper reassembly and component retention. Never attempt to heat or straighten any suspension or steering components. Instead, replace any bent or damaged part with a new one.*

2   Strut assembly (front) - removal, inspection and installation

Removal

1   Loosen the wheel lug nuts, raise the vehicle and support it securely on jackstands. Remove the wheel.

2   Unbolt the brake hose bracket from the strut (see illustration). If the vehicle is equipped with ABS, detach the speed sensor wiring harness from the strut by removing the clamp bracket bolt.

3   Disconnect the stabilizer bar link from the strut (see Section 4).

4   Mark the relationship of the strut to the steering knuckle (see illustrations).

5   Separate the strut from the steering knuckle. Be careful not to overextend the inner CV joint. Also, don't let the steering knuckle fall outward and strain the brake hose.

6   Loosen the three upper nuts (see illustration). With the nuts loose, support the strut and spring assembly with one hand and remove the three strut-to-body nuts. Now remove the assembly out from the fenderwell.
Warning: *Do NOT remove the center top nut of the strut assembly when removing the assembly from the vehicle. Remove this nut ONLY when a spring compressor is properly attached and is compressed enough to relieve the spring pressure off of the upper strut bearing. Follow the strut/spring assembly removal procedures in Section 3.*

Inspection

7   Check the strut body for leaking fluid, dents, cracks and other obvious damage which would warrant repair or replacement.

8   Check the coil spring for chips or cracks in the spring coating (this will cause premature spring failure due to corrosion). Inspect the spring seat for cuts, hardness and general deterioration.

2.4a Mark the relationship of the strut to the steering knuckle (to preserve the camber setting when reassembling)

2.4b To detach the strut from the steering knuckle, remove the two nuts, then knock out the bolts with a hammer and punch

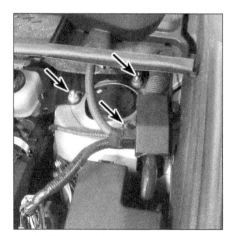

2.6 To detach the upper end of the strut from the body, remove the upper mounting nuts

9   If any undesirable conditions exist, proceed to the strut disassembly procedure (see Section 3).

Installation

10   Guide the strut assembly up into the fenderwell and insert the upper mounting studs through the holes in the body. Once the studs protrude, install the nuts so the strut won't fall back through. This is most easily accomplished with the help of an assistant, as the strut is quite heavy and awkward.

11   Slide the steering knuckle into the strut flange and insert the two bolts. Install the nuts, align the previously made matchmarks and tighten them to the torque listed in this Chapter's Specifications.

12   Connect the brake hose bracket to the strut and tighten the bolt securely. If the vehicle is equipped with ABS, install the speed sensor wiring harness bracket.

13   Install the wheel and lug nuts, then lower the vehicle and tighten the lug nuts to the torque listed in the Chapter 1 Specifications.

14   Tighten the upper mounting nuts to the torque listed in this Chapter's Specifications.

15   Have the front end alignment checked, and if necessary, adjusted.

---

3   Strut/spring assembly - replacement

---

1   If the struts or coil springs exhibit the telltale signs of wear (leaking fluid, loss of damping capability, chipped, sagging or cracked coil springs) explore all options before beginning any work. The strut/shock absorber assemblies are not serviceable and must be replaced if a problem develops. However, strut assemblies complete with springs may be available on an exchange basis, which eliminates much time and work. Whichever route you choose to take, check on the cost and availability of parts before disassembling

your vehicle.

Warning: *Disassembling a strut is potentially dangerous and utmost attention must be directed to the job, or serious injury may result. Use only a high-quality spring compressor and carefully follow the manufacturer's instructions furnished with the tool. After removing the coil spring from the strut assembly, set it aside in a safe, isolated area.*

Disassembly

2   Remove the strut assembly following the procedure described in the previous Section. Mount the strut assembly in a vise. Line the vise jaws with wood or rags to prevent damage to the unit and don't tighten the vise excessively.

3   Following the tool manufacturer's instructions, install the spring compressor (which can be obtained at most auto parts stores or equipment yards on a daily rental basis) on the spring and compress it sufficiently to

relieve all pressure from the upper spring seat (see illustration). This can be verified by wiggling the spring.

4   Loosen the damper shaft nut with a socket wrench (see illustration).

Note: *Inspect the bearing in the suspension support for smooth operation. If it doesn't turn smoothly, replace the suspension support. Check the rubber portion of the suspension support for cracking and general deterioration. If there is any separation of the rubber, replace it.*

5   Remove the nut and suspension support (see illustration).

6   Lift the spring seat and upper insulator from the damper shaft (see illustration). Check the rubber spring seat for cracking and hardness, replacing it if necessary.

7   Carefully lift the compressed spring from the assembly (see illustration) and set it in a safe place.

Warning: *Never place your head near the end of the spring!*

3.3 Install the spring compressor according to the tool manufacturer's instructions and compress the spring until all pressure is relieved from the upper spring seat

3.4 Remove the damper shaft nut - if the upper spring seat turns while loosening the nut, immobilize it with a chain wrench or strap wrench

3.5 Lift the suspension support off the damper shaft

3.6 Remove the spring seat from the damper shaft

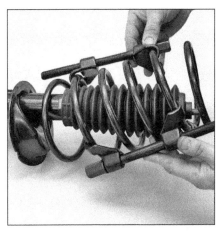

3.7 Remove the compressed spring assembly - keep the ends of the spring pointed away from your body

3.11 When installing the spring, make sure the end fits into the recessed portion of the lower seat

3.12 The flats on the damper shaft must match up with the flats in the spring seat

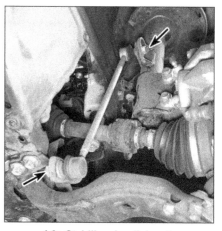

4.2a Stabilizer bar link nuts

8   Slide the rubber bumper off the damper shaft.

9   Check the lower insulator (if equipped) for wear, cracking and hardness and replace it if necessary.

Reassembly

10   If the lower insulator is being replaced, set it into position with the dropped portion seated in the lowest part of the seat. Extend the damper rod to its full length and install the rubber bumper.

11   Carefully place the coil spring onto the lower insulator, with the end of the spring resting in the lowest part of the insulator (see illustration).

12   Install the upper insulator and spring seat, making sure that the flats in the hole in the seat match up with the flats on the damper shaft (see illustration).

13   Install the dust seal and suspension support to the damper shaft.

14   Install the nut and tighten it to the torque listed in this Chapter's Specifications.

15   Install the strut assembly (see Section 2).

---

4   Stabilizer bar and bushings (front) - removal and installation

Removal

1   Loosen the front wheel lug nuts. Raise the front of the vehicle and support it securely on jackstands. Apply the parking brake and block the rear wheels to keep the vehicle from rolling off the stands. Remove the front wheels.

2   Detach the stabilizer bar link from the bar (see illustrations). If the ballstud turns with the nut, use an Allen wrench to hold the stud.

3   Unbolt and lower the subframe (see Section 19). Remove the stabilizer bar from the subframe.

4   Check the stabilizer bar bushings; if they're cracked, worn or deteriorated, replace them. It's also a good idea to inspect the

stabilizer bar links. To check them, flip each balljoint stud side-to-side five or six times, then install the nut. Using an inch-pound torque wrench, turn the nut continuously one turn every two to four seconds and note the torque reading on the fifth turn. It should be no less than 0.4 in-lbs. If it is, it's too loose and the link should be replaced.

5   Clean the bushing area of the stabilizer bar with a stiff wire brush to remove any rust or dirt.

Installation

6   Lubricate the inside and outside of the new bushing with vegetable oil (used in cooking) to simplify reassembly.

Note: *Don't use petroleum or mineral-based lubricants or brake fluid - they will lead to deterioration of the bushings. The slits of the bushings must face the rear of the vehicle. Also, the bushings must be positioned to the outside of the bushing stop.*

7   Installation is the reverse of removal. Tighten the fasteners to the torque values listed in this Chapter's Specifications.

---

5   Control arm - removal, inspection and installation

Removal

1   Loosen the wheel lug nuts on the side to be dismantled, raise the front of the vehicle, support it securely on jackstands and remove the wheel. Disconnect the stabilizer bar link from the control arm (see Section 4).

2   Remove the bolt and two nuts securing the balljoint to the control arm. Use a prybar to disconnect the control arm from the steering knuckle (see illustrations).

3   Support the subframe with a floor jack, then refer to Section 19 to unbolt and lower the subframe. Keep in mind that it isn't necessary to completely remove the subframe, but lower it just enough to allow removal of the control arm fasteners.

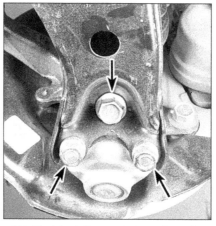

4.2b If the ballstud spins while loosening the nut, hold it with a hex bit

5.2a To detach the control arm from the steering knuckle balljoint, remove this bolt and these two nuts . . .

5.2b . . . and pry the control arm and balljoint apart with a large prybar or screwdriver

5.4a Control arm front mounting fastener location

5.4b Control arm rear mounting fastener

4   Slowly lower the subframe with the floor jack just enough to remove the front and rear control arm mounting fasteners (see illustrations).

5   Remove the control arm.

Inspection

6   Check the control arm for distortion and the bushings for wear, replacing parts as necessary. Do not attempt to straighten a bent control arm.

Installation

Note: *Before tightening the pivot bolt, raise the outer end of the control arm with a floor jack to simulate normal ride height.*

7   Installation is the reverse of removal. Tighten all of the fasteners to the torque values listed in this Chapter's Specifications.

8   Install the wheel and lug nuts, lower the vehicle and tighten the lug nuts to the torque listed in the Chapter 1 Specifications.

9   It's a good idea to have the front wheel alignment checked and, if necessary, adjusted after this job has been performed.

---

6   Balljoint - replacement

---

1   Loosen the wheel lug nuts, raise the vehicle and support it securely on jackstands. Remove the wheel.

2   Remove the cotter pin from the balljoint stud and loosen the nut (but don't remove it yet).

3   Separate the balljoint from the steering knuckle with a picklefork-type balljoint separator (see illustration). Remove the balljoint stud nut. The clearance between the balljoint stud and the CV joint is very tight. To remove the stud nut, you'll have to alternately back off the nut a turn or two, pull down the stud, turn the nut another turn or two, etc., until the nut is off.

4   Remove the bolt and nuts securing the balljoint to the control arm. Separate the balljoint from the control arm with a prybar

(see Section 5).

5   To install the balljoint, insert the balljoint stud through the hole in the steering knuckle and install the nut.

6   Attach the balljoint to the control arm and install the bolt and nuts, tightening them to the torque listed in this Chapter's Specifications.

7   Tighten the balljoint stud nut to the torque listed in this Chapter's Specifications and install a new cotter pin. If the cotter pin hole doesn't line up with the slots on the nut, tighten the nut additionally until it does line up - don't loosen the nut to insert the cotter pin.

8   Install the wheel and lug nuts. Lower vehicle and tighten the lug nuts to the torque listed in the Chapter 1 Specifications.

---

7   Steering knuckle and hub - removal and installation

Warning: *Dust created by the brake system is harmful to your health. Never blow it out with compressed air and don't inhale any of it. Do not, under any circumstances, use petroleum-based solvents to clean brake parts. Use brake system cleaner only.*

Removal

1   Remove the wheel cover and loosen, but don't remove, the driveaxle/hub nut. Loosen the wheel lug nuts, raise the vehicle and support it securely on jackstands, then remove the wheel.

2   Remove the brake caliper (don't disconnect the hose) and the brake disc (see Chapter 9), and disconnect the brake hose from the strut. Hang the caliper from the coil spring with a piece of wire - don't let it hang by the brake hose.

3   Remove the ABS wheel speed sensor (see Chapter 9).

4   Mark the relationship of the strut to the steering knuckle. Loosen, but don't remove the strut-to-steering knuckle nuts and bolts (see Section 2).

5   Separate the tie-rod end from the steering knuckle arm (see Section 16).

6   Remove the balljoint-to-control arm bolt and nuts (see Section 5).

7   Remove the driveaxle/hub nut and push the driveaxle from the hub as described in Chapter 8. Support the end of the driveaxle with a piece of wire.

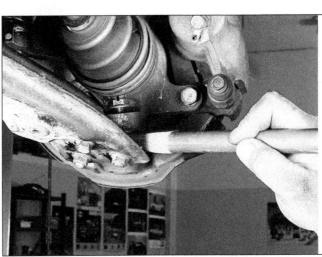

6.3 Separate the balljoint from the steering knuckle with a picklefork-type balljoint separator

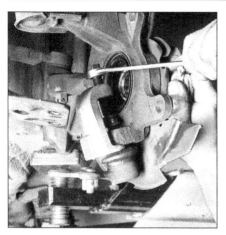

7.8 Use a balljoint removal tool or small puller to force the balljoint stud from the steering knuckle

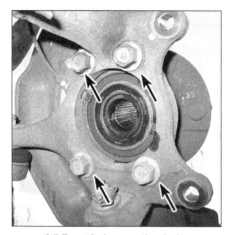

8.7 Front hub mounting bolts

9.2 Left subframe brace bolts

8   Using a balljoint removal tool (see illustration) or a small puller, remove the balljoint from the steering knuckle.
Note: *If you're removing the steering knuckle to replace the hub bearings, and the balljoint is in good condition, the balljoint can remain attached.*
9   The strut-to-knuckle bolts can now be removed.
10   Carefully separate the steering knuckle from the strut.

Installation
11   Guide the knuckle and hub assembly into position, inserting the driveaxle into the hub.
12   Push the knuckle into the strut flange and install the bolts and nuts, but don't tighten them yet.
13   If you removed the balljoint from the old knuckle, and are planning to use it with the new knuckle, connect the balljoint to the knuckle and tighten the balljoint stud nut to the torque listed in this Chapter's Specifications. Install a new cotter pin.
14   Attach the balljoint to the control arm, but don't tighten the bolt and nuts yet.
15   Attach the tie-rod to the steering knuckle arm (see Section 16). Tighten the strut bolt nuts, the balljoint-to-control arm bolt and nuts

and the tie-rod nut to the torque listed in this Chapter's Specifications.
16   Place the brake disc on the hub and install the caliper as outlined in Chapter 9.
17   Install the driveaxle/hub nut and tighten it securely, but not completely yet.
18   Install the wheel and lug nuts. Lower the vehicle and tighten the lug nuts to the torque listed in the Chapter 1 Specifications.
19   Tighten the driveaxle/hub nut to the torque listed in the Chapter 8 Specifications. Install the wheel cover.
20   Have the front-end alignment checked and, if necessary, adjusted.

---

8   Hub and bearing assembly (front) - removal and installation

Warning: *Dust created by the brake system is harmful to your health. Never blow it out with compressed air and don't inhale any of it. Do not, under any circumstances, use petroleum-based solvents to clean brake parts. Use brake system cleaner only.*

Removal
1   Loosen the driveaxle/hub nut (see Chapter 8).

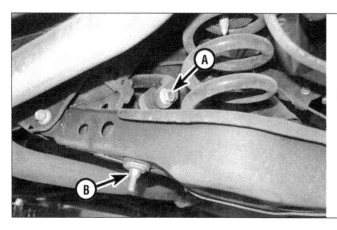

9.3 Stabilizer bar link nut (A) and link-to-lower arm nut (B)

2   Loosen the wheel lug nuts, raise the vehicle and support it securely on jackstands. Block the rear wheels to prevent the vehicle from rolling. Remove the wheel.
3   Remove the brake caliper and disc (see Chapter 9).
4   Remove the front speed sensor (see Chapter 9).
5   Remove the balljoint-to-lower arm bolt and nuts (see illustrations 5.2a and 5.2b).
6   Pull the steering knuckle outward and remove the driveaxle from the hub splines (see Chapter 8). Support the driveaxle with a piece of wire or rope.
7   Remove the four hub-to-knuckle bolts from the back side of the steering knuckle (see illustration).
8   Remove the hub and bearing assembly.

Installation
9   Position the hub and bearing assembly on the knuckle. Install the bolts. After all four bolts have been installed, tighten them to the torque listed in this Chapter's Specifications.
10   Install the driveaxle into the hub splines (see Chapter 8).
11   Install the brake disc, caliper, speed sensor and the wheel.
12   Lower the vehicle and tighten the lug nuts to the torque listed in the Chapter 1 Specifications. Tighten the driveaxle/hub nut to the torque listed in the Chapter 8 Specifications, then install a new cotter pin.

---

9   Stabilizer bar and bushings (rear) - removal and installation

1   Raise the rear of the vehicle and support it securely on jackstands. Block the front wheels to prevent the vehicle from rolling.
2   Remove the left subframe brace (see illustration).
3   Remove the stabilizer bar-to-link nuts (see illustration). If the ballstud turns with the nut, use a hex bit to hold the stud. Also loosen

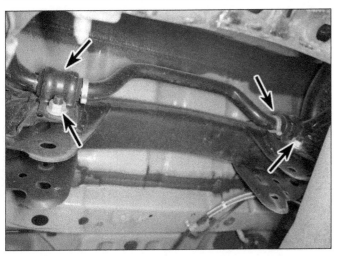

9.4 Rear stabilizer bar bushing clamp nuts (upper nuts not visible)

10.3 Remove these two bolts from the shock absorber bracket

the left link-to-lower arm nut, then separate the ballstud from the bar.

4   Unbolt the stabilizer bar bushing clamps from the subframe (see illustration).

5   The stabilizer bar can now be removed from the vehicle. Pull the bushing clamps off the stabilizer bar using a rocking motion.

6   Check the bushings for wear, hardness, distortion, cracking and other signs of deterioration, replacing them if necessary.

7   Check the stabilizer bar links for wear as described in Section 4.

8   Using a wire brush, clean the areas of the bar where the bushings ride. Install the bushings on the bar with the slits facing down. Also, the bushings must be installed to the inside of the paint line on the bar. If necessary, use a light coat of vegetable oil to ease bushing and U-bracket installation (don't use petroleum-based products or brake fluid, as

these will damage the rubber).

9   Installation is the reverse of removal. Tighten the fasteners to the torque listed in this Chapter's Specifications.

---

10   Shock absorber (rear) - removal, inspection and installation

1   Loosen the wheel lug nuts, raise the vehicle and support it securely on jackstands. Block the front wheels to prevent the vehicle from rolling. Remove the wheel.

2   Support the rear suspension arm "B" with a floor jack and a board (to protect the suspension arm from any damage). Raise the jack slightly to take the spring pressure off the shock absorber lower mount (roughly 1.5 to 2 inches is sufficient).

Note: *The jack must remain in this position throughout the entire procedure.*

3   Working from the underside the vehicle, remove the lower two bolts from the shock absorber bracket (see illustration).

4   Remove the upper mounting bolt (see illustration).

5   Remove the shock absorber from the vehicle, then remove the lower mounting bracket, and separate the shock absorber from the bracket (see illustration).

6   Installation is the reverse of the removal procedure. Tighten the mounting fasteners to the torque listed in this Chapter's Specifications.

Note: *Before tightening the shock absorber-to-lower mounting bracket fasteners, raise the lower arm with a floor jack to simulate normal ride height.*

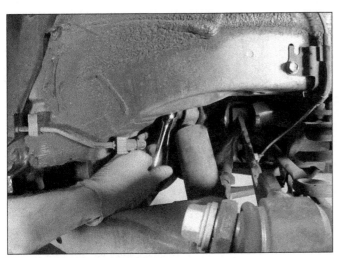

10.4 Rear shock absorber upper mounting bolt

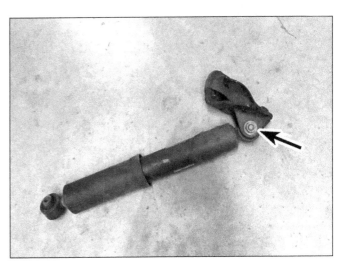

10.5 Shock absorber lower mounting bracket fastener. Install the bracket on the replacement shock absorber

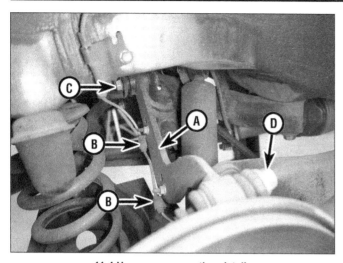

11.4 Upper arm mounting details

A   *Upper arm*
B   *Harness fasteners*
C   *Upper arm inner pivot bolt*
D   *Upper arm-to-rear knuckle bolt*

11.9 Make matchmarks on the toe adjuster cam and subframe before loosening the nut/bolt

11   Suspension arms and knuckle (rear) - removal and installation

1   Loosen the wheel lug nuts, raise the vehicle and support it securely on jackstands. Block the front wheels to prevent the vehicle from rolling. Remove the wheel.

Upper suspension arm

2   Remove the ABS wheel speed sensor harness mounting fasteners.
3   Support the lower arm B with a floor jack placed under the arm, between the coil spring seat and the outer end of the arm.

Note: *Do not remove the jack while it is supporting the lower arm B.*
4   Remove the fasteners at each end of the upper arm and remove the arm from the vehicle (see illustration).
5   Check the arm for bending and cracks, and the bushing for wear, hardness or deterioration.
6   Installation is the reverse of removal. Raise the lower arm with the floor jack to simulate normal ride height, then tighten the fasteners to the torque listed in this Chapter's Specifications.

Note: *Before tightening the fasteners, use the floor jack to raise the suspension to simulate normal ride height.*

Lower suspension arms
Lower arm A

7   Support the lower arm B with a floor jack placed under the arm, between the coil spring seat and the outer end of the arm.

Warning: *Do not remove the jack while it is supporting the lower arm B.*

Note: *When removing lower arm A, it may be necessary to remove the shock absorber for tool clearance.*
8   If equipped, remove the cover from lower arm B.
9   Mark the position of the adjuster cam on the inner pivot bolt to the subframe (see illustration).
10   Remove the inner end mounting fastener from lower arm A (see illustration).
11   Remove the outer end mounting nut-to-knuckle. Using a balljoint removal tool or a small puller, detach the ballstud from the knuckle and remove the arm from the vehicle.
12   Installation is the reverse of removal. Be sure to align the matchmarks made in Step 9, and tighten the fasteners to the torque listed in this Chapter's Specifications.

Note: *Before tightening the fasteners, use the floor jack to raise the suspension to simulate normal ride height.*

Lower arm B

13   Support the lower arm B with a floor jack placed under the arm at the end where the knuckle is connected to the arm.

Warning: *Do not remove the jack while it is supporting the lower arm B.*
14   Remove the wheel speed sensor harness mounting fasteners.
15   Disconnect the rear stabilizer bar link from the arm (see Section 9).

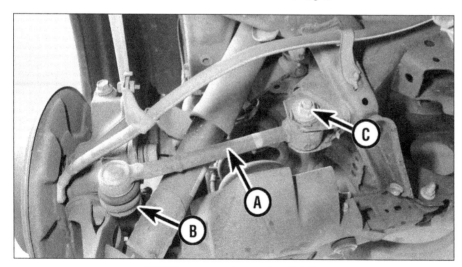

11.10 Lower arm A mounting details

A   *Lower arm A*
B   *Lower arm A-to-knuckle nut (not visible)*
C   *Lower arm A-to-subframe fastener and adjuster cam*

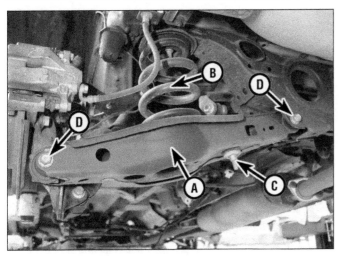

11.16 Lower arm B mounting details

A	Lower arm B	D	Lower arm
B	Coil spring		mounting fasteners
C	Stabilizer bar link		

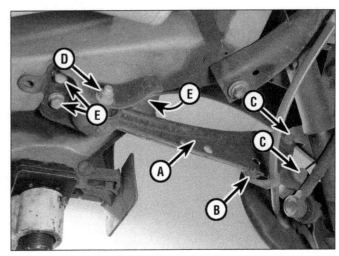

11.24 Trailing arm mounting details

A	Trailing arm	D	Trailing arm pivot bolt/nut
B	Parking brake cable	E	Trailing arm bracket bolts
	mounting bracket		
C	Trailing arm-to-knuckle		
	fasteners		

16   Loosen the lower arm inner mounting fastener (see illustration).

17   Remove the lower arm outer mounting fastener and disconnect the knuckle from the arm.

18   Slowly lower the floor jack until the coil spring is extended, then remove the coil spring (see Section 12).

19   Remove the lower arm inner mounting fastener and remove the arm.

20   Installation is the reverse of removal. Be sure to position the lower end of the coil spring in the depressed area of the lower arm. Tighten all fasteners to the proper torque specifications.

Note: *Before tightening the fasteners, use the floor jack to raise the suspension to simulate normal ride height*

Trailing arm

21   Support the lower arm B with a floor jack placed under the arm at the end where the knuckle is connected to the arm.

Warning: *Do not remove the jack while it is supporting the lower arm B.*

22   Disconnect the parking brake cable bracket from the trailing arm.

23   Remove the rear shock absorber (see Section 10).

24   Remove the trailing arm-to-knuckle fasteners, then loosen the trailing arm pivot bolt/ nut (see illustration).

25   Remove the trailing arm bracket bolts and detach the arm from the chassis, then remove the bolt and detach the bracket from the arm.

26   Inspect the trailing arm pivot bushing for signs of deterioration. If it is in need of replacement, take the trailing arm to an automotive machine shop to have the bushing replaced.

27   Installation is the reverse of removal, noting the following points:

a) *Before fully tightening the trailing arm fasteners, raise the suspension with a floor jack to simulate normal ride height.*

b) *Tighten all fasteners to the torque values listed in this Chapter's Specifications.*

c) *Have the rear wheel alignment checked and, if necessary, adjusted.*

Rear knuckle

28   On AWD models, remove the driveaxle/ hub nut and discard it.

29   Remove the rear brake disc caliper and brake disc (see Chapter 9).

30   Remove the hub and bearing assembly (see Section 13).

31   Disconnect the parking brake cable and the rear wheel speed sensor.

32   Support the lower arm B with a floor jack placed under the arm at the end where the knuckle is connected to the arm.

Warning: *Do not remove the jack while it is supporting the lower arm B.*

33   Detach the lower arm B from the knuckle (see Step 16).

34   Slowly lower the floor jack until the coil spring is extended, then remove the coil spring (see Section 12).

35   Detach the upper arm from the knuckle (see Step 9).

36   Detach the lower arm A from the knuckle (see illustration 11.10).

37   Detach the trailing arm from the knuckle (see illustration 11.24).

38   On AWD models, carefully push the driveaxle while removing the knuckle. Use a puller if the driveaxle is stuck in the hub (see Chapter 8).

Note: *Do not pull on the driveaxle because the inner CV joint could separate.*

39   Installation is the reverse of removal, noting the following points:

a) *Place the spring into the lower arm B spring seat and raise the arm with a floor jack, then tighten the fasteners to the torque listed in this Chapter's Specifications.*

b) *Before fully tightening the pivot bolts of the suspension arms, raise the suspension with a floor jack to simulate normal ride height.*

c) *Tighten all suspension fasteners to the torque values listed in this Chapter's Specifications.*

d) *Tighten the brake fasteners to the torque listed in the Chapter 9 Specifications.*

e) *Install a new driveaxle/hub nut and tighten the nut to the torque listed in the Chapter 8 Specifications.*

f) *It won't be necessary to bleed the brakes unless a hydraulic fitting was loosened.*

g) *Have the rear wheel alignment checked and, if necessary, adjusted.*

12   Coil spring (rear) - removal and installation

Warning: *Always replace the springs as a set - never replace just one of them.*

1   Loosen the wheel lug nuts, raise the vehicle and support it securely on jackstands. Block the front wheels to prevent the vehicle from rolling. Remove the wheel.

2   Detach the stabilizer bar link from the lower suspension arm B.

3   Support the lower suspension arm B with a floor jack.

4   Unbolt the outer end of lower suspension arm B from the knuckle, then slowly lower the arm until the coil spring is extended. Remove the coil spring.

5   Check the spring for cracks and chips,

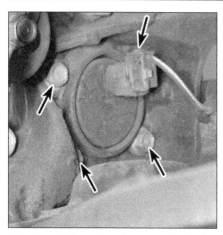

13.6 Rear hub and bearing mounting bolts

14.3 Pry this trim cover off to gain access to the airbag module right-side locking claw

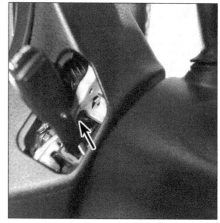

14.4 Use the Torx bit screwdriver to dislodge the three retaining springs. One from the lower steering column cover area, one on the opposite side, and one at the 6 o'clock position of the steering wheel.

replacing the springs as a set if any defects are found. Also check the upper insulator for damage and deterioration, replacing it if necessary.

6   Installation is the reverse of removal. Be sure to position the lower end of the coil spring in the depressed area of the arm. Tighten all fasteners to the torque values listed in this Chapter's Specifications.

Note: *Before tightening the fasteners, use the floor jack to raise the suspension to simulate normal ride height.*

---

13   Hub and bearing assembly (rear) - removal and installation

---

Warning: *Dust created by the brake system is harmful to your health. Never blow it out with compressed air and don't inhale any of it. Do not, under any circumstances, use petroleum-based solvents to clean brake parts. Use brake system cleaner only.*

Note: *The strut assembly should not be disassembled by the home mechanic. The assembly can be removed, however, and taken to a dealer service department or other repair shop to have the bearing replaced.*

1   On AWD models, loosen the rear driveaxle/hub nut (see Chapter 8).

2   Loosen the wheel lug nuts, raise the vehicle and support it securely on jackstands. Block the front wheels to prevent the vehicle from rolling. Remove the wheel.

Removal

3   If you're working on an AWD model, remove the driveaxle/hub nut (see Chapter 8).

4   Remove the brake caliper and disc (see Chapter 9).

5   Unplug the electrical connector from the ABS wheel speed sensor (2WD models) or remove the rear wheel speed sensor (AWD

models) (see Chapter 9).

6   Remove the hub and bearing mounting bolts from the back side of the knuckle (see illustration).

7   Remove the hub and bearing assembly. If you're working on an AWD model, it might be necessary to tap the driveaxle through the hub splines as the hub and bearing is withdrawn. If the driveaxle sticks in the hub, use a puller to push it out of the hub splines.

Installation

8   If you're working on an AWD model, lubricate the splines of the driveaxle with multi-purpose grease.

9   Position the hub and bearing assembly on the rear knuckle (guiding the driveaxle into the hub splines on AWD models) and install the bolts. After all four bolts have been installed, tighten them to the torque listed in this Chapter's Specifications.

10   If you're working on an AWD model, install a new driveaxle/hub nut.

11   Install the brake disc, caliper, speed sensor and the wheel.

12   Lower the vehicle and tighten the lug nuts to the torque listed in the Chapter 1 Specifications. On AWD models tighten the driveaxle/hub nut to the torque listed in the Chapter 8 Specifications, then stake the collar of the nut into the slot in the driveaxle.

---

14   Steering wheel - removal and installation

Warning: *These models are equipped with a Supplemental Restraint System (SRS), more commonly known as airbags. Always disable the airbag system before working in the vicinity of any airbag system component to avoid the possibility of accidental deployment of the airbag(s), which could cause personal injury (see Chapter 12).*

Warning: *Do not use a memory saving device to preserve the PCM or radio memory when working on or near airbag system components.*

Removal

1   Turn the ignition key to Off, then disconnect the cable from the negative battery terminal (see Chapter 5). Wait at least two minutes before proceeding.

2   Turn the steering wheel so the wheels are pointing straight ahead.

3   Use a small screwdriver to release the trim cover from around the cruise control switch (see illustration).

4   A T30H Torx bit screwdriver is the right size needed to push in the three torsion spring retainers out of the way to release the airbag from the steering wheel (see illustration).

5   Pull the airbag module off the steering wheel, disengage the safety connector lock, then disconnect the main electrical connectors (see illustrations). Set the airbag module in a safe and isolated area.

Warning: *Carry the airbag module with the trim side facing away from you, and set the airbag module down with the trim side facing up. Don't place anything on top of the airbag module.*

6   Unplug the electrical connector for the horn and, if equipped, the cruise control switch (see illustration).

7   Remove the steering wheel retaining nut, then mark the relationship of the steering shaft to the hub (if marks don't already exist or don't line up) to simplify installation and ensure steering wheel alignment (see illustration).

8   Use a puller to disconnect the steering wheel from the shaft (see illustration).

Warning: *Do not hammer on the shaft or the puller in an attempt to loosen the wheel from the shaft. Also, don't allow the steering shaft to turn with the steering wheel removed. If the*

14.5a Lift the airbag module off the steering wheel ("A" shows the posts that engage with the retaining springs) . . .

14.5b . . . disengage the safety connector locks by pulling them up slightly with a small screwdriver, then unplug the airbag electrical connectors

14.6 Unplug the electrical connectors for the horn and cruise control switches

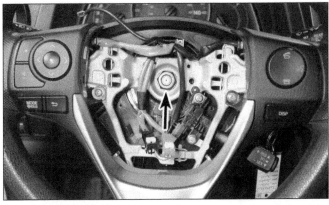

14.7 After removing the steering wheel nut, mark the relationship of the steering wheel to the shaft before removing the wheel

14.8 Use a steering wheel puller to remove the steering wheel from the steering shaft

shaft turns, the airbag spiral cable will become uncentered, which may cause the wire inside to break when the vehicle is returned to service.

9   Installation is the reverse of removal. Be sure to torque all fasteners to the proper specifications listed in this Chapter's Specifications.

Spiral cable (clockspring) removal

10   Remove the steering wheel, if not already done.

11   Remove the steering column upper and lower covers (see Chapter 11).

12   Apply a small piece of tape to the spiral cable housing to keep it from turning once it has been removed from the steering column (see illustration).

13   Disconnect the electrical connection to the spiral cable (see illustration).

14   Gently pry the claw-type clips away from

14.12 Add a bit of duct tape to keep the spiral cable from rotating once it has been removed

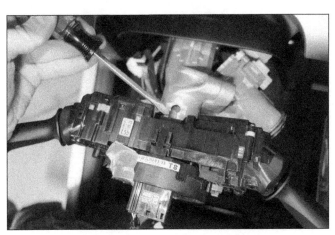

14.13 Spiral cable electrical connector

14.14a The arrow shows the upper claw-type clip retaining the spiral cable sub-assembly.

14.14b One clip is on the top, one on this side, and one on the opposite side (not shown)

the multi-function switch to release the spiral cable sub-assembly. There are three claw like clips holding it in place (see illustrations).

Note: *The spiral cable (clockspring) can be removed separately from the multi-function switch or as one entire component.*

Installation

Note: *If for any reason you are unsure that your tape has held the spiral cable in the correct orientation, follow the next step to ensure it has been installed correctly.*

Warning: *Do NOT force the spiral cable to turn. It should gently turn in both directions. If it fails to turn, replace the unit.*

15   Make sure that the front wheels are facing straight ahead, then center the spiral cable. Depress the interlock in the center of the spiral cable and turn the spiral cable counterclockwise by hand until it stops (but don't force it, as the cable could break). Rotate the spiral cable clockwise about 2-1/2 turns and align the two pointers shaped like arrowheads so they are pointed directly at each other.

Note: *When the unit is properly centered and the marks are aligned, the yellow roller will be visible in the window.*

16   Snap the spiral cable into place on the multi-function switch.

17   Reconnect the spiral cable electrical connections.

18   Reinstall the steering wheel covers.

19   Install the steering wheel by aligning the mark on the steering wheel hub with the mark on the shaft and slip the wheel onto the shaft. Install the nut and tighten it to the torque listed in this Chapter's Specifications.

20   Plug in the cruise control and horn connectors.

21   Plug in the electrical connectors for the airbag module and push the yellow safety locks down to secure the airbag connections to the airbag.

22   Make sure the airbag module electrical wiring is positioned so that it does not interfere with installing the airbag onto the steering wheel, then push the airbag into the retaining springs in the steering wheel. It should snap into place and not be able to be removed again

(unless you release the retaining springs).

23   Connect the negative battery cable, start vehicle and watch the airbag light. The light should remain on for no more than 5 seconds and go off. With the light off, the airbag system is functioning normally.

---

15   Steering column and power steering motor assembly - removal and installation

Warning: *These models are equipped with airbags. Always disable the airbag system before working in the vicinity of any airbag system component to avoid the possibility of accidental deployment of the airbag(s), which could cause personal injury (see Chapter 12).*

Steering column

Removal

1   Park the vehicle with the wheels pointing straight ahead. Disconnect the cable from the negative battery terminal (see Chapter 5).

2   Remove the steering wheel (see Section 14), then turn the ignition key to the Lock position to prevent the steering shaft from turning.

Warning: *If this is not done, the airbag spiral cable could be damaged.*

3   Remove the driver's knee bolster and the reinforcement behind it (see Chapter 11).

4   Remove the steering column covers (see Chapter 11).

5   Remove the upper instrument panel trim (see Chapter 11).

6   Remove the combination switch and clockspring assembly from the steering column (see Chapter 12).

7   Remove the column cover from the firewall by folding back the carpet and removing the two retaining clips (see illustration).

8   Mark the relationship of the U-joint to the steering shaft, then remove the pinch bolt (see illustration).

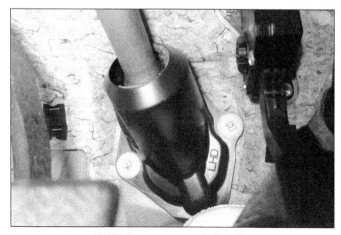

15.7 Column cover retaining clips

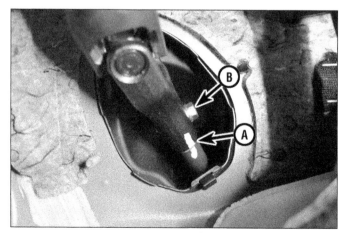

15.8 Steering shaft with column cover removed

A   Matchmarks   B   Pinch bolt

15.9 Clamps and electrical locations for power steering sub-assembly

A   *Electrical connections*
B   *Ground lead*
C   *Clamp*

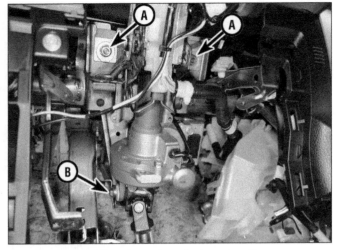

15.10a Steering column fastener locations

A   *Upper mounting nuts*   B   *Lower mounting bolt*

9   Remove the electrical connections to the power steering column behind the dash (see illustration).
10   Remove the steering column mounting fasteners (see illustration), then lower the column and pull it to the rear, making sure nothing is still connected. Separate the intermediate shaft from the steering column shaft (see illustration).

Installation

11   Guide the steering column into position, connect the intermediate shaft back to the match marks you made earlier, then install the mounting fasteners.
12   Tighten the column mounting fasteners to the torque listed in this Chapter's Specifications.
13   Install the pinch bolt, tightening it to the torque listed in this Chapter's Specifications.

14   The remainder of installation is the reverse of removal.

Power steering ECU

15   Disconnect the cable from the negative terminal of the battery (see Chapter 5).
16   Remove the upper instrument panel trim (see Chapter 11).
17   Remove the steering column.
18   Remove the wire harness fasteners to the power steering ECU.
19   Release the two claw like clips from the ECU dust cover.
20   Release the four claws securing the ECU protector to the ECU.
21   Remove the three terminal bolts and two bolts securing the ECU assembly to the steering column.
22   Remove the ECU.

23   Installation is the reverse of removal.
24   The ECU must now be calibrated. Using a jumper wire, bridge terminals 4 and 12 of the Data Link Connector (DLC), located under the dash, to the left of the steering column (see illustration).
25   Turn the ignition On for at least five seconds, then turn it off.
26   Check for the presence of any stored Diagnostic Trouble Codes (see Chapter 6, Section 2).

Power steering motor
Removal
27   Remove the steering column.
28   Remove the power steering ECU.
29   Remove the two bolts securing the power steering motor to the steering column.
30   Lift the motor off of the steering column.

15.10b Mark the intermediate shaft orientation with the column before removing the bolt. Then remove the pinch bolt and detach the intermediate shaft from the column.

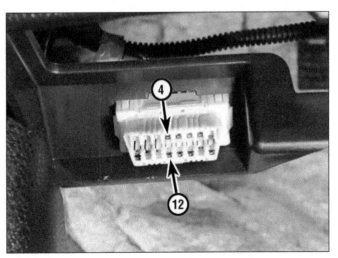

15.24 Terminals 4 and 12 of the DLC

16.2a Loosen the jam nut . . .

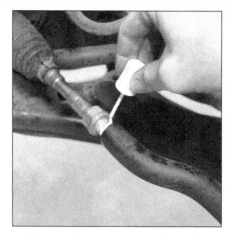

16.2b . . . then mark the position of the tie-rod end in relation to the threads

16.3 Straighten the cotter pin out, then pull the cotter pin out of the castle nut

Installation

31   Install a new O-ring around the motor shaft.

32   Apply grease to the motor spline, then guide the motor shaft back into the steering column.

33   Align the two bolt holes with the holes in the column, then install the bolts. Tighten the bolts to the torque listed in this Chapter's Specifications.

34   The remainder of the installation is the reverse of removal.

16   Tie-rod ends - removal and installation

Removal

1   Loosen the wheel lug nuts. Apply the parking brake, raise the front of the vehicle and support it securely on jackstands. Remove the front wheel.

2   Hold the tie-rod with a pair of locking pliers or a wrench and loosen the jam nut enough to mark the position of the tie-rod end in relation to the threads (see illustrations).

3   Remove the cotter pin from the tie-rod castle nut (see illustration). Loosen the nut a few turns, but don't remove it yet.

4   Disconnect the tie-rod end from the steering knuckle arm with a puller (see illustration), then remove the castle nut.

5   Unscrew the tie-rod end from the tie-rod.

Note: *As an extra insurance when reinstalling the tie rod, count the number of turns you make as you remove the old tie-rod end. Record this number and install the new tie rod using the same number of turns.*

Installation

6   Thread the tie-rod end on to the marked position and insert the tie-rod stud into the steering knuckle arm. Tighten the jam nut securely.

7   Install the castle nut on the stud and tighten it to the torque listed in this Chapter's Specifications. Install a new cotter pin. If the

hole for the cotter pin doesn't line up with one of the slots in the nut, tighten the nut an additional amount until it does.

8   Install the wheel and lug nuts. Lower the vehicle and tighten the lug nuts to the torque listed in the Chapter 1 Specifications.

9   Have the alignment checked and, if necessary, adjusted.

17   Steering gear boots - replacement

1   Loosen the lug nuts, raise the vehicle and support it securely on jackstands. Remove the wheel.

2   Remove the tie-rod end and jam nut (see Section 16).

3   Remove the outer steering gear boot clamp with a pair of pliers (see illustration). Cut off the inner boot clamp with a pair of diagonal cutters (see illustration). Slide off the boot.

4   Before installing the new boot, wrap

16.4 Disconnect the tie-rod end from the steering knuckle arm with a puller or balljoint separator

17.3a The outer ends of the steering gear boots are secured by spring-type clamps; they're easily released with a pair of pliers

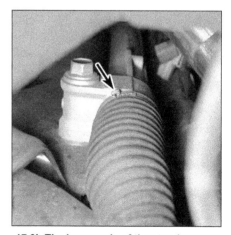

17.3b The inner ends of the steering gear boots are retained by boot clamps which must be cut off and discarded

19.9 Subframe reinforcement member bolts

19.10 Subframe reinforcement brackets

the threads and serrations on the end of the steering rod with a layer of tape so the small end of the new boot isn't damaged.

5   Slide the new boot into position on the steering gear until it seats in the groove in the steering rod and install new clamps.

6   Remove the tape and install the tie-rod end (see Section 16).

7   Install the wheel and lug nuts. Lower the vehicle and tighten the lug nuts to the torque listed in the Chapter 1 Specifications.

18   Steering gear - removal and installation

Warning: *These models are equipped with airbags. Always disable the airbag system before working in the vicinity of any airbag system component to avoid the possibility of accidental deployment of the airbag(s), which could cause personal injury (see Chapter 12).*

Removal

1   Remove the subframe (see Section 19).

2   Mark the relationship of the steering intermediate shaft to the steering gear input shaft, then loosen the pinch bolt and detach the intermediate shaft.

3   Remove the steering gear mounting bolts/nuts and detach the steering gear from the subframe.

Note: *Unscrew the bolts, not the nuts (the nuts have rotation-prevention tabs).*

Installation

4   Installation is the reverse of removal, noting the following points:

a)   *Make sure the steering gear is centered before installing it.*

b)   *When connecting the steering gear input shaft to the intermediate shaft U-joint, be sure to align the matchmarks.*

c)   *Tighten all suspension and steering fasteners to the torque values listed in this Chapter's Specifications.*

d)   *Tighten the lug nuts to the torque listed in the Chapter 1 Specifications.*

e)   *Set the front wheels in the straight-ahead position, then center the spiral cable and install the steering wheel and airbag module (see Section 14).*

19   Subframe, front - removal and installation

Front

1   Pass the seat belt through the steering wheel and clip it into its latch.

Warning: *This will prevent the steering wheel from rotating when the intermediate shaft is disconnected from the steering gear, which could damage the airbag spiral cable.*

2   Remove the steering column intermediate shaft universal joint cover. Mark the relationship of the lower universal joint to the steering gear input shaft and remove the pinch bolt (see illustrations 15.7 and 15.8 in Section 15).

3   Loosen the wheel lug nuts. Apply the parking brake, raise the front of the vehicle and support it securely on jackstands. Remove the front wheel.

4   Remove the lower engine cover and the rear lower engine covers. (The LH side uses either a half cover type which has three clips, or a full cover type which uses two clips. RH side uses two clips.)

5   Remove the smaller lower engine cover located to the front of the engine comnpartment.

6   Disconnect the stabilizer bar links from the stabilizer bar (see Section 4).

7   Detach the tie-rod ends from the steering knuckles (see Section 16).

8   Remove the fasteners and separate the balljoints from the control arms (see Section 5).

9   Remove the subframe reinforcement members from the right and left sides (see illustration).

10   Remove the rear subframe reinforcement brackets (see illustration).

11   Remove the rear engine mount-to-subframe fasteners (see illustration).

12   Support the subframe with a floor jack, then remove the subframe mounting bolts.

13   Very carefully lower the subframe with the floor jack, making sure nothing is still attached.

14   Installation is the reverse of removal. Tighten all fasteners to the torque values listed in this Chapter's Specifications.

15   Have the wheel alignment checked and, if necessary, adjusted.

19.11 Rear engine mount bracket fasteners

20.3 Use a press tool to push the stud out of the flange

1 Hub flange   3 Press tool
2 Lug nut on stud

20.4 Install a spacer and a lug nut on the stud, then tighten the nut to draw the stud into place

1   Hub flange   2   Spacer

20   Wheel studs - replacement

Note: *This procedure applies to both the front and rear wheel studs.*

1   Loosen the wheel lug nuts, raise the vehicle and support it securely on jackstands. Remove the wheel.

2   Remove the brake disc or drum (see Chapter 9).

3   Install a lug nut part way onto the stud being replaced. Push the stud out of the hub flange with a press tool (see illustration).

4   Insert the new stud into the hub flange from the back side and install some flat washers and a lug nut on the stud (see illustration).

5   Tighten the lug nut until the stud is seated in the flange.

6   Reinstall the brake drum or disc. Install the wheel and lug nuts. Lower the vehicle and tighten the lug nuts to the torque listed in the Chapter 1 Specifications.

21   Wheels and tires - general information

1   All vehicles covered by this manual are equipped with metric-sized fiberglass or steel belted radial tires (see illustration). Use of other size or type of tires may affect the ride and handling of the vehicle. Don't mix different types of tires, such as radials and bias belted, on the same vehicle as handling may be seriously affected. It's recommended that tires be replaced in pairs on the same axle, but if only one tire is being replaced, be sure it's the same size, structure and tread design as the other.

2   Because tire pressure has a substantial effect on handling and wear, the pressure

on all tires should be checked at least once a month or before any extended trips (see Chapter 1).

3   Wheels must be replaced if they are bent, dented, leak air, have elongated bolt holes, are heavily rusted, out of vertical symmetry or if the lug nuts won't stay tight. Wheel repairs that use welding or peening are not recommended.

4   Tire and wheel balance is important in the overall handling, braking and performance of the vehicle. Unbalanced wheels can adversely affect handling and ride characteristics as well as tire life. Whenever a tire is installed on a wheel, the tire and wheel should be balanced by a shop with the proper equipment.

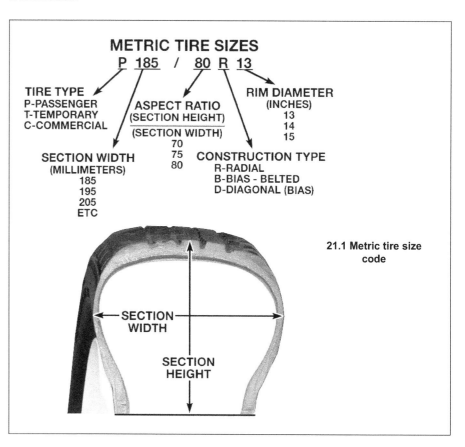

21.1 Metric tire size code

22   Wheel alignment - general information

1   A wheel alignment refers to the adjustments made to the wheels so they are in proper angular relationship to the suspension and the ground. Wheels that are out of proper alignment not only affect vehicle control, but also increase tire wear. The front end angles normally measured are camber, caster and toe-in (see illustration). Toe-in and camber are adjustable; if the caster is not correct, check for bent components. Rear camber and toe-in are also adjustable.

2   Getting the proper wheel alignment is a very exacting process, one in which complicated and expensive machines are necessary to perform the job properly. Because of this, you should have a technician with the proper equipment perform these tasks.

3   Toe-in is the turning in of the wheels. The purpose of a toe specification is to ensure parallel rolling of the wheels. In a vehicle with zero toe-in, the distance between the front edges of the wheels will be the same as the distance between the rear edges of the wheels. The actual amount of toe-in is normally only a fraction of an inch. Incorrect toe-in will cause the tires to wear improperly by making them scrub against the road surface.

4   Camber is the tilting of the wheels from vertical when viewed from one end of the vehicle. When the wheels tilt out at the top, the camber is said to be positive (+). When the wheels tilt in at the top the camber is negative (-). The amount of tilt is measured in degrees from vertical and this measurement is called the camber angle. This angle affects the amount of tire tread which contacts the road and compensates for changes in the suspension geometry when the vehicle is cornering or traveling over an undulating surface.

5   Caster is the tilting of the front steering axis from vertical. A tilt toward the rear is positive caster and a tilt toward the front is negative caster.

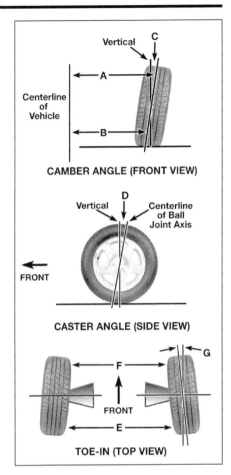

22.1 Camber, caster and toe-in angles

A minus B = C (degrees camber)
D = degrees caster
E minus F = toe-in (measured in inches)
G = toe-in (expressed in degrees)

Notes

Chapter 11
Body

Contents

Specifications

Torque specifications

	Ft-lbs (unless otherwise indicated)	Nm
Seat belt fasteners		
Front		
Belt buckle to seat	31	42
Shoulder belt anchor to body	31	42
Floor anchor	31	42
Retractor	75 in-lbs	8.5
Rear		
Inner belt assembly to seat assembly	31	42
Outer anchor bolt	31	42
Center seat lap belt anchor bolts	31	42
Upper retractor bolt	75 in-lbs	8.5
Lower retractor bolt	31	42
Seat mounting bolts		
Front seat to body	27	37
Rear seat(s) to body	31	42

---

1   General information

Warning: *The models covered by this manual are equipped with a Supplemental Restraint System (SRS), more commonly known as airbags. Always disable the airbag system before working in the vicinity of any airbag system components to avoid the possibility of accidental deployment of the airbags, which could cause personal injury (see Chapter 12).*

1   Certain body components are particularly vulnerable to accident damage and can be unbolted and repaired or replaced. Among these parts are the hood, doors, tailgate, liftgate, bumpers and front fenders.

2   Only general body maintenance practices and body panel repair procedures within the scope of the do-it-yourselfer are included in this Chapter.

Make sure the damaged area is perfectly clean and rust free. If the touch-up kit has a wire brush, use it to clean the scratch or chip. Or use fine steel wool wrapped around the end of a pencil. Clean the scratched or chipped surface only, not the good paint surrounding it. Rinse the area with water and allow it to dry thoroughly

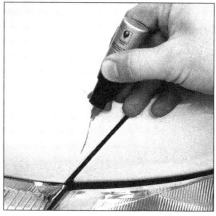

Thoroughly mix the paint, then apply a small amount with the touch-up kit brush or a very fine artist's brush. Brush in one direction as you fill the scratch area. Do not build up the paint higher than the surrounding paint

2   Repairing minor paint scratches

1   No matter how hard you try to keep your vehicle looking like new, it will inevitably be scratched, chipped or dented at some point. If the metal is actually dented, seek the advice of a professional. But you can fix minor scratches and chips yourself. Buy a touch-up paint kit from a dealer service department or an auto parts store. To ensure that you get the right color, you'll need to have the specific make, model and year of your vehicle and, ideally, the paint code, which is located on a special metal plate under the hood or in the door jamb.

3   Body repair - minor damage

Plastic body panels

1   The following repair procedures are for minor scratches and gouges. Repair of more serious damage should be left to a dealer service department or qualified auto body shop. Below is a list of the equipment and materials necessary to perform the following repair procedures on plastic body panels.

Wax, grease and silicone removing solvent
Cloth-backed body tape
Sanding discs
Drill motor with three-inch disc holder
Hand sanding block
Rubber squeegees
Sandpaper
Non-porous mixing palette
Wood paddle or putty knife
Wood paddle or putty knife
Curved-tooth body file
Flexible parts repair material

Flexible panels (bumper trim)

2   Remove the damaged panel, if necessary or desirable. In most cases, repairs can be carried out with the panel installed.
3   Clean the area(s) to be repaired with a

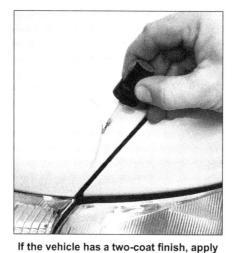

If the vehicle has a two-coat finish, apply the clear coat after the color coat has dried

wax, grease and silicone removing solvent applied with a water-dampened cloth.
4   If the damage is structural, that is, if it extends through the panel, clean the backside of the panel area to be repaired as well. Wipe dry.
5   Sand the rear surface about 1-1/2 inches beyond the break.
6   Cut two pieces of fiberglass cloth large enough to overlap the break by about 1-1/2 inches. Cut only to the required length.
7   Mix the adhesive from the repair kit according to the instructions included with the kit, and apply a layer of the mixture approximately 1/8-inch thick on the backside of the panel. Overlap the break by at least 1-1/2 inches.
8   Apply one piece of fiberglass cloth to the adhesive and cover the cloth with additional adhesive. Apply a second piece of fiberglass cloth to the adhesive and immediately cover the cloth with additional adhesive in sufficient quantity to fill the weave.
9   Allow the repair to cure for 20 to 30 min-

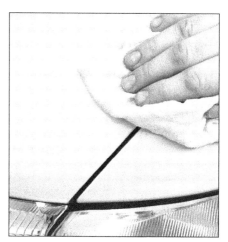

Wait a few days for the paint to dry thoroughly, then rub out the repainted area with a polishing compound to blend the new paint with the surrounding area. When you're happy with your work, wash and polish the area

utes at 60-degrees to 80-degrees F.
10   If necessary, trim the excess repair material at the edge.
11   Remove all of the paint film over and around the area(s) to be repaired. The repair material should not overlap the painted surface.
12   With a drill motor and a sanding disc (or a rotary file), cut a "V" along the break line approximately 1/2-inch wide. Remove all dust and loose particles from the repair area.
13   Mix and apply the repair material. Apply a light coat first over the damaged area; then continue applying material until it reaches a level slightly higher than the surrounding finish.
14   Cure the mixture for 20 to 30 minutes at 60-degrees to 80-degrees F.
15   Roughly establish the contour of the area being repaired with a body file. If low areas or

pits remain, mix and apply additional adhesive.

16   Block sand the damaged area with sandpaper to establish the actual contour of the surrounding surface.

17   If desired, the repaired area can be temporarily protected with several light coats of primer. Because of the special paints and techniques required for flexible body panels, it is recommended that the vehicle be taken to a paint shop for completion of the body repair.

Steel body panels

Repairing simple dents

18   When repairing dents, the first job is to pull the dent out until the affected area is as close as possible to its original shape. There is no point in trying to restore the original shape completely as the metal in the damaged area will have stretched on impact and cannot be restored to its original contours. It is better to bring the level of the dent up to a point that is about 1/8-inch below the level of the surrounding metal. In cases where the dent is very shallow, it is not worth trying to pull it out at all.

19   If the backside of the dent is accessible, it can be hammered out gently from behind using a soft-face hammer. While doing this, hold a block of wood firmly against the opposite side of the metal to absorb the hammer blows and prevent the metal from being stretched.

20   If the dent is in a section of the body which has double layers, or some other factor makes it inaccessible from behind, a different technique is required. Drill several small holes through the metal inside the damaged area, particularly in the deeper sections. Screw long, self-tapping screws into the holes just enough for them to get a good grip in the metal. Now pulling on the protruding heads of the screws with locking pliers can pull out the dent.

21   The next stage of repair is the removal of paint from the damaged area and from an inch or so of the surrounding metal. This is easily done with a wire brush or sanding disk in a drill motor, although it can be done just as effectively by hand with sandpaper. To complete the preparation for filling, score the surface of the bare metal with a screwdriver or the tang of a file or drill small holes in the affected area. This will provide a good grip for the filler material. To complete the repair, see the Section on filling and painting.

Repair of rust holes or gashes

22   Remove all paint from the affected area and from an inch or so of the surrounding metal using a sanding disk or wire brush mounted in a drill motor. If these are not available, a few sheets of sandpaper will do the job just as effectively.

23   With the paint removed, you will be able to determine the severity of the corrosion and decide whether to replace the whole panel, if possible, or repair the affected area. New body panels are not as expensive as most people think and it is often quicker to install a new panel than to repair large areas of rust.

24   Remove all trim pieces from the affected area except those which will act as a guide to the original shape of the damaged body, such as headlight shells, etc. Using metal snips or a hacksaw blade, remove all loose metal and any other metal that is badly affected by rust. Hammer the edges of the hole in to create a slight depression for the filler material.

25   Wire-brush the affected area to remove the powdery rust from the surface of the metal. If the back of the rusted area is accessible, treat it with rust inhibiting paint.

26   Before filling is done, block the hole in some way. This can be done with sheet metal riveted or screwed into place, or by stuffing the hole with wire mesh.

27   Once the hole is blocked off, the affected area can be filled and painted. See the following subsection on filling and painting.

Filling and painting

28   Many types of body fillers are available, but generally speaking, body repair kits which contain filler paste and a tube of resin hardener are best for this type of repair work. A wide, flexible plastic or nylon applicator will be necessary for imparting a smooth and contoured finish to the surface of the filler material. Mix up a small amount of filler on a clean piece of wood or cardboard (use the hardener sparingly). Follow the manufacturer's instructions on the package, otherwise the filler will set incorrectly.

29   Using the applicator, apply the filler paste to the prepared area. Draw the applicator across the surface of the filler to achieve the desired contour and to level the filler surface. As soon as a contour that approximates the original one is achieved, stop working the paste. If you continue, the paste will begin to stick to the applicator. Continue to add thin layers of paste at 20-minute intervals until the level of the filler is just above the surrounding metal.

30   Once the filler has hardened, the excess can be removed with a body file. From then on, progressively finer grades of sandpaper should be used, starting with a 180-grit paper and finishing with 600-grit wet-or-dry paper. Always wrap the sandpaper around a flat rubber or wooden block, otherwise the surface of the filler will not be completely flat. During the sanding of the filler surface, the wet-or-dry paper should be periodically rinsed in water. This will ensure that a very smooth finish is produced in the final stage.

31   At this point, the repair area should be surrounded by a ring of bare metal, which in turn should be encircled by the finely feathered edge of good paint. Rinse the repair area with clean water until all of the dust produced by the sanding operation is gone.

32   Spray the entire area with a light coat of primer. This will reveal any imperfections in the surface of the filler. Repair the imperfections with fresh filler paste or glaze filler and once more smooth the surface with sandpaper. Repeat this spray-and-repair procedure until you are satisfied that the surface of the filler and the feathered edge of the paint are perfect. Rinse the area with clean water and allow it to dry completely.

33   The repair area is now ready for painting. Spray painting must be carried out in a warm, dry, windless and dust free atmosphere. These conditions can be created if you have access to a large indoor work area, but if you are forced to work in the open, you will have to pick the day very carefully. If you are working indoors, dousing the floor in the work area with water will help settle the dust that would otherwise be in the air. If the repair area is confined to one body panel, mask off the surrounding panels. This will help minimize the effects of a slight mismatch in paint color. Trim pieces such as chrome strips, door handles, etc., will also need to be masked off or removed. Use masking tape and several thickness of newspaper for the masking operations.

34   Before spraying, shake the paint can thoroughly, then spray a test area until the spray painting technique is mastered. Cover the repair area with a thick coat of primer. The thickness should be built up using several thin layers of primer rather than one thick one. Using 600-grit wet-or-dry sandpaper, rub down the surface of the primer until it is very smooth. While doing this, the work area should be thoroughly rinsed with water and the wet-or-dry sandpaper periodically rinsed as well. Allow the primer to dry before spraying additional coats.

35   Spray on the top coat, again building up the thickness by using several thin layers of paint. Begin spraying in the center of the repair area and then, using a circular motion, work out until the whole repair area and about two inches of the surrounding original paint is covered. Remove all masking material 10 to 15 minutes after spraying on the final coat of paint. Allow the new paint at least two weeks to harden, then use a very fine rubbing compound to blend the edges of the new paint into the existing paint. Finally, apply a coat of wax

4   Body repair - major damage

1   Major damage must be repaired by an auto body shop specifically equipped to perform body and frame repairs. These shops have the specialized equipment required to do the job properly.

2   If the damage is extensive, the frame must be checked for proper alignment or the vehicle's handling characteristics may be adversely affected and other components may wear at an accelerated rate.

3   Due to the fact that all of the major body components (hood, fenders, etc.) are separate and replaceable units, any seriously damaged components should be replaced rather than repaired. Sometimes the components can be found in a wrecking yard that specializes in used vehicle components, often at considerable savings over the cost of new parts.

These photos illustrate a method of repairing simple dents. They are intended to supplement *Body repair - minor damage* in this Chapter and should not be used as the sole instructions for body repair on these vehicles.

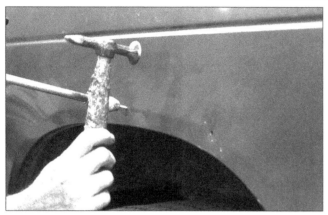

1 If you can't access the backside of the body panel to hammer out the dent, pull it out with a slide-hammer-type dent puller. Tap with a hammer near the edge of the dent to help 'pop' the metal back to its original shape, about 1/8-inch below the surface of the surrounding metal

2 Using coarse-grit sandpaper, remove the paint down to the bare metal. Clean the repair area with wax/silicone remover.

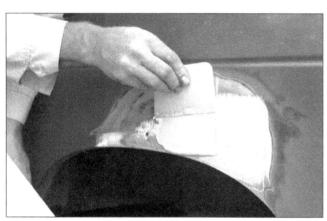

3 Following label instructions, mix up a batch of plastic filler and hardener, then quickly press it into the metal with a plastic applicator. Work the filler until it matches the original contour and is slightly above the surrounding metal

4 Let the filler harden until you can just dent it with your fingernail. File, then sand the filler down until it's smooth and even. Work down to finer grits of sandpaper - always using a board or block - ending up with 360 or 400 grit

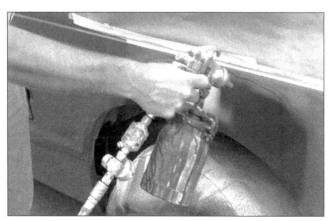

5 When the area is smooth to the touch, clean the area and mask around it. Apply several layers of primer to the area. A professional-type spray gun is being used here, but aerosol spray primer works fine

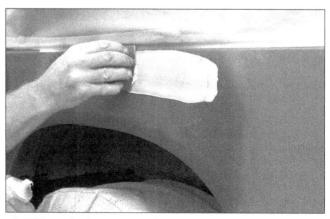

6 Fill imperfections or scratches with glazing compound. Sand with 360 or 400-grit and re-spray. Finish sand the primer with 600 grit, clean thoroughly, then apply the finish coat. Don't attempt to rub out or wax the repair area until the paint has dried completely (at least two weeks)

5   Fastener and trim removal

1   There is a variety of plastic fasteners used to hold trim panels, splash shields and other parts in place in addition to typical screws, nuts and bolts. Once you are familiar with them, they can usually be removed without too much difficulty.

2   The proper tools and approach can prevent added time and expense to a project by minimizing the number of broken fasteners and/or parts.

3   The following illustration shows various types of fasteners that are typically used on most vehicles and how to remove and install them (see illustration). Replacement fasteners are commonly found at most auto parts stores, if necessary.

4   Trim panels are typically made of plastic and their flexibility can help during removal. The key to their removal is to use a tool to pry the panel near its retainers to release it without damaging surrounding areas or breaking-off any retainers. The retainers will usually snap out of their designated slot or hole after force is applied to them. Stiff plastic tools designed for prying on trim panels are available at most auto parts stores (see illustration). Tools that are tapered and wrapped in protective tape, such as a screwdriver or small pry tool, are also very effective when used with care.

6   Upholstery, carpets and vinyl trim - maintenance

Upholstery and carpets

1   Every three months remove the floormats and clean the interior of the vehicle (more frequently if necessary). Use a stiff whiskbroom to brush the carpeting and loosen dirt and dust, then vacuum the upholstery and carpets thoroughly, especially along seams and crevices.

2   Dirt and stains can be removed from carpeting with basic household or automotive carpet shampoos available in spray cans. Follow the directions and vacuum again, then

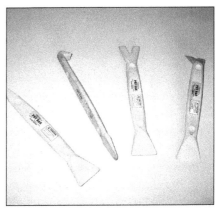

5.4 These small plastic pry tools are ideal for prying off trim panels

use a stiff brush to bring back the "nap" of the carpet.

3   Most interiors have cloth or vinyl upholstery, either of which can be cleaned and maintained with a number of material-specific cleaners or shampoos available in

Fasteners

This tool is designed to remove special fasteners. A small pry tool used for removing nails will also work well in place of this tool

A Phillips head screwdriver can be used to release the center portion, but light pressure must be used because the plastic is easily damaged. Once the center is up, the fastener can easily be pried from its hole

Here is a view with the center portion fully released. Install the fastener as shown, then press the center in to set it

This fastener is used for exterior panels and shields. The center portion must be pried up to release the fastener. Install the fastener with the center up, then press the center in to set it

This type of fastener is used commonly for interior panels. Use a small blunt tool to press the small pin at the center in to release it . . .

. . . the pin will stay with the fastener in the released position

Reset the fastener for installation by moving the pin out. Install the fastener, then press the pin flush with the fastener to set it

This fastener is used for exterior and interior panels. It has no moving parts. Simply pry the fastener from its hole like the claw of a hammer removes a nail. Without a tool that can get under the top of the fastener, it can be very difficult to remove

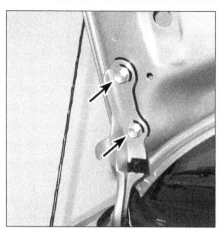

9.3 Draw alignment marks around the hood hinges to ensure proper alignment of the hood when it's reinstalled - arrows indicate the hinge-to-hood retaining bolts

9.9 To adjust the hood latch horizontally or vertically, loosen these bolts

9.10 To adjust the vertical height of the leading edge of the hood so that it's flush with the fenders, turn each edge cushion (arrow indicates one) clockwise (to lower the hood) or counterclockwise (to raise the hood)

auto supply stores. Follow the directions on the product for usage, and always spot-test any upholstery cleaner on an inconspicuous area (bottom edge of a backseat cushion) to ensure that it doesn't cause a color shift in the material.

4   After cleaning, vinyl upholstery should be treated with a protectant.

Note: *Make sure the protectant container indicates the product can be used on seats - some products may make a seat too slippery.*

Caution: *Do not use protectant on vinyl-covered steering wheels.*

5   Leather upholstery requires special care. It should be cleaned regularly with saddle-soap or leather cleaner. Never use alcohol, gasoline, nail polish remover or thinner to clean leather upholstery.

6   After cleaning, regularly treat leather upholstery with a leather conditioner, rubbed in with a soft cotton cloth. Never use car wax on leather upholstery.

7   In areas where the interior of the vehicle is subject to bright sunlight, cover leather seating areas of the seats with a sheet if the vehicle is to be left out for any length of time.

Vinyl trim

8   Don't clean vinyl trim with detergents, caustic soap or petroleum-based cleaners. Plain soap and water works just fine, with a soft brush to clean dirt that may be ingrained. Wash the vinyl as frequently as the rest of the vehicle.

9   After cleaning, application of a high-quality rubber and vinyl protectant will help prevent oxidation and cracks. The protectant can also be applied to weather-stripping, vacuum lines and rubber hoses, which often fail as a result of chemical degradation, and to the tires.

7   Hinges and locks - maintenance

1   Once every 3000 miles, or every three months, the hinges and latch assemblies on the doors, hood and liftgate should be given a few drops of light oil or lock lubricant. The door latch strikers should also be lubricated with a thin coat of grease to reduce wear and ensure free movement. Lubricate the door and liftgate locks with spray-on graphite lubricant.

8   Windshield and fixed glass - replacement

1   Replacement of the windshield and fixed glass requires the use of special fast-setting adhesive/caulk materials and some specialized tools and techniques. These operations should be left to a dealer service department or a shop specializing in glass work.

9   Hood - removal, installation and adjustment

Note: *The hood is somewhat awkward to remove and install; at least two people should perform this procedure.*

Removal and installation

1   Open the hood, then place blankets or pads over the fenders and cowl area of the body. This will protect the body and paint as the hood is lifted off.

2   Disconnect any cables or wires that will interfere with removal. Disconnect the windshield washer tubing from the nozzles on the hood.

3   Make marks around the hood hinge to ensure proper alignment during installation (see illustration).

4   Have an assistant support one side of the hood. Take turns removing the hinge-to-hood bolts and lift off the hood.

5   Installation is the reverse of removal. Align the hinge bolts with the marks made in Step 3.

Adjustment

6   Fore-and-aft and side-to-side adjustment of the hood is done by moving the hinge plate slot after loosening the bolts or nuts.

Note: *The factory bolts are a centering type that will not allow adjustment. To adjust the hood in relation to the hinges, these bolts must be replaced with standard bolts with flat washers and lock washers.*

7   Mark around the entire hinge plate so you can determine the amount of movement.

8   Loosen the bolts and move the hood into correct alignment. Move it only a little at a time. Tighten the hinge bolts and carefully lower the hood to check the position.

9   If necessary after installation, the entire hood latch assembly can be adjusted up-and-down as well as from side-to-side on the radiator support so the hood closes securely and flush with the fenders. Scribe a line or mark around the hood latch mounting bolts to provide a reference point, then loosen them and reposition the latch assembly, as necessary (see illustration). Following adjustment, retighten the mounting bolts.

10   Finally, adjust the hood bumpers on the radiator support so the hood, when closed, is flush with the fenders (see illustration).

11   The hood latch assembly, as well as the hinges, should be periodically lubricated with white lithium-base grease to prevent binding and wear.

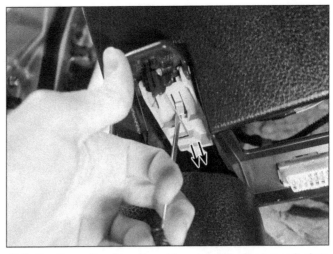

10.2 Pry out the cable retainer from the backside of the hood latch assembly, then disengage the cable

10.5a Gently pry the claw clip out far enough to disengage it, then slide the lever assembly free from the dash

10 Hood latch and release cable - removal and installation

Warning: *The models covered by this manual are equipped with a Supplemental Restraint System (SRS), more commonly known as airbags. Always disable the airbag system before working in the vicinity of any airbag system component to avoid the possibility of accidental deployment of the airbag, which could cause personal injury (see Chapter 12).*

Latch

1   Scribe a line around the latch to aid alignment when installing, then remove the retaining bolts securing the hood latch to the radiator support (see illustration 9.9). Remove the latch.
2   Disconnect the hood release cable by disengaging the cable from the latch assembly (see illustration).
3   Installation is the reverse of removal.
Note: *Adjust the latch so the hood engages securely when closed and the hood bumpers are slightly compressed.*

Hood release lever

4   Working inside the vehicle, remove the driver's side kick panel (see Section 25).
5   Lift the hood release lever upward, then pull down on the cable housing end and disengage the cable from the hood release lever handle. If the handle lever needs to be replaced, pull outward on the handle retaining tab and push downward to release the claw like clips from the instrument panel. The cable and lever can now be removed (see illustrations).
6   Installation is the reverse of removal.

Hood release cable

7   Remove the cable from the inside release lever (see previous Steps), then

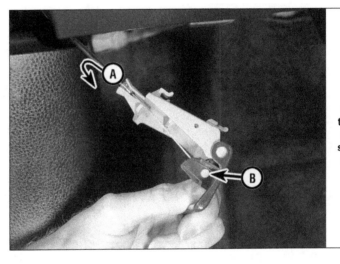

10.5b To remove the cable from the release lever, slide the barrel end of the cable (A) out of the slot, then detach the cable end (B) from the handle

attach a piece of thin wire or string to the end of the cable.
Note: *The thin wire or string needs to be long enough to reach from the inside lever to the hood latch.*
8   Working in the engine compartment, disconnect the hood release cable from the latch assembly as described in Steps 1 and 2. Unclip all the cable retaining clips on the radiator support and the inner fenderwell.
9   Pull the cable forward into the engine compartment until you can see the wire or string, then remove the wire or string from the old cable and fasten it to the new cable.
10   Working inside the vehicle, with the new cable attached to the wire or string, pull the wire or string back through the firewall until the new cable reaches the inside handle.
11   Install the new cable into the hood release lever, making sure the cable housing (barrel end) fits snugly into the notch in the lever bracket.
12   Pull on the cable until the cable stop seats in the grommet on the firewall.

13   The remainder of the installation is the reverse of removal.

11 Bumpers and bumper covers - removal and installation

Warning: *The models covered by this manual are equipped with a Supplemental Restraint System (SRS), more commonly known as airbags. Always disable the airbag system before working in the vicinity of any airbag system component to avoid the possibility of accidental deployment of the airbag, which could cause personal injury (see Chapter 12).*

Front bumper components

Front bumper cover

1   Apply the parking brake, raise the front of the vehicle and support it securely on jackstands. Open the hood.
2   Disconnect the cable from the negative battery terminal (see Chapter 5), and wait two

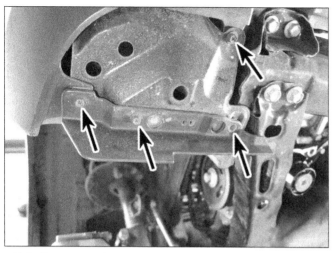

11.3a Remove the fasteners for the front wheel mud guards and bumper cover . . .

11.3b . . . then remove the remaining bumper cover lower fasteners

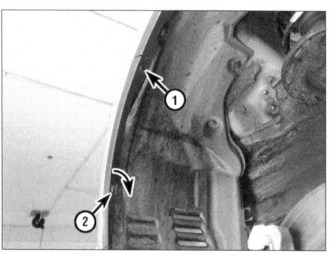

11.4 Remove the upper side fastener (1), then turn the lower side fastener 90-degrees (2) to disengage it from the bumper cover

11.5 Bumper cover upper fasteners

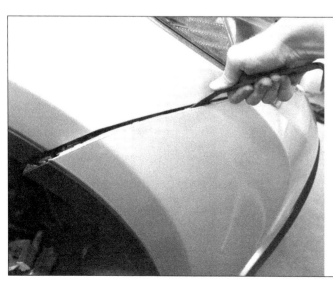

11.6 Carefully pry the bumper and fender seam apart

minutes before proceeding any further.

3　　Working below the vehicle, remove the front wheel mud guards and lower bumper cover fasteners (see illustrations).

4　　Turn the steering wheel as necessary to move the wheels left/right for access to the side fasteners (see illustration).

5　　Remove the bumper cover upper fasteners (see illustration).

6　　Using a plastic trim tool, detach the claw-like clips securing the bumper cover to the fender and headlight area (see illustration).

7　　Disconnect the fog lights (if equipped).

8　　With the aid of an assistant, remove the bumper cover from the vehicle.

9　　Installation is the reverse of removal. When installing, make sure the tabs on the back of the bumper cover snap securely and evenly into the corresponding clips on the body before installing the bolts and screws.

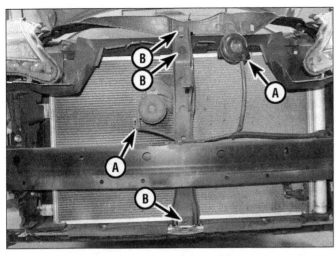

11.12 Disconnect the horn connectors (A) and remove the mounting bolts and clip (B) from the center support

11.14 Remove the upper support fasteners (A) and the upper radiator splash shield fasteners (B)

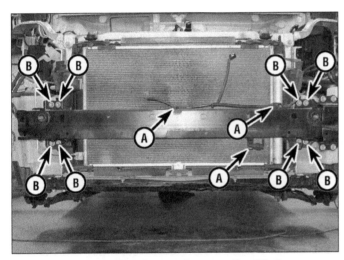

11.16 Disengage the front bumper harness clips (A), then remove the mounting bolts (B) and bumper

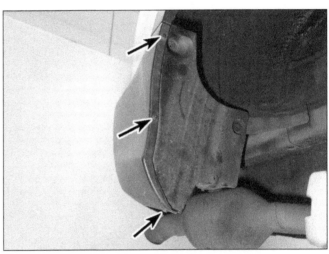

11.18a Rear bumper cover fenderwell fasteners

An assistant would be helpful at this point.

Front bumper center and upper supports

10   Remove the front bumper cover.

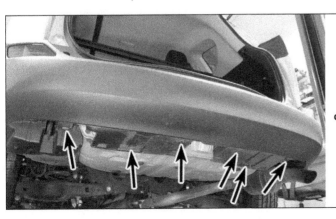

11.18b Rear bumper cover lower fasteners

11   Remove the bumper impact foam piece.
12   To remove the bumper center support- disconnect the horn electrical connectors from the center support, then remove the center support fasteners (see illustration).

13   To remove the upper support- remove the headlight housings first (see Chapter 12).
14   Remove the upper support fasteners, then, if necessary, the upper radiator splash shield fasteners (see illustration).

Front bumper

15   Remove the front bumper cover and center support.
16   Disengage the wiring harness clips from the bumper, then remove the bumper mounting bolts (see illustration).

Rear bumper components
Rear bumper cover

17   Open the liftgate.
18   Remove the rear bumper cover fasteners, in order of the following illustrations (see illustrations).
19   Once the fasteners have all been

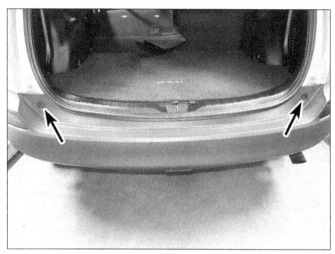

11.18c Rear bumper cover upper fasteners

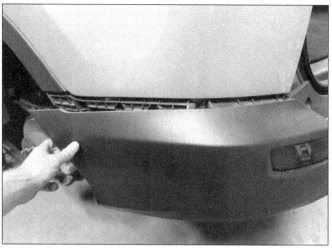

11.19 Gently separate the quarter panel and bumper cover apart

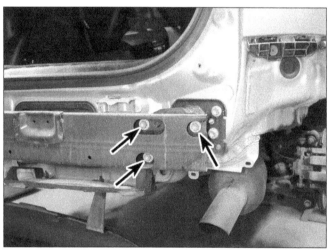

11.23 Rear bumper mounting fasteners

12.6a Fenderwell splash shield fastener locations (rear fenderwell splash shield similar)

12.6b Remove the fender end cap screws . . .

removed, carefully separate the bumper cover claw-like clips (see illustration).
20   With aid of an assistant, remove the rear bumper cover from the vehicle.
21   Installation is the reverse of removal.

Rear bumper

22   Remove the rear bumper cover.
23   Remove the rear bumper mounting fasteners (see illustration).
24   If necessary, remove the rear bumper mounting brackets to vehicle body.
25   Installation is the reverse of removal.

12   Front fender - removal and installation

Warning: *The models covered by this manual are equipped with a Supplemental Restraint System (SRS), more commonly known as airbags. Always disable the airbag system before working in the vicinity of any airbag system component to avoid the possibility of accidental deployment of the airbag, which could cause personal injury (see Chapter 12).*
1   Place the vehicle on a smooth flat surface with plenty of room to maneuver the fender, bumper, and inner liner off the body of the car. Raise and support the hood.
2   Loosen the front wheel lug nuts. Raise the vehicle, support it securely on jackstands and remove the front wheels.
3   Remove the front bumper cover (see Section 11).
4   Remove the headlight housing (see Chapter 12).
5   Remove the cowl paneling (see Section 13).
6   The remaining Steps for the removal of the fender should be followed in sequence of the photos (see illustrations 12.6a through 12.6n).

12.6c . . . then, at the inner end, disengage the square clips with a screwdriver to remove the cap

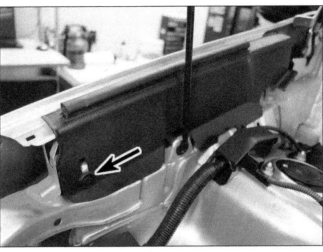

12.6d Remove the pushpin clip . . .

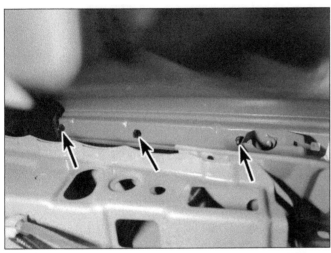

12.6e . . . disengage the 3 clips inside the fenderwell with needle-nose pliers . . .

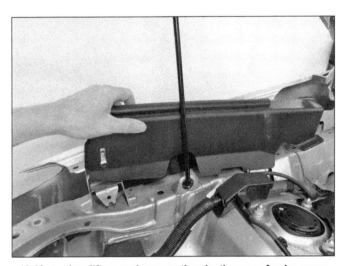

12.6f . . . then lift up and remove the plastic upper fender cover

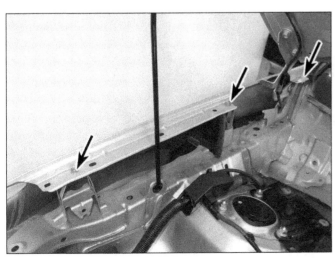

12.6g Remove the fender upper bolts

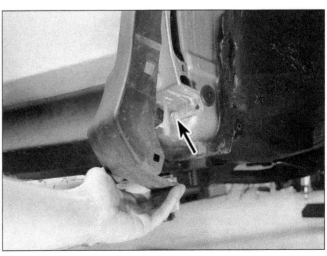

12.6h Pry out the plastic fender guard enough to remove the lower fender bolt

12.6i Open the front door and remove the exposed bolt

12.6j With the door still open, disengage the clips for the inner plastic shield (1 of 3 shown) . . .

12.6k . . . then guide the shield out from the fenderwell . . .

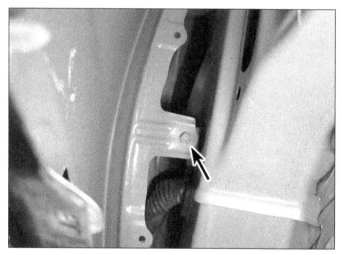

12.6l . . . and remove the exposed inner fenderwell bolt

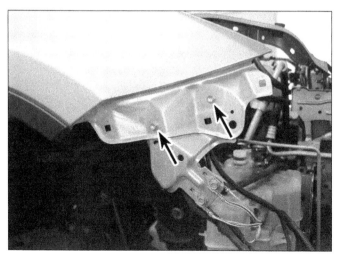

12.6m Remove the fender end bolts

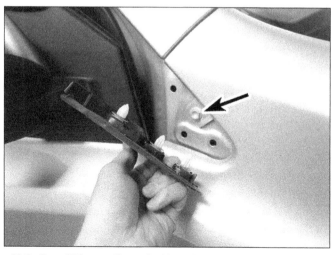

12.6n Pry off the small panel with a plastic trim tool, remove the bolt and carefully guide the fender off the vehicle

13.2 With the hood open, carefully pry out the fender-to-cowl seal

13.3a Remove these fasteners at each end by pushing in the center of the clip with a small screwdriver, then pulling the clip out . . .

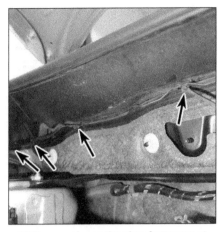

13.3b . . . release the claw fasteners at the underside . . .

13.3c . . . then remove the cowling upper trim

13.6 Vent tray mounting fasteners

7   With the aid of an assistant, support the fender while it's being moved away from the vehicle to prevent damage to the surrounding body panels.

8   Installation is the reverse of removal. Check the alignment of the fender to the hood, door, and the rest of the surrounding body before final tightening of the fender fasteners. It also helps to align the bolt head indent markings cast into the fender when tightening the bolts.

13   Cowl cover and vent tray - removal and installation

Note: *Disconnect the cable from the negative battery terminal before beginning*

Cowl cover

1   Remove the wiper arms (see Chapter 12).

2   With a plastic trim tool, pry up and remove both sides of the fender-to-cowl trim seals (see illustration).

3   Remove the two corner clips and disengage the remaining clips at the underside to remove the upper cowl panel cover (see illustrations).

4   Installation is the reverse of removal.

Vent tray

5   Remove the cowl cover (see previous Steps) and windshield wiper motor assembly (see Chapter 12).

6   Disengage the clip retaining the wiper motor wiring harness, then remove the vent tray mounting fasteners along the tray (see illustration).

7   Release the clip and remove the vent tray baffle, then guide the metal vent tray out of the engine compartment (see illustrations).

8   Installation is the reverse of removal.

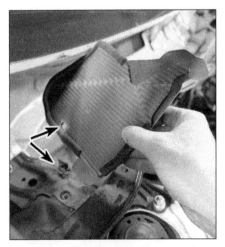

13.7a Release the clip and remove the foam baffle . . .

13.7b ... then remove the metal vent tray

14.1 Pry up the power window switch plate and disconnect the electrical connector

14.2a Pry off the outside mirror trim cover . . .

14.2b ... the door handle trim cover . . .

14.2c ... and the trim behind the inner release handle

14.3a Door trim panel retaining screws

14.3b Use a flat trim or equivalent to gently release the retaining clips from around the outer edge of the door panel

14   Door trim panels - removal and installation

Warning: *The models covered by this manual are equipped with a Supplemental Restraint System (SRS), more commonly known as airbags. Always disable the airbag system before working in the vicinity of any airbag system component to avoid the possibility of accidental deployment of the airbag, which could cause personal injury (see Chapter 12).*

Removal

Note: *When using a screwdriver to pry trim panels, be sure to apply electrical tape to the tip to prevent scratching the panels. Alternatively, a plastic trim tool should be used.*

Note: *The removal of the rear door panel is similar to the front door panel removal*

Warning: *Disconnect the cable from the negative battery terminal and wait two minutes before beginning work.*

Door panel

1   Pry up the window switch plate and disconnect the electrical connector to remove it (see illustration).

2   Pry out the outside mirror cover (sail trim) and door handle trim panels (see illustrations).

3   Remove the door trim panel retaining screws, then carefully pry the panel out along the perimeter until the clips disengage (see illustrations). Work slowly and carefully around the outer edge of the trim panel until it's free.

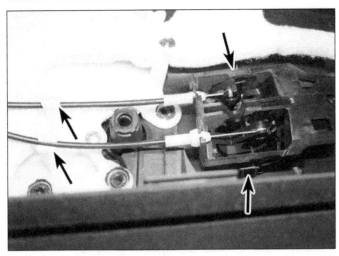

14.4 Separate the cables from the clips, then disengage the side tabs and separate the inside handle from the door panel

14.5 Working around the perimeter, pull the watershield off of the door, using a razor blade to separate the butyl tape along the way

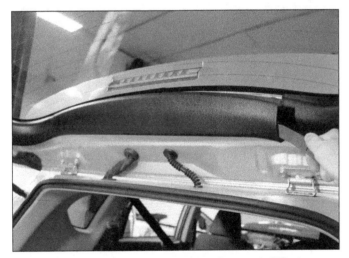

14.6a Start by prying the top center trim panel off first . . .

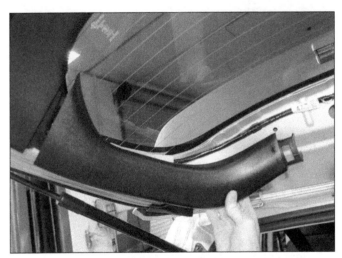

14.6b . . . then the upper side trim panels . . .

4   Once all of the clips are disengaged, pull the trim panel up and away from the door enough to gain access to release the inside handle. Release the cables from the panel, then detach the inside handle assembly by disengaging the side clips (see illustration).

5   For access to the components inside the door, remove the door watershield (see illustration). Note the position of the wiring harness(es) that are routed through the watershield for installation. Also, to allow for complete removal of the watershield, the ground bolt and airbag side impact sensor electrical connector will need to be removed/disconnected.

Liftgate trim paneling

Note: *When prying off these panels (especially the two side ones), the plastic retaining clips have a tendency to separate from the panels and fly-off into the inside of the liftgate body, making them nearly impossible to retrieve without removing/tilting the entire liftgate. Use caution when prying off these panels, working in a meticulous manner.*

6   Open the liftgate. Use a flat trim tool to release the trim paneling (see illustrations).

Installation

7   Prior to installation, be sure to reinstall any clips into the paneling which may have fallen off when you removed the panel.

8   The remainder of installation is the reverse of removal.

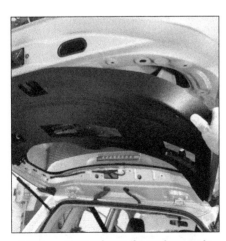

14.6c . . . then release the various main trim panel clips and disconnect any electrical connectors (where equipped)

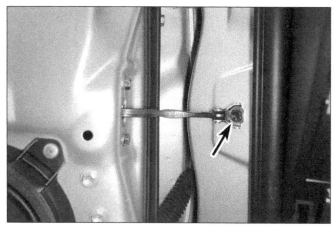

15.6 Remove the bolt retaining the door stop strut

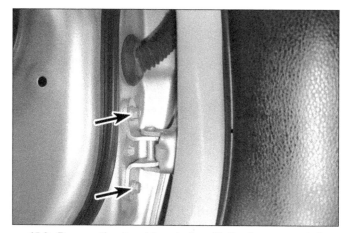

15.8a Remove the door hinge bolts with the door supported
(bottom hinge shown, top hinge similar)

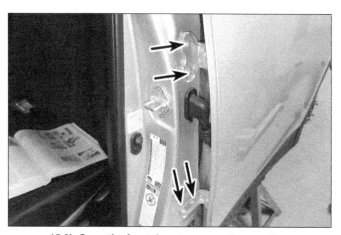

15.8b Open the front door to access the rear door
hinge-to-body bolts

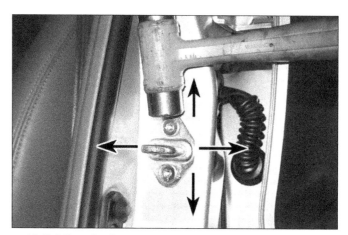

15.13 Adjust the door lock striker by loosening the mounting
screws and gently tapping the striker in the desired direction

15   Door - removal, installation and adjustment

Warning: *The models covered by this manual are equipped with a Supplemental Restraint System (SRS), more commonly known as airbags. Always disable the airbag system before working in the vicinity of any airbag system component to avoid the possibility of accidental deployment of the airbag, which could cause personal injury (see Chapter 12).*

Note: *The door is heavy and somewhat awkward to remove and install - at least two people should perform this procedure.*

Removal and installation

1   Raise the window completely in the door and disconnect the negative cable from the battery (see Chapter 5).

2   Open the door all the way and support it from the ground on jacks or blocks covered with rags to prevent damaging the paint.

3   Remove the door trim panel and watershield as described in Section 14.

4   Disconnect all electrical connections, ground wires and harness retaining clips from the door.

5   From the door side, detach the rubber conduit between the body and the door. Pull the wiring harness through the conduit hole and remove it from the door.

Note: *It is a good idea to label all connections to aid the reassembly process.*

6   Remove the door stop strut bolt (see illustration).

7   Mark around the door hinges with a pen or a scribe to facilitate realignment during reassembly.

8   With an assistant holding the door, remove the hinge-to-door bolts (see illustrations) and lift the door off.

9   Installation is the reverse of removal.

Adjustment

10   Having proper door-to-body alignment is a critical part of a well-functioning door assembly. First check the door hinge pins for excessive play. Fully open the door and lift up and down on the door without lifting the body. If a door has 1/16-inch or more excessive play, the hinges should be replaced.

11   Door-to-body alignment adjustments are made by loosening the hinge-to-body bolts or hinge-to-door bolts and moving the door. Proper body alignment is achieved when the top of the doors are parallel with the roof section, the front door is flush with the fender, the rear door is flush with the rear quarter panel and the bottom of the doors are aligned with the lower rocker panel. If these goals can't be reached by adjusting the hinge-to-body or hinge-to-door bolts, body alignment shims may have to be purchased and inserted behind the hinges to achieve correct alignment.

12   To adjust the door-closed position, scribe a line or mark around the striker plate to provide a reference point, then check that the door latch is contacting the center of the latch striker. If not, adjust the up-and-down position first.

13   Finally adjust the latch striker sideways position, so that the door panel is flush with the center pillar or rear quarter panel and provides positive engagement with the latch mechanism (see illustration).

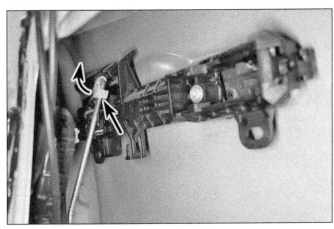

16.2 Detach the plastic clips on the actuating rods (by rotating them 90-degrees) leading to the outside handle and the door lock cylinder, then disengage the rods

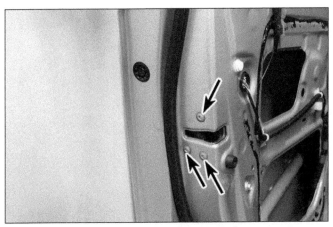

16.4 Door latch mounting bolts

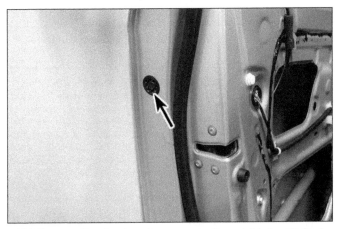

16.6 Remove the plug to gain access to the outside handle/lock cylinder retaining screw

16.7 Pull the lock cylinder straight out to remove it

16   Door latch, lock cylinder and handles - removal and installation

Warning: *The models covered by this manual are equipped with a Supplemental Restraint System (SRS), more commonly known as airbags. Always disable the airbag system before working in the vicinity of any airbag system component to avoid the possibility of accidental deployment of the airbag, which could cause personal injury (see Chapter 12).*

Warning: *Wear gloves when working inside the door openings to protect against cuts from sharp metal edges.*

Door latch

1   Raise the window, then remove the door trim panel and watershield (see Section 14).
2   Working through the large access hole, disengage the outside door handle-to-latch rod and outside door lock-to-latch rod (see illustration).
3   All door lock rods are attached by plas-

tic clips. The plastic clips can be removed by unsnapping the portion engaging the connecting rod and pulling the rod out of its locating hole. On models with power door locks, disconnect the electrical connectors at the latch.
4   Remove the screws securing the latch to the door (see illustration). Remove the latch assembly through the door opening.
Note: *If the latch still cannot be easily removed, it may help to remove the outside door handle inner housing (see illustration 16.2) and also unbolt the outer-end door window glass track, repositioning it to allow for removal*
5   Installation is the reverse of removal.

Outside handle and door lock cylinder

6   Remove the plastic plug from the edge of the door (see illustration) to gain access to the retaining bolt for the lock cylinder and outside handle. Loosen the bolt until the lock cylinder can be pulled out of the door.
7   Withdraw the lock cylinder from the door handle (detaching the lock cylinder clips, if necessary) (see illustration).

8   Remove the outside handle by sliding it rearward and pulling it off of the door (see illustration). If equipped, disconnect the electrical connector from the handle.

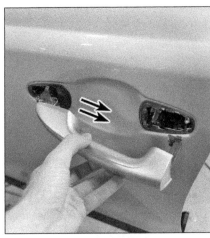

16.8 Slide the handle rearward, then pull it off the door to remove it

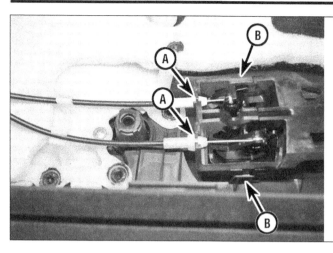

16.12 Detach the cable housings from the slots (A) and the cable ends, then remove the assembly by disengaging the side clips (B)

9   The trim pads for the outside handle can now be removed as well. Release the claw-type clips securing the pads to the door and remove them.
10   Installation is the reverse of removal.

Inside door handle

11   Remove the door trim panel (see Section 14).
Note: *Make a note of the cables' locations to ensure correct installation*
12   Disconnect the cables from the inside handle, then release the side clips to remove the assembly (see illustration).
13   Installation is the reverse of removal.

17   Door window glass - removal and installation

Warning: *The models covered by this manual are equipped with a Supplemental Restraint System (SRS), more commonly known as airbags. Always disable the airbag system before working in the vicinity of any airbag system component to avoid the possibility of accidental deployment of the airbag, which could cause personal injury (see Chapter 12).*

Warning: *Wear gloves when working inside the door openings to protect against cuts from sharp metal edges.*
Caution: *Disconnect the cable from the negative battery terminal before beginning*

Front door glass

1   Remove the door trim panel and the plastic watershield (see Section 14).
2   Raise the window just enough to access the window retaining bolts through the holes in the door frame (see illustration). The negative battery cable and window switch can be reconnected again to accomplish this. Disconnect the cable once the window has been raised to the proper height.
3   Remove the two glass mounting bolts.
4   Remove the glass by pulling it up and carefully maneuvering it out of the door.
5   Installation is the reverse of removal.

Rear door glass (four-door models)

6   Remove the door panel and watershield (see Section 14).
7   Lower the window glass all the way down into the door.
8   Remove the outer weatherstripping. Pry

upward on the clips to release it from the door.
9   Remove the bolt and screw securing the division bar to the door frame, then remove the division bar and the fixed glass.
10   Raise the window just enough to access the window retaining bolts through the holes in the door frame. The battery can be reconnected to accomplish this - see Step 2. Remove the bolts.
11   Remove the glass by pulling it up and guiding it out.
12   Installation is the reverse of the removal procedure.

Back door glass

13   Replacement of the back door glass requires the use of special fast-setting adhesive/caulk materials and some specialized tools and techniques. These operations should be left to a dealer service department or a shop specializing in glass work.

18   Door window glass regulator - removal and installation

Warning: *The models covered by this manual are equipped with a Supplemental Restraint System (SRS), more commonly known as airbags. Always disable the airbag system before working in the vicinity of any airbag system component to avoid the possibility of accidental deployment of the airbag, which could cause personal injury (see Chapter 12).*
Warning: *Wear gloves when working inside the door openings to protect against cuts from sharp metal edges.*

Front

1   Remove the door trim panel and the plastic watershield (see Section 14).
2   Remove the window glass (see Section 17).
3   The regulator assembly is a scissors-type. Disconnect the electrical connector and remove the bolts holding the window regulator/motor to the door (see illustration).

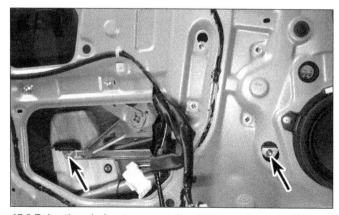

17.2 Raise the window to access the glass retaining bolts through the holes in the door frame

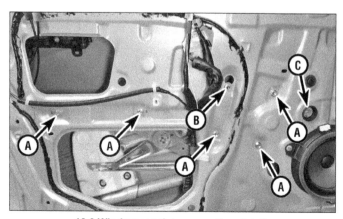

18.3 Window regulator mounting details

A   Remove these regulator mounting bolts completely
B   Only loosen this mounting bolt - don't remove it
C   Disconnect the regulator electrical connector

4   Pull the equalizer arm and regulator assembly through the service hole in the door frame to remove it.

5   Installation is the reverse of removal. If a new window motor was purchased individually from the regulator, transfer the new motor onto the regulator. Lubricate the rollers and wear points on the regulator with white grease before installation (see illustration).

Rear

6   Remove the door trim panel and the plastic watershield (see Section 14).

7   Remove the window glass assembly (see Section 17).

8   Disconnect the electrical connector from the window regulator motor.

9   Remove the regulator/motor assembly mounting bolts.

10   Remove the equalizer bar and raise the regulator assembly through the service hole in the door frame to remove it.

11   Installation is the reverse of removal.

Lubricate the rollers and wear points on the regulator with white grease before installation.

---

19   Mirrors - removal and installation

Outside mirrors

1   Pry off the mirror trim cover on the inside of the door.

2   Remove the door trim panel (see Section 14).

3   Disconnect the mirror electrical connector.

4   Remove the three mirror retaining bolts and detach the mirror from the door (see illustration).

5   Installation is the reverse of removal.

Inside mirror

6   On auto-dimming mirrors, disengage the

mirror mount cover clips from each side of the cover and slide the cover down.

7   Depress the clip at the bottom of the base and slide the mirror towards the top of the windshield and remove the mirror (see illustrations).

8   To install the mirror, slide the mirror onto the base, pushing downward until the mirror is secured.

9   On auto-dimming mirrors, slide the cover back up over the base and engage the clips.

Note: *If the mount plate itself has come off the windshield, adhesive kits are available at auto parts stores to resecure it. Follow the instructions included with the kit.*

---

20   Liftgate - removal and installation

Warning: *The liftgate is heavy and somewhat awkward to remove and install - at least two people should perform this procedure.*

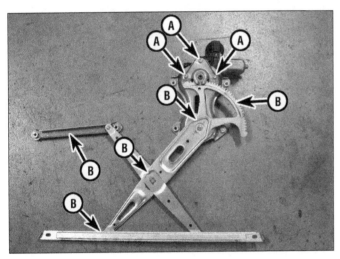

18.5 Window regulator details

A   *Window motor mounting bolts*
B   *Make sure these areas are well lubricated with white lithium based grease before installing*

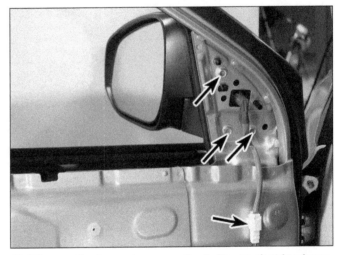

19.4 Remove the three mirror mounting bolts - on electric mirrors, disconnect the electrical connector

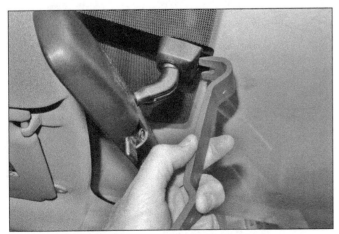

19.7a Using a plastic trim tool, pry up the clip at the base . . .

19.7b . . . and slide the mirror up and off of its base

Removal and installation

1   Open the liftgate all the way and remove the liftgate trim panel (see Section 14).

2   Disconnect all electrical connections, ground wires and harness retaining clips from the liftgate.

Note: *It is a good idea to label all connections to aid the reassembly process.*

3   Detach the rubber conduits between the body and the liftgate. Pull the wiring harnesses through the conduit holes.

4   Mark around the door hinges with a pen or a scribe to facilitate realignment during reassembly.

5   While an assistant holds the liftgate, Unbolt the liftgate support strut brackets.

6   Remove the hinge-to-liftgate bolts (see illustration) and remove the liftgate.

7   Installation is the reverse of removal.

21 Liftgate latch - removal and installation

1   Open the liftgate and remove the trim panel (see Section 14).

2   Working through the access hole, disconnect the electrical connector.

3   Remove the bolts securing the latch to the door (see illustration). Detach the clamp and remove the latch assembly through the access hole.

4   Installation is the reverse of removal.

22 Center console - removal and installation

Warning: *The models covered by this manual are equipped with a Supplemental Restraint System (SRS), more commonly known as airbags. Always disable the airbag system before working in the vicinity of any airbag system component to avoid the possibility of accidental deployment of the airbag, which could cause personal injury (see Chapter 12).*

Note: *When using a screwdriver to pry trim panels, be sure to tape the tip of the screwdriver tip to prevent scratching the panels.*

1   Disconnect the cable from the negative battery terminal (see Chapter 5).

2   Raise the parking brake handle.

3   Follow the illustrations, in order, to remove the center console.

4   Lift the center console out of the vehicle.

5   Installation is the reverse of removal.

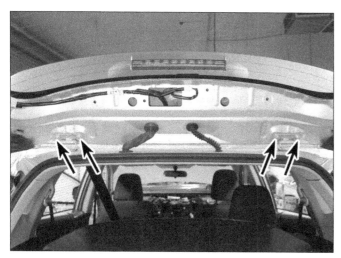

20.6 Liftgate hinge bolts

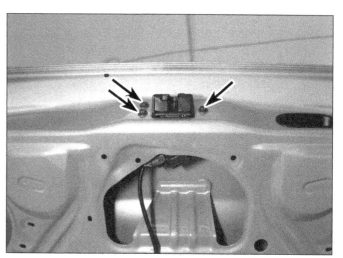

21.3 Liftgate latch fasteners

22.3a Rotate the shift knob counterclockwise and remove it

22.3b Remove the parking brake lever trim

22.3c Pry up the console trim plate . . .

22.3d . . . and disconnect the electrical connector

22.3e Pry the trim plate off of the tray trim panel

22.3f Remove the screws for the tray trim, then remove the trim

22.3g Remove the front side panel on the LH side

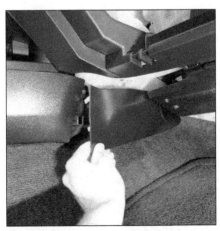

22.3h Remove the front side panel on the RH side

22.3i Remove the center console carpet, then remove the four bolts

22.3j Use a flat trim tool to remove the center instrument panel trim

23   Dashboard trim panels and cover

Warning: *The models covered by this manual are equipped with a Supplemental Restraint System (SRS), more commonly known as airbags. Always disable the airbag system before working in the vicinity of any airbag system component to avoid the possibility of accidental deployment of the airbag, which could cause personal injury (see Chapter 12).*

Note: *When using a screwdriver to pry trim panels, be sure to tape the tip to prevent scratching the panels.*

1   Disconnect the cable from the negative battery terminal (see Chapter 5).

Instrument cluster bezel

2   On tilt steering columns, move the column to the lowest position.

3   Use a flat trim tool to release the clips securing the bezel to the instrument panel (see illustration).

4   Installation is the reverse of the removal procedure. Make sure the clips are engaged properly before pushing the bezel firmly into place.

Center instrument panel trim

5   Use a flat trim panel tool and release the clips to remove the center instrument panel trim (see illustration).

6   Installation is the reverse of removal. Be sure the clips and fasteners are aligned before pressing into place.

Instrument panel vent registers

7   Remove the center instrument panel trim (see illustration 23.5).

8   Pry the top of the vent grille to free the clips, then detach the two claws at the bottom and remove the register (see illustration).

9   Installation is the reverse of removal, making sure to have the clips aligned before pressing into place.

Center lower trim panel (except for automatic HVAC equipped vehicles)

10   Using a flat trim tool, pry the upper corners of the trim panel off. Then remove any electrical connections.

11   Installation is the reverse of removal. Be sure to have all clips aligned before pressing into place.

Lower instrument panel garnish trim panel

12   Using a flat trim tool, pry the clips free to remove the trim panel (see illustration).

13   Installation is the reverse of removal. Be sure to have all the clips aligned before pressing into place.

23.3 Carefully pry the bezel from the instrument panel

23.5 Pry the center trim loose with a flat trim tool. Hold the tool as close as possible to each of the clips to avoid cracking the trim panel

23.8 Pry out the top of the vent register

23.12 The lower garnish trim panel is secured by 13 clips

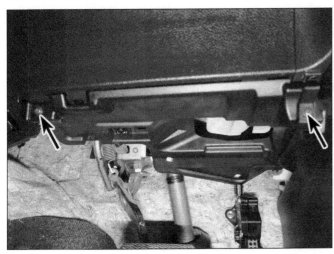

23.14a Remove these two screws, then slide the rear tabs out of their slots

23.14b Disconnect the electrical connection to the lower under cover panel (if equipped)

Knee bolster

14   Remove the lower under cover panel (see illustrations).

15   Pry the knee bolster trim panel off with a flat trim tool (see illustration).

16   Vehicles equipped with a knee bolster airbag, remove the airbag retaining bolts and electrical connection (see illustrations).

17   Installation is the reverse of removal.

23.15 Pry from the outer edges to remove the trim panel

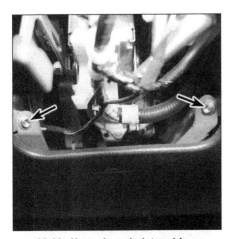

23.16a Upper knee bolster airbag retaining bolts

23.16b Lower knee bolster airbag retaining bolts

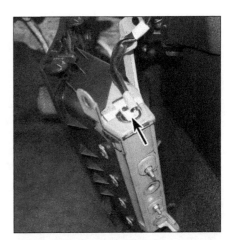

23.16c With the retaining bolts removed, disconnect the electrical connection to the knee bolster airbag

23.18a Pry the trim off with a flat trim tool

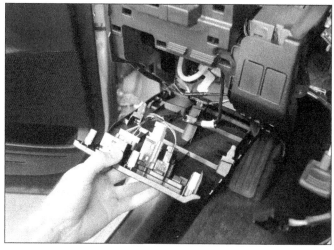

23.18b With trim removed, disconnect the electrical connections

Driver's side trim panel

18   Carefully pry the panel off with a flat trim tool (see illustrations).

19   Installation is the reverse of removal. Make sure the clips are engaged properly before pushing the panels firmly into place.

Glove box underside panel

20   Pull the panel free by grasping the top edge (see illustration).

Glove box

21   Squeeze the post and detach the damper from the right side of the glove box.

22   Squeeze the sides of the glove box inward, lower the glove box and detach the hinges.

23   Installation is the reverse of removal.

Dash upper cover

24   Remove the center trim panel (see illustration 23.5).

25   Remove the dash vent registers (see illustration 23.8).

26   Remove the steering column covers (see Section 24).

27   Remove the combination switch assembly (see Chapter 12).

28   Remove the instrument cluster bezel (see illustration 23.3) and the instrument cluster (see Chapter 12).

29   Remove the lower center trim panel.

30   Remove the radio (see Chapter 12).

31   Remove the HVAC control panel (see Chapter 3, Section 10).

32   Remove the center dash garnish trim panel (see illustration 23.12).

33   Remove the upper driver's side finish panel (see illustrations 23.18a and 23.18b).

34   Remove the glove box (see Section 23).

35   Remove the speakers from the dash (see Chapter 12).

36   Remove the right-hand and left-hand "A" pillar trim (see illustration).

37   Remove the dash speakers (see Chapter 12).

38   Remove the main bolts securing the dash to the vehicle located behind the corner speaker in the dash: One in the upper corner - driver's side, one in the upper corner - passenger's side (see illustrations).

39   Remove the two bolts securing the passenger's airbag to the instrument panel support beam behind the glove box area (see illustration).

40   Pull upwards on the dash cover to release the clips on the underside.

41   Disconnect any remaining electrical leads and/or wire harness fasteners as you proceed with the removal.

42   Lift the dash cover out of the vehicle with care. At this point it is very fragile and can be easily cracked.

43   Installation is the reverse of removal.

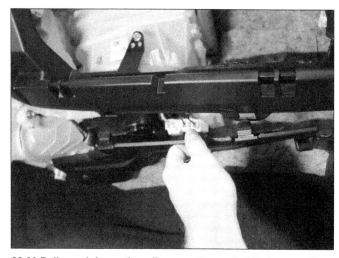

23.20 Pull panel down, then disconnect any electrical connections

23.36 Start at the top edge to release the clips to the A pillar trim

23.38a Main upper dash bolt location, driver's side

23.38b Main upper dash bolt location, passenger's side

23.39 Passenger's airbag bolts

24 Steering column covers - removal and installation

Warning: *The models covered by this manual are equipped with a Supplemental Restraint System (SRS), more commonly known as airbags. Always disable the airbag system before working in the vicinity of any airbag system component to avoid the possibility of accidental deployment of the airbag, which could cause personal injury (see Chapter 12).*

1   Remove the steering wheel (see Chapter 10).
2   The steering column covers are held in place by two screws just behind the steering wheel and pressure clips on the sides (see illustrations).
3   The rear part of the upper cover is hooked into the lower cover. Guide the hooked end out of the lower cover as you remove the upper cover.
4   Installation is the reverse of removal.

25 Instrument panel - removal and installation

Warning: *The models covered by this manual are equipped with a Supplemental Restraint System (SRS), more commonly known as airbags. Always disable the airbag system before working in the vicinity of any airbag system component to avoid the possibility of accidental deployment of the airbag, which could cause personal injury (see Chapter 12).*

Note: *When using a screwdriver to pry trim panels, be sure to tape the tip of the screwdriver tip to prevent scratching the panels.*

1   Disconnect the cable from the negative battery terminal (see Chapter 5).
2   Remove the dashboard trim panels (see Section 23) and the center floor console (see Section 22).
3   Remove the steering wheel (see Chapter 10).
4   Remove the steering column covers (see

Section 24).
5   Remove the instrument cluster (see Chapter 12).
6   Remove the radio and, if equipped, the navigation unit (see Chapter 12).
7   Remove the A-pillar trim panels (see Section 23).
8   Remove the glove box (see Section 23).
9   Remove the dash cover (see Section 23).
10   Remove the steering column (see Chapter 10).
11   Remove the heater and air conditioning control panel (see Chapter 3).
12   Remove the knee bolster (see Section 23) and lower insulator cover.
13   Remove the power steering ECU assembly fasteners and disconnect the electrical connectors.
14   Remove the hood latch release cable from the instrument panel (see Section 10).
15   Mark the electrical connectors with masking tape and a marking pen before disconnecting them.

24.2a With the steering wheel removed the two screws can be reached

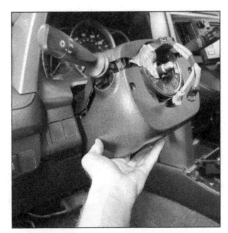

24.2b Tilt the steering column to its highest point, then release the claw type fasteners on each side of the covers

24.2c Release the three claw type fasteners, four clips, and two pins from the upper steering column cover to remove it

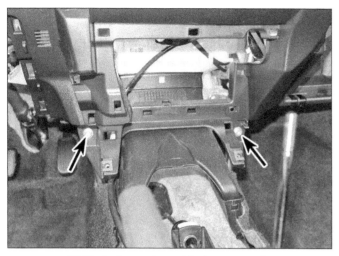

25.16a Instrument panel fasteners - center

25.16b Instrument panel fasteners - left side

16   Remove all of the fasteners from the ends and middle (bolts, screws, nuts and clips) holding the instrument panel to the body (see illustrations). Lift the panel, pull it away from the windshield and take it out through the driver's door opening.

17   Installation is the reverse of removal.

---

26   Seats - removal and installation

Warning: *The models covered by this manual are equipped with a Supplemental Restraint System (SRS), more commonly known as airbags. Always disable the airbag system before working in the vicinity of any airbag system component to avoid the possibility of ac-*cidental deployment of the airbag, which could cause personal injury (see Chapter 12).

Front seat

Warning: *The front seat belts are equipped with pre-tensioners, which are pyrotechnic (explosive) devices designed to retract the seat belts in the event of a collision. On models equipped with pre-tensioners, do not remove the front seat belt retractor assemblies, and do not disconnect the electrical connectors leading to the assemblies. Problems with the pre-tensioners will turn on the SRS (airbag) warning light on the dash. If any pre-tensioner problems are suspected, take the vehicle to a dealer service department.*

1   Disconnect the cable from the nega-tive terminal of the battery (see Chapter 5). Pry out the plastic covers to access the seat tracks and their mounting bolts (see illustration).

2   Remove the retaining bolts (see illustration).

3   Tilt the seat upward to access the underside, then disconnect any electrical connectors and lift the seat from the vehicle.

4   Installation is the reverse of removal.

Rear seat

5   Fold the seat down.

6   Remove the retaining bolts (see illustrations).

7   Installation is the reverse of removal.

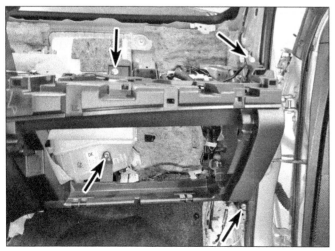

25.16c Instrument panel fasteners - right side

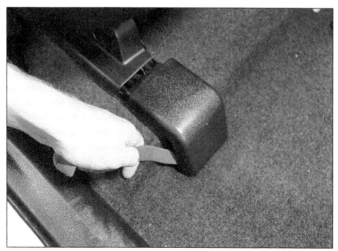

26.1 Use a flat trim tool to remove the trim covers

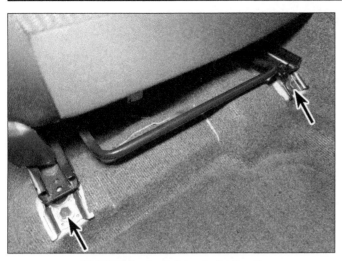

26.2 Seat track retaining bolts (front)

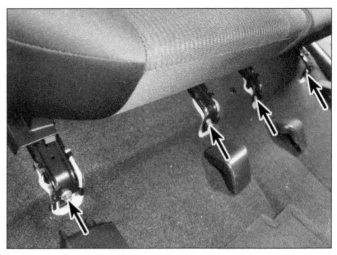

26.6a Remove the front retaining bolts . . .

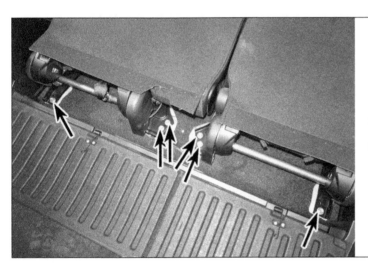

26.6b . . . and the rear retaining bolts

NOTES

Chapter 12
Chassis electrical system

Contents

1 General information

1 The electrical system is a 12-volt, negative ground type. Power for the lights and all electrical accessories is supplied by a lead/acid-type battery that is charged by the alternator.

2 This Chapter covers repair and service procedures for the various electrical components not associated with the engine. Information on the battery, alternator, ignition system and starter motor can be found in Chapter 5.

3 It should be noted that when portions of the electrical system are serviced, the negative cable should be disconnected from the battery to prevent electrical shorts and/or fires.

2 Electrical troubleshooting - general information

Refer to illustrations 2.5a, 2.5b, 2.6 and 2.9

1 A typical electrical circuit consists of an electrical component, any switches, relays, motors, fuses, fusible links or circuit breakers related to that component and the wiring and connectors that link the component to both the battery and the chassis. To help you pinpoint an electrical circuit problem, wiring diagrams are included at the end of this Chapter.

2 Before tackling any troublesome electrical circuit, first study the appropriate wiring diagrams to get a complete understanding of what makes up that individual circuit. Trouble spots, for instance, can often be narrowed down by noting if other components related to the circuit

are operating properly. If several components or circuits fail at one time, chances are the problem is in a fuse or ground connection, because several circuits are often routed through the same fuse and ground connections.

3 Electrical problems usually stem from simple causes, such as loose or corroded connections, a blown fuse, a melted fusible link or a failed relay. Visually inspect the condition of all fuses, wires and connections in a problem circuit before troubleshooting the circuit.

4 If test equipment and instruments are going to be utilized, use the diagrams to plan ahead of time where you will make the necessary connections in order to accurately pinpoint the trouble spot.

5 The basic tools needed for electrical troubleshooting include a circuit tester or voltmeter (a 12-volt bulb with a set of test

leads can also be used), a continuity tester, which includes a bulb, battery and set of test leads, and a jumper wire, preferably with a circuit breaker incorporated, which can be used to bypass electrical components (see illustrations). Before attempting to locate a problem with test instruments, use the wiring diagram(s) to decide where to make the connections.

Voltage checks

6   Voltage checks should be performed if a circuit is not functioning properly. Connect one lead of a circuit tester to either the negative battery terminal or a known good ground. Connect the other lead to a connector in the circuit being tested, preferably nearest to the battery or fuse (see illustration). If the bulb of the tester lights, voltage is present, which means that the part of the circuit between the connector and the battery is problem free. Continue checking the rest of the circuit in the same fashion. When you reach a point at which no voltage is present, the problem lies between that point and the last test point with voltage. Most of the time the problem can be traced to a loose connection. **Note:** *Keep in mind that some circuits receive voltage only when the ignition key is in the Accessory or Run position.*

Finding a short

7   One method of finding shorts in a circuit is to remove the fuse and connect a test light or voltmeter in place of the fuse terminals. There should be no voltage present in the circuit. Move the wiring harness from side-to-side while watching the test light. If the bulb goes on, there is a short to ground somewhere in that area, probably where the insulation has rubbed through. The same test can be performed on each component in the circuit, even a switch.

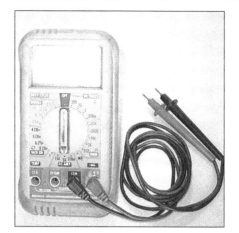

2.5a The most useful tool for electrical troubleshooting is a digital multimeter that can check volts, amps, and test continuity

Ground check

8   Perform a ground test to check whether a component is properly grounded. Disconnect the battery and connect one lead of a continuity tester or multimeter (set to the ohms scale), to a known good ground. Connect the other lead to the wire or ground connection being tested. If the resistance is low (less than 5 ohms), the ground is good. If the bulb on a self-powered test light does not go on, the ground is not good.

Continuity check

9   A continuity check is done to determine if there are any breaks in a circuit - if it is passing electricity properly. With the circuit off (no power in the circuit), a self-powered continuity tester or multimeter can be used to check the circuit. Connect the test leads to both ends of the circuit (or to the power end and a good ground), and if the test light comes

2.5b A test light is a very handy tool for checking voltage

on the circuit is passing current properly (see illustration). If the resistance is low (less than 5 ohms), there is continuity; if the reading is 10,000 ohms or higher, there is a break somewhere in the circuit. The same procedure can be used to test a switch, by connecting the continuity tester to the switch terminals. With the switch turned On, the test light should come on (or low resistance should be indicated on a meter).

Finding an open circuit

10   When diagnosing for possible open circuits, it is often difficult to locate them by sight because the connectors hide oxidation or terminal misalignment. Merely wiggling a connector on a sensor or in the wiring harness may correct the open circuit condition. Remember this when an open circuit is indicated when troubleshooting a circuit. Intermittent problems may also be caused by oxidized

2.6 In use, a basic test light's lead is clipped to a known good ground, then the pointed probe can test connectors, wires or electrical sockets - if the bulb lights, the part being tested has battery voltage

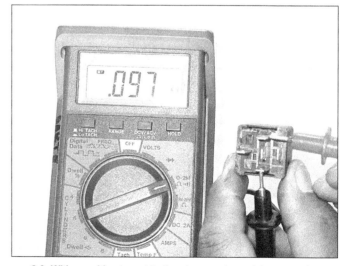

2.9 With a multimeter set to the ohms scale, resistance can be checked across two terminals - when checking for continuity, a low reading indicates continuity, a high reading indicates lack of continuity

3.1a The interior fuse box is located at the left (driver's) side of the instrument panel, under a cover just above the diagnostic port

3.1b The engine compartment fuse/relay box is located at the left side of the engine compartment

3.1c The number 2 fuse/relay box is on the right-hand side of the engine compartment

3.1d The relay block number 5 is behind this access panel

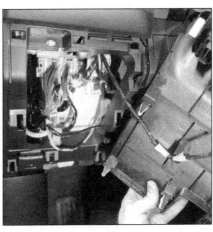

3.1e There are some fuse circuits on the relay block number 5 as well

3.3 When a fuse blows, the element between the terminals melts

or loose connections.

11   Electrical troubleshooting is simple if you keep in mind that all electrical circuits are basically electricity running from the battery, through the wires, switches, relays, fuses and fusible links to each electrical component (light bulb, motor, etc.) and to ground, from which it is passed back to the battery. Any electrical problem is an interruption in the flow of electricity to and from the battery.

3   Fuses, fusible links and circuit breakers - general information

Fuses

1   The electrical circuits of the vehicle are protected by a combination of fuses, circuit breakers and fusible links. The main fuse/relay panel is in the engine compartment (see illustration), while the interior fuse/relay panel is located inside the passenger compartment (see illustrations). Each of the fuses is designed to protect a specific circuit, and the various circuits are identified on the fuse

panel itself.

2   Several sizes of fuses are employed in the fuse blocks. There are small, medium and large sizes of the same design, all with the same blade terminal design. The medium and large fuses can be removed with your fingers, but the small fuses require the use of pliers or the small plastic fuse-puller tool found in most fuse boxes.

3   If an electrical component fails, always check the fuse first. The best way to check the fuses is with a test light. Check for power at the exposed terminal tips of each fuse. If power is present at one side of the fuse but not the other, the fuse is blown. A blown fuse can also be identified by visually inspecting it (see illustration).

4   Be sure to replace blown fuses with the correct type. Fuses (of the same physical size) of different ratings may be physically interchangeable, but only fuses of the proper rating should be used. Replacing a fuse with one of a higher or lower value than specified is not recommended. Each electrical circuit needs a specific amount of protection. The amperage value of each fuse is molded into

the top of the fuse body.

5   If the replacement fuse immediately fails, don't replace it again until the cause of the problem is isolated and corrected. In most cases, this will be a short circuit in the wiring caused by a broken or deteriorated wire.

Fusible links

6   Some circuits are protected by fusible links. The links are used in circuits which are not ordinarily fused, such as the ignition circuit, or which carry high current.

7   Cartridge type fusible links are located in the engine compartment fuse/relay box and are similar to large fuses. After disconnecting the negative battery cable, simply unplug and replace a fusible link of the same amperage.

Circuit breakers

8   Circuit breakers protect certain circuits, such as the power windows or heated seats. Depending on the vehicle's accessories, there may be one or two circuit breakers, located in the fuse/relay box in the engine compartment.

4.1a The cover for the engine compartment fuse/relay box (left-hand side of the engine compartment) has the relay and fuse location chart printed on the inside of the lid

4.1b The number 2 relay box (right-hand side of the engine compartment) is marked under the lid as well

9   Because the circuit breakers reset automatically, an electrical overload in a circuit breaker-protected system will cause the circuit to fail momentarily, then come back on. If the circuit does not come back on, check it immediately.

10   For a basic check, pull the circuit breaker up out of its socket on the fuse panel, but just far enough to probe with a voltmeter. The breaker should still contact the sockets. With the voltmeter negative lead on a good chassis ground, touch each end prong of the circuit breaker with the positive meter probe. There should be battery voltage at each end. If there is battery voltage only at one end, the circuit breaker must be replaced.

11   Some circuit breakers must be reset manually.

4   Relays - general information and testing

General information

1   Several electrical accessories in the vehicle, such as the fuel injection system, horns, starter, and fog lamps use relays to transmit the electrical signal to the component. Relays use a low-current circuit (the control circuit) to open and close a high-current circuit (the power circuit). If the relay is defective, that component will not operate properly. Most relays are mounted in the engine compartment fuse/relay boxes, with some specialized relays located above the interior fuse box under the dash (see illustrations). If a faulty relay is suspected, it can be removed and tested using the procedure below or by

a dealer service department or a repair shop. Defective relays must be replaced as a unit.

Testing

2   Refer to the wiring diagrams for the circuit to determine the proper connections for the relay you're testing. If you can't determine the correct connection from the wiring diagrams, however, you may be able to determine the test connections from the information that follows.

3   There are four basic types of relays used on these models (see illustrations). Some are normally open type and some normally closed, while others include a circuit of each type.

4   On most relays, two of the terminals are the relay control circuit (they connect to the relay coil which, when energized, closes the large contacts to complete the circuit). The other terminals are the power circuit (they are connected together within the relay when the control-circuit coil is energized).

5   Some relays may be marked as an aid to help you determine which terminals are the control circuit and which are the power circuit. If the relay is not marked, refer to the wiring diagrams at the end of this Chapter to determine the proper hook-ups for the relay you're testing.

6   To test a relay, connect an ohmmeter across the two terminals of the power circuit, continuity should not be indicated (see illustration). Now connect a fused jumper wire between one of the two control circuit terminals and the positive battery terminal. Connect another jumper wire between the other control circuit terminal and ground. When the connections are made, the relay should click and continuity should be indicated on the meter. On some relays, polarity may be critical, so, if the relay doesn't click, try swapping

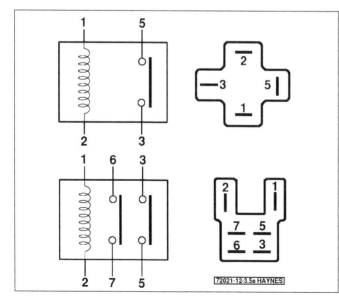

4.3a These two relays are typical normally open types; the one above completes a single circuit (terminal 5 to terminal 3) when energized - the lower relay type completes two circuits (6 and 7, and 3 and 5) when energized

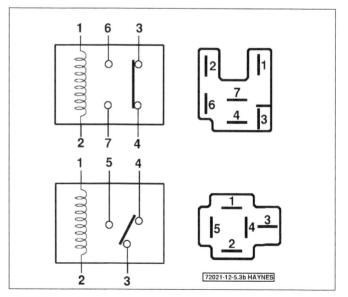

4.3b These relays are normally closed types, where current flows though one circuit until the relay is energized, which interrupts that circuit and completes the second circuit

the jumper wires on the control circuit terminals.

7   If the relay fails the above test, replace it.

5   Electrical connectors - general information

1   Most electrical connections on these vehicles are made with multiwire plastic connectors. The mating halves of many connectors are secured with locking clips molded into the plastic connector shells. The mating halves of some large connectors, such as some of those under the instrument panel,

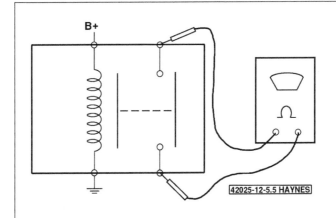

4.6 To test a typical four-terminal normally open relay, connect an ohmmeter to the two terminals of the power circuit - the meter should indicate continuity with the relay energized and no continuity with the relay not energized

Electrical connectors

Most electrical connectors have a single release tab that you depress to release the connector

Some electrical connectors have a retaining tab which must be pried up to free the connector

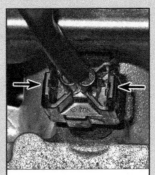

Some connectors have two release tabs that you must squeeze to release the connector

Some connectors use wire retainers that you squeeze to release the connector

Critical connectors often employ a sliding lock (1) that you must pull out before you can depress the release tab (2)

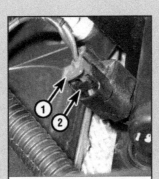

Here's another sliding-lock style connector, with the lock (1) and the release tab (2) on the side of the connector

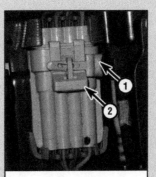

On some connectors the lock (1) must be pulled out to the side and removed before you can lift the release tab (2)

Some critical connectors, like the multi-pin connectors at the Powertrain Control Module employ pivoting locks that must be flipped open

are held together by a bolt through the center of the connector.

2   To separate a connector with locking clips, use a small screwdriver to pry the clips apart carefully, then separate the connector halves. Pull only on the shell, never pull on the wiring harness as you may damage the individual wires and terminals inside the connectors. Look at the connector closely before trying to separate the halves. Often the locking clips are engaged in a way that is not immediately clear. Additionally, many connectors have more than one set of clips.

3   Each pair of connector terminals has a male half and a female half. When you look at the end view of a connector in a diagram, be sure to understand whether the view shows the harness side or the component side of the connector. Connector halves are mirror images of each other, and a terminal shown on the right side end-view of one half will be on the left side end-view of the other half.

4   It is often necessary to take circuit voltage measurements with a connector connected. Whenever possible, carefully insert a small straight pin (not your meter probe)

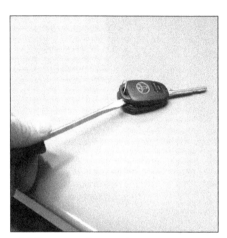

6.7 Use a small screwdriver to carefully separate the key halves

into the rear of the connector shell to contact the terminal inside, then clip your meter lead to the pin. This kind of connection is called "backprobing." When inserting a test probe into a terminal, be careful not to distort the terminal opening. Doing so can lead to a poor connection and corrosion at that terminal later. Using the small straight pin instead of a meter probe results in less chance of deforming the terminal connector.

---

6   Remote keyless entry fob - battery replacement

1   The keyless entry system consists of a remote control transmitter that sends a coded infrared signal to a receiver which then operates the door lock system. On models so equipped, the transmitter may also engage the alarm system and provide a panic button which flashes the lights and blows the horn for emergencies.

2   Replace the key battery when the smart key or wireless remote does not work properly. As the batteries deteriorate with age, the distance at which the remote transmitter operates will diminish.

With Smart Key System

3   Remove the mechanical key from the fob body.

4   Insert and twist a small screwdriver carefully between the case halves and separate the two halves for battery replacement.

5   Replace the lithium battery with the same type as originally installed, observing the polarity diagram on the case.

6   Snap the case halves together.

Without Smart Key System

7   Insert and twist a small screwdriver carefully between the case halves and separate the two halves for battery replacement (see illustration).

8   Remove the module from the key and separate the cover from the module.

9   Replace the lithium battery with the same type as originally installed, observing the polarity diagram on the case (see illustration).

10   Snap the cover onto the module and insert the module back into the case.

11   Snap the case halves together.

---

7   Steering column switches - replacement

Warning: *The models covered by this manual are equipped with a Supplemental Restraint System (SRS), more commonly known as airbags. Always disable the airbag system before working in the vicinity of any airbag system components to avoid the possibility of accidental deployment of the airbag(s), which could cause personal injury (see Section 24).*

Combination switch

1   Disconnect the cable from the negative battery terminal (see Chapter 5).

2   Remove the steering wheel (see Chapter 10).

3   Remove the steering column covers (see Chapter 11).

4   Unplug the electrical connectors from the combination switch (see illustration).

5   Lift the claw-like clips upward and pull the combination switch assembly from the column (see illustration).

6   Installation is the reverse of removal.

Combination switch disassembly

7   The next series of photos will guide you through the removal of the individual components of the combination switch assembly.

8   Reconnect the components as needed utilizing the claw clips. The components should snap into place. Be sure that the individual components align up with their per-

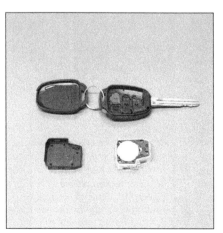

6.9 Remove the lithium battery, and install a new one with the positive (+) side of the battery facing up

7.4 Carefully unplug all of the electrical connectors. They only fit in one place so you won't need to mark them

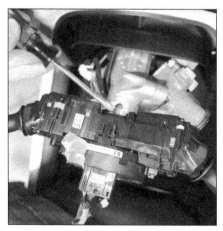

7.5 Do not put excessive pressure on the claw clips or you could damage them

7.7a The wiper switch and combination switch are held on with claw-type clips. Lift the clip slightly and slide the switch off

7.7b The spiral cable (clockspring) can be removed by lifting the claw clips slightly, then pull the unit off

7.11 Remove the two screws holding the switch in place

spective slots and guides before snapping into place. Otherwise, installation is the reverse of removal.

9   When installing the combination switch/ clockspring, be sure to center the clockspring as described in Chapter 10, Section 14.

Steering wheel switches

Cruise control switch

10   Remove the steering wheel see Chapter 11.

11   Remove the two screws securing the switch to the steering wheel (see illustration).

12   Guide the wire and connector out of the steering wheel as you remove the switch.

13   Installation is the reverse of removal.

Audio and function buttons

14   Remove the steering wheel (see Chapter 11).

15   Remove the cruise control switch.

16   Remove the two screws securing the switch trim plate onto the steering wheel (see illustrations).

17   With the trim cover removed, remove the individual switches.

18   Installation is the reverse of removal.

Paddle shift levers

19   Remove the steering wheel see Chapter 11.

20   Remove the cruise control switch and audio switches.

21   Remove the trim from the back of the steering wheel by releasing the claw clips.

22   Release the paddle switches by prying the claw clips open.

23   Remove the two wire straps securing the wire leads to the steering wheel.

24   Installation is the reverse of removal.

8   Ignition switch/key lock cylinder and steering lock actuator - replacement (without Smart Key System)

Warning: *The models covered by this manual are equipped with a Supplemental Restraint System (SRS), more commonly known as airbags. Always disable the airbag system before working in the vicinity of any airbag system components to avoid the possibility of accidental deployment of the airbag(s), which could cause personal injury (see Section 24).*

Note: *These models are equipped with an integration relay. The integration relay works in conjunction with the ignition switch to activate the key unlock warning system and anti-theft system.*

7.16a These two screws will need to be removed . . .

7.16b . . . then use a flat trim tool to release the claw clips and remove the switch trim plate

8.7a Lift the clip up slightly in order to slide the transponder coil off of the ignition switch assembly

8.7b Disconnect the electrical connection, then set aside in a safe place to avoid any damage while it is off of the vehicle

8.8 Ignition switch electrical connections

General steering column repair procedures

1   Disconnect the cable from the negative battery terminal (see Chapter 5).
2   Place the ignition key in the ACC position.
3   Remove the steering column covers (see Chapter 11).
4   If needed, depending on the type of repair, remove the steering column (see Chapter 10).
5   Remove the steering column switches (see Section 7).
6   Remove steering lock actuator assembly (see Section 8).

Transponder coil

7   Remove the transponder coil from the ignition switch assembly (see illustrations).

Ignition switch

8   Unplug the ignition switch wiring harness connector (see illustration) and key light con-

nector (if equipped).
9   Remove the lock actuator assembly.
10   Remove the switch retaining screws (see illustration), then pull the switch from the lock cylinder housing.
11   Installation is reverse of the removal.

Key lock cylinder

Note: *This applies to models without push-button starting systems.*
12   Remove the transponder coil.
13   Turn the ignition switch to ON (ACC).
14   Use a small flat screwdriver with a fine tip, inserted into the slot on the top of the steering column bracket to depress the locking claw. Pull the ignition switch lock cylinder out until its claw comes into contact with the stopper of the steering column bracket.
15   Insert the tip of a screwdriver into the hole in the side of the steering column bracket and tilt it downward to disengage the claw. Pull out the ignition switch lock cylinder.

16   Installation is the reverse of removal.

Steering lock actuator

Note: *This applies to vehicles with or without the Smart Key System.*
17   Remove the steering column by following the general steering column repair procedures in Section 8. Then place the steering column shaft in a soft-jaw vise. Tighten firmly without crushing or damaging the steering column.
18   Use a center punch and mark the center of the tapered-head bolts (see illustration).
19   Use a 3 or 4 mm drill bit (0.12 to 0.16 in) and drill a hole in the center of the bolt.
20   Using a screw extractor that will fit into the hole you just created, turn the bolt counterclockwise and remove the bolt.
21   Install the steering lock actuator and steering column clamp with new shear bolts, then tighten the bolts until their heads break off.

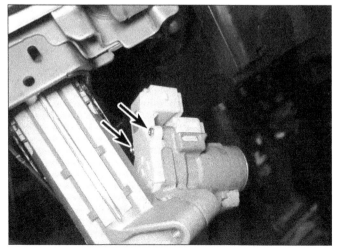

8.10 Ignition switch screws

8.18 Mark the centers of the shear-head bolts in order to keep the drill bit centered

9.2a Gently pry the switch assembly off
with a flat trim tool...

9.2b ... then turn it over to disconnect the
electrical connector

9.4a Remove the access panel

9.4b Turn the panel over, disconnect the
electrical connection, then press in on the
claw clips to remove the switch

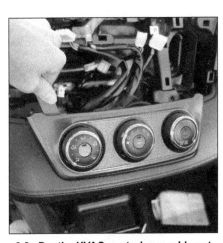

9.8a Pry the HVAC control assembly out
with a flat trim tool

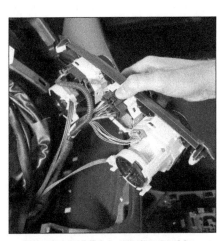

9.8b Pull the HVAC control assembly out
far enough to remove the connectors but
not so far as to stretch or damage the
wire harnesses

9   Instrument panel switches - replacement

Warning: *The models covered by this manual are equipped with a Supplemental Restraint System (SRS), more commonly known as airbags. Always disable the airbag system before working in the vicinity of any airbag system components to avoid the possibility of accidental deployment of the airbag(s), which could cause personal injury (see Section 24).*

Hazard warning switch

1   Remove the center trim side finish panel from the instrument panel (see Chapter 11).
2   Gently pry the clips free to remove the switch housing, then disconnect the electrical connector at the rear of the switch (see illustrations).
3   Installation is the reverse of removal.

Power mirror control switch

4   Remove the fuse box cover (see illustrations).

5   Disconnect the electrical connector and remove the switch. Depress the tabs and push the switch out of the panel.
6   Installation is the reverse of removal.

HVAC controls and defogger switch

Note: *The defogger control switch and air conditioning control switch is one unit and must be replaced as an assembly.*

7   Carefully remove the center cluster finish panel (see Chapter 11) from the instrument panel.
8   Pry the clips free from the HVAC control assembly and lift it out until you can disconnect the electrical connectors at the rear of the switch (see illustrations).
9   Installation is the reverse of removal.

Instrument panel illumination rheostat

10   Carefully pry the switch from the instrument panel, using tape on the screwdriver tip

to prevent scratching the instrument panel.
11   Disconnect the electrical connector and remove the switch.
12   Installation is the reverse of removal.

Console switches and power outlet

Switches

13   Carefully pry the switch panel from the instrument panel (see Chapter 11).
14   Disconnect the electrical connector and remove the switch. Depress the tabs and push the switch(es) out of the panel.
15   Installation is the reverse of removal.

Power outlet

16   Carefully pry the switch panel from the instrument panel (see Chapter 11).
17   Disconnect the electrical connector. Depress the tabs and push the outlet through the panel.
18   Installation is the reverse of removal.

10.5a Remove the two mounting screws from the instrument cluster

A   *Mounting screws*
B   *Clips (pull outward to remove the cluster)*

10.5b Guide the cluster out, then disconnect the electrical connectors

Engine start/stop switch (Smart Key Models)

19   Disconnect the cable from the negative battery terminal (see Chapter 5).
20   Remove the heater and air conditioning control assembly (see Chapter 3).
21   Unplug the engine switch wiring harness connectors.
22   Pull the two retaining clips back and pull the switch from the control assembly.
23   Installation is reverse of the removal.

10   Instrument cluster - removal and installation

Warning: *The models covered by this manual are equipped with a Supplemental Restraint System (SRS), more commonly known as airbags. Always disable the airbag system before working in the vicinity of any airbag system components to avoid the possibility of accidental deployment of the airbag(s), which could cause personal injury (see Section 24).*
Note: *Before disconnecting the battery, allow the car to sit (undisturbed) for approximately one minute. This will allow some of the various modules to go into their sleep mode.*
1   Disconnect the cable from the negative battery terminal (see Chapter 5).
2   Remove the center trim panel from the instrument panel (see Chapter 11).
3   Remove the instrument panel vent registers (see Chapter 11).
4   Remove the instrument cluster bezel trim (see Chapter 11).
5   Remove the cluster mounting screws, thenpull outward to release the instrument

cluster from the lower clips (see illustrations).
6   Installation is the reverse of removal.

11   Wiper systems

Front wiper system

Wiper arms

1   Disconnect the cable from the negative battery terminal (see Chapter 5).
2   Remove the cap from the wiper arm retaining nut (see illustration).
3   Mark the positions of the wiper arm(s) to the wiper shaft, then remove the wiper arm(s) (see illustration).
4   Use a small puller, if necessary, to remove the wiper arm from the wiper shaft.

11.2 Use a pocket screwdriver to remove the cap

11.3 Mark the position of the arm to the wiper shaft before removing

11.7 Two bolts (A) and one electrical connector (B) need to be removed

11.10 Pop the wiper arm cover up to expose the retaining nut

Wiper motor

5   Remove the wiper arms.

6   Remove the cowl cover see Chapter 11.

7   Disconnect the wiper motor electrical connector and remove the two bolts securing the windshield wiper motor/linkage assembly to the cowling (see illustration). Detach the linkage from the motor crank by prying it off with a flat screwdriver. Do not remove the attachment arm from the motor. Then remove the motor mounting bolts and detach the motor from the linkage frame.

8   Installation is the reverse of removal.

Rear wiper motor

9   Remove the liftgate trim panel (see Chapter 11).

10   Remove the wiper arm cover to expose the wiper arm retaining nut (see illustration).

11   From inside the vehicle, disconnect the wiper motor harness connector and remove the windshield wiper motor mounting bolts (see illustration).

12   Lift the windshield wiper motor assembly from the liftgate.

13   Installation is the reverse of removal.

Washer fluid reservoir and washer pump

14   Remove front bumper cover (see Chapter 11).

15   Remove the right-hand inner fender liner (see Chapter 11).

Note: *It's not necessary to remove the entire inner fender liner. Just the front section needs to be pulled back for access.*

Note: *Before removing the pump, drain any remaining washer fluid into a suitable container for reuse.*

Washer pump

16   To remove the washer pump, disconnect the electrical connection and the fluid hose, then pull the pump out of the reservoir.

17   Installation is the reverse of removal.

Reservoir

18   Remove the fastener securing the reservoir neck to the radiator support.

19   Now pull the bottle neck upwards to remove it.

20   Remove the six retaining clips for the front and rear hoses, and the electrical connections to the washer pump(s).

21   Remove the three reservoir mounting bolts, then remove the reservoir (see illustration).

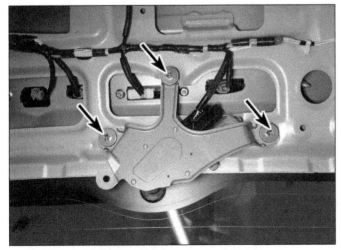

11.11 Rear wiper motor mounting bolts

11.21 Washer reservoir shown with fender and bumper cover removed for clarity

12.3a Remove the radio mounting bolts

12.3b Disconnect the electrical and antenna leads

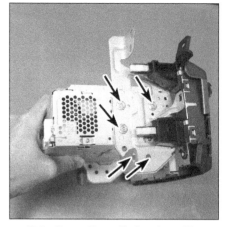

12.3c Reuse the radio brackets. The navigation/radio assembly utilizes five fasteners on each side (arrows show locations of all five fasteners)

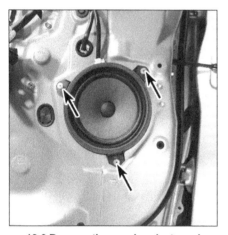

12.6 Remove the speaker rivets and disconnect the electrical connector to remove the speaker from the vehicle

12.8a Pry the speaker cover off with a flat trim tool

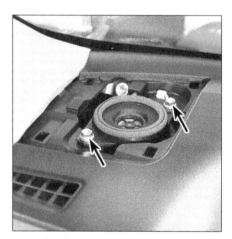

12.8b Tweeter mounting bolts

12   Radio and speakers - removal and installation

Warning: *The models covered by this manual are equipped with a Supplemental Restraint System (SRS), more commonly known as airbags. Always disable the airbag system before working in the vicinity of any airbag system components to avoid the possibility of accidental deployment of the airbag(s), which could cause personal injury (see Section 24).*

1   Disconnect the cable from the negative battery terminal (see Chapter 5).

Radio

Note: *After replacing a radio system containing a satellite receiver, in order to receive "pay-type" broadcast stations, the radio's ID needs to be registered with the appropriate satellite provider.*

2   Access the radio by removing the dash center trim surrounding the radio by gently prying the trim free with a flat trim tool (see

Chapter 11). (Work around the outside edges near the corners to release the clips.)

3   Remove the four bolts retaining the radio, then pull outwards to gain access to the electrical connections. Disconnect the electrical connectors and the antenna lead (see illustrations).

4   Installation is the reverse of removal.

Door speakers

5   Remove the door trim panel (see Chapter 11). All models have speakers in both front and rear doors. Rear door speakers are mounted similarly to the front door speakers.

6   Drill out the rivets, then disconnect the electrical connector (see illustration).

7   Installation is the reverse of removal.

Tweeters

8   In addition to the standard door and rear speakers, some models are equipped with tweeters mounted in the upper corners of the dash (see illustrations).

9   Remove the mounting fasteners, pull the tweeter out and disconnect the electrical connector.

10   Installation is the reverse of removal.

Rear woofer assembly

11   Remove the carpet and left-hand storage bin.

12   Remove rear hatch weatherstrip trim.

13   Remove the left-hand quarter panel trim (see Chapter 11).

14   Remove the four bolts securing the woofer to the left-hand quarter panel area. Disconnect the electrical connector, and remove the woofer assembly.

15   Installation is the reverse of removal.

13   Antenna - replacement

Warning: *The models covered by this manual are equipped with a Supplemental Restraint System (SRS), more commonly known as airbags. Always disable the airbag system before working in the vicinity of any airbag system components to avoid the possibility of accidental deployment of the airbag(s), which could cause personal injury (see Section 24).*

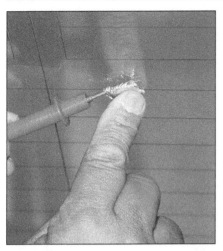

14.4 When measuring the voltage at the rear window defogger grid, wrap a piece of aluminum foil around the positive probe of the voltmeter and press the foil against the wire with your finger

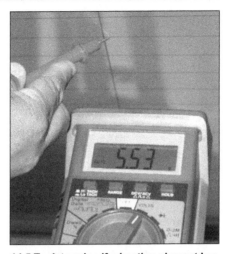

14.5 To determine if a heating element has broken, check the voltage at the center of each element - if the voltage is 5- or 6-volts, the element is unbroken; if the voltage is 10- or 12-volts, the element is broken between the center and the ground side; if there is no voltage, the element is broken between the center and the positive side

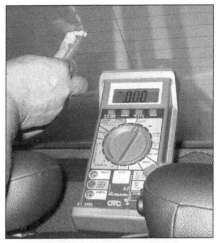

14.7 To find the break, place the voltmeter negative lead against the defogger ground terminal, place the voltmeter positive lead with the foil strip against the heating element at the positive terminal end and slide it toward the negative terminal end - the point at which the voltmeter reading changes abruptly is the point at which the element is broken

Removal

Fixed antenna

1   Unscrew the antenna mast from the base mount.
2   From inside the vehicle, remove the access panel at the rear of the headliner (if equipped).
Note: *If there is no access panel, the end of the headliner must be lowered.*
3   Disconnect the antenna cord from the base and unscrew the base from the roof panel.
4   On fender-mounted antennas, raise the vehicle and secure it on jackstands.
5   On fender-mounted antennas, remove the front wheel.
6   Remove the antenna assembly from the vehicle body.

 a) *On front top-mounted antennas, remove the driver's side door pillar trim to access the antenna cable.*
 b) *On fender-mounted antennas, working in the fenderwell area, remove the antenna mounting bolt*

7   On fender-mounted antennas, push the antenna mount into the fender to separate the assembly from the grommet.
8   Remove the radio (see Section 12) and disconnect the antenna cable.
9   On fender mounted antennas, remove the glovebox (see Chapter 11) and pull the cable from behind the dash.
10   On automatic power antennas, disconnect the antenna cable from the motor.

Installation

Fixed antenna

11   Thread the antenna mast onto the base mount and tighten it securely.

12   Working in the dash area, feed the antenna cable behind the glovebox and into the radio.
13   Install the radio (see Section 12).
14   Install the glove box (see Chapter 11).
15   The remainder of installation is the reverse of removal.

---

14   Rear window defogger - check and repair

1   The rear window defogger consists of a number of horizontal elements baked onto the glass surface.
2   Small breaks in the element can be repaired without removing the rear window.

Check

3   Turn the ignition switch and defogger system switches to the On position. Using a voltmeter, place the positive probe against the defogger grid positive terminal and the negative probe against the ground terminal. If battery voltage is not indicated, check the fuse, defogger switch and related wiring. If voltage is indicated, but all or part of the defogger doesn't heat, proceed with the following tests.
4   When measuring voltage during the next two tests, wrap a piece of aluminum foil around the tip of the voltmeter positive probe and press the foil against the heating element with your finger (see illustration). Place the negative probe on the defogger grid ground terminal.
5   Check the voltage at the center of each

heating element (see illustration). If the voltage is 5- or 6-volts, the element is okay (there is no break). If the voltage is 0-volts, the element is broken between the center of the element and the positive end. If the voltage is 10- to 12-volts the element is broken between the center of the element and ground. Check each heating element.
6   Connect the negative lead to a good body ground. The reading should stay the same. If it doesn't, the ground connection is bad.
7   To find the break, place the voltmeter negative probe against the defogger ground terminal. Place the voltmeter positive probe with the foil strip against the heating element at the positive terminal end and slide it toward the negative terminal end. The point at which the voltmeter deflects from several volts to zero is the point at which the heating element is broken (see illustration).

Repair

8   Repair the break in the element using a repair kit specifically recommended for this purpose, available at most auto parts stores. Included in this kit is plastic conductive epoxy.
9   Prior to repairing a break, turn off the system and allow it to cool off for a few minutes.
10   Lightly buff the element area with fine steel wool, then clean it thoroughly with rubbing alcohol.
11   Use masking tape to mask off the area being repaired.
12   Thoroughly mix the epoxy, following the instructions provided with the repair kit.
13   Apply the epoxy material to the slit in the masking tape, overlapping the undamaged

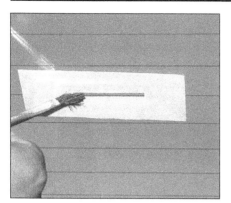

14.13 To use a defogger repair kit, apply masking tape to the inside of the window at the damaged area, then brush on the special conductive coating

15.2 Turn the weather sealing cap counterclockwise to detach it from the headlight housing

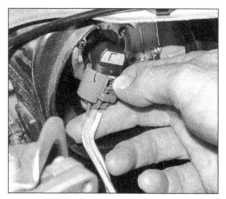

15.3 Lift the tab and disconnect the electrical connector

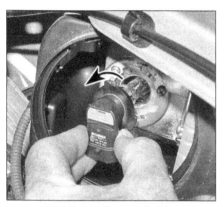

15.4 Rotate the bulb holder counterclockwise and remove it from the housing

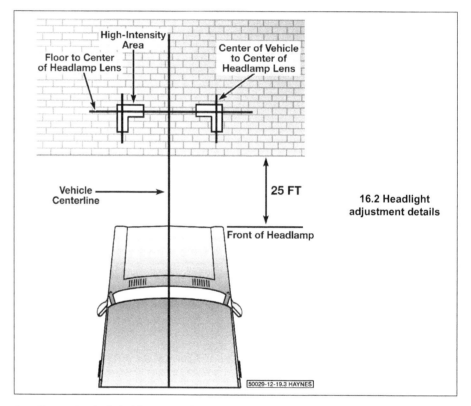

16.2 Headlight adjustment details

High-Intensity Area
Floor to Center of Headlamp Lens
Center of Vehicle to Center of Headlamp Lens
Vehicle Centerline
25 FT
Front of Headlamp
50029-12-19.3 HAYNES

area about 3/4-inch on either end (see illustration).

14   Allow the repair to cure for 24 hours before removing the tape and using the system.

15   Headlight bulb - replacement

Warning: *Halogen gas-filled headlight bulbs which are under pressure and may shatter if the surface is damaged or the bulb is dropped. Wear eye protection and handle the bulbs carefully, grasping only the base whenever possible. Do not touch the surface of the bulb with your fingers because the oil from your skin could cause it to overheat and fail prematurely. If you do touch the bulb surface, clean it with rubbing alcohol.*

Note: *These vehicles are equipped with either Halogen or LED (Light Emitting Diode) type headlamps. The procedures for bulb replacement are for the halogen type only.*

1   If you're replacing a bulb on the right side, remove the plastic retainer and reposition the windshield washer fluid reservoir neck.

2   If you're replacing a low-beam bulb, remove the weather sealing cap (see illustration).

3   Unplug the electrical connector from the bulb (see illustration).

4   Grasp the bulb holder securely and rotate it counterclockwise to remove it from the housing (see illustration).

5   Without touching the glass with your bare fingers, insert the new bulb assembly into the headlight housing, aligning the three tabs with the cutouts in the housing.

6   Plug in the electrical connector and, if you replaced a low-beam bulb, install the weather sealing cap.

16   Headlights - adjustment

Note: *The headlights must be aimed correctly. If adjusted incorrectly they could blind the driver of an oncoming vehicle and cause a serious accident or seriously reduce your ability to see the road. The headlights should be checked*

for proper aim every 12 months and any time a new headlight is installed or front end body work is performed. It should be emphasized that the following procedure is only an interim step which will provide temporary adjustment until the headlights can be adjusted by a properly equipped shop.

1   There are several methods of adjusting the headlights. The simplest method requires masking tape, a blank wall and a level floor.

2   Position masking tape vertically on the wall in reference to the vehicle centerline and the centerlines of both headlights (see illustration).

3   Position a horizontal tape line in reference to the centerline of all the headlights.

Note: *It may be easier to position the tape on the wall with the vehicle parked only a few inches away.*

4   Adjustment should be made with the

16.5 Insert a long Phillips screwdriver into the gear-drive mechanism to adjust the headlight

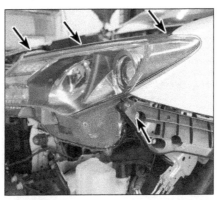

17.2a With the bumper cover removed, remove the remaining two bolts at the top of the housing . . .

17.2b . . . and the bolt from the side

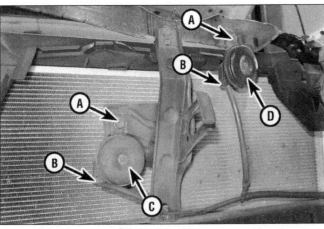

18.2 Horn details

A	Mounting bolts	C	Hi-tone horn
B	Electrical connectors	D	Lo-tone horn

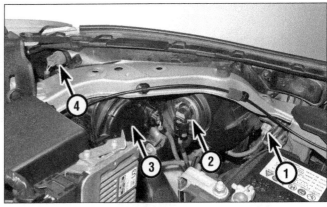

19.1 Front lighting details

1	Park/turn signal light	3	Low-beam headlight (behind weather sealing cap)
2	High-beam headlight	4	Side marker light

vehicle parked 25 feet from the wall, sitting level, the gas tank half-full and no unusually heavy load in the vehicle.

5   Insert a Phillips screwdriver into the gear-drive mechanism to turn the screw (see illustration).

6   Starting with the low beam adjustment, position the high intensity zone so it is two inches below the horizontal line and two inches to the side of the headlight vertical line, away from oncoming traffic. Adjustment is made by turning the vertical adjusting screw to raise or lower the beam. The horizontal adjusting screw should be used in the same manner to move the beam left or right.

7   With the high beams on, the high intensity zone should be vertically centered with the exact center just below the horizontal line.

Note: *It may not be possible to position the headlight aim exactly for both high and low beams. If a compromise must be made, keep in mind that the low beams are the most used and have the greatest effect on driver safety.*

8   Have the headlights adjusted by a dealer service department or service station at the earliest opportunity.

17   Headlight housing - replacement

Warning: *These vehicles are equipped with halogen gas-filled headlight bulbs which are under pressure and may shatter if the surface is damaged or the bulb is dropped. Wear eye protection and handle the bulbs carefully, grasping only the base whenever possible. Do not touch the surface of the bulb with your fingers because the oil from your skin could cause it to overheat and fail prematurely. If you do touch the bulb surface, clean it with rubbing alcohol.*

1   Remove the bumper cover (see Chapter 11).

2   Remove the retaining bolts, detach the housing and withdraw it from the vehicle (see illustrations). Unplug the electrical connectors.

3   Installation is the reverse of removal. Be sure to check headlight adjustment (see Section 16).

18   Horn - replacement

Warning: *The models covered by this manual are equipped with a Supplemental Restraint System (SRS), more commonly known as airbags. Always disable the airbag system before working in the vicinity of any airbag system components to avoid the possibility of accidental deployment of the airbag(s), which could cause personal injury (see Section 24).*

1   To access the horns, remove the front bumper cover (see Chapter 11).

2   Disconnect the electrical connectors and remove the mounting bolts (see illustration).

3   Installation is the reverse of removal.

19   Bulb replacement

Front park/turn signal lights and side marker lights

1   Open the hood and locate the park/turn signal or side marker light bulb holder (see illustration). Turn the bulb holder counter-

19.4a Remove the two screws . . .

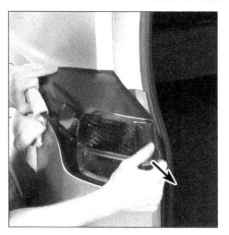

19.4b . . . then pull the housing straight back

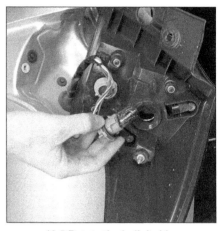

19.5 Rotate the bulb holder counterclockwise to remove it from the housing then, pull the bulb out of the socket and install the new one

19.7a Remove the access cover

19.7b With access cover removed, turn the bulb holder counterclockwise to remove the bulb and socket, then pull the bulb straight out of the socket and install the new one

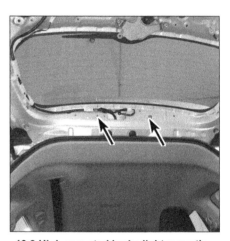

19.9 High-mounted brake light mounting screw locations

clockwise and pull it from the housing. Pull the bulb from the socket and install the new one.

2   Rotate the bulb holder counterclockwise and pull the bulb out. Remove the bulb from the holder.

3   Installation is the reverse of removal.

Rear tail light/brake light/turn signal

4   To access the tail light bulbs, remove the two screws securing the lens assembly to the vehicle (see illustrations), then grasp the housing and pull it straight back to release the pins from their retainers.

5   Rotate the bulb holders counterclockwise and pull the bulbs out to remove them (see illustration).

6   Installation is the reverse of removal.

Liftgate bulb replacement

7   Remove the access cover with a flat trim tool to gain access to the bulb holders (see illustrations).

High-mounted brake light

Note: *High-mounted brake lights are the LED type. Replace the assembly, not a bulb.*

8   Remove the liftgate trim panel (see Chapter 11).

9   Remove the high-mounted brake light mounting screws and pull the light assembly out from the rear spoiler (see illustration).

10   Installation is the reverse of removal.

Interior lights

11   Basically, all the interior light fixtures are removed in the same manner. Start by prying the lens off of the light housing (see illustrations), then replace the bulb.

12   Installation is the reverse of removal.

License plate lights

13   Carefully pry out the access panel from the liftgate trim panel, then turn the bulb holder(s) counterclockwise to remove (see illustration). Pull the bulb straight out of the

holder and install the new one.

14   Installation is the reverse of removal.

Fog light

15   Park the vehicle with the wheel pointing inward on the side of the burned-out bulb.

16   Remove the fasteners from the front of the inner fender splash shield (see Chapter 11), then peel back the splash shield for access to the fog light.

17   Unplug the electrical connector, then twist the bulb holder to remove it from the light housing.

18   Installation is the reverse of removal.

Outside mirror turn signal light

19   Carefully pry the outer cover off of the side view mirror(s) housing.

Note: *On mirrors equipped with a heating element and/or blind spot monitor systems, disconnect the heating element and/or the blind spot monitor connector.*

19.11a Pry off the front reading light lens using a flat-bladed screwdriver or trim tool

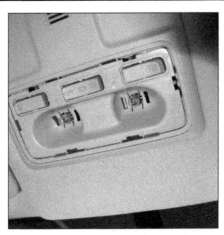

19.11b Replace the bulbs by pulling straight out from their terminals

19.11c Center overhead dome light lens can be pried off with a small flat screwdriver

19.11d Use the screwdriver again to work one end of the dome light bulb out, then grasp the bulb, pull it out the rest of the way and replace it

Warning: *Don't pry on the glass!*

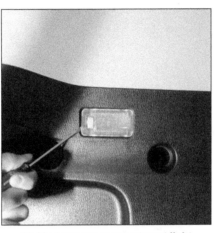

19.11e Luggage compartment light fixtures can be removed with a pocket screwdriver

19.11f Slide the fixture out partially to disconnect it

19.11g With the fixture removed, separate the lens and housing. Replace the bulb and reinstall

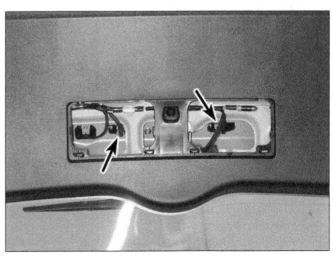

19.13 Remove the access panel to gain access to the license plate bulb holders

Bulb removal

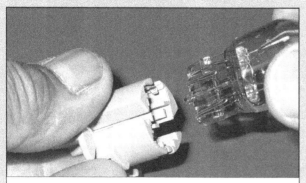

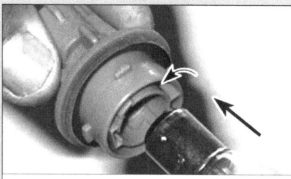

To remove many modern exterior bulbs from their holders, simply pull them out

On bulbs with a cylindrical base ("bayonet" bulbs), the socket is spring-loaded; a pair of small posts on the side of the base hold the bulb in place against spring pressure. To remove this type of bulb, push it into the holder, rotate it 1/4-turn counterclockwise, then pull it out

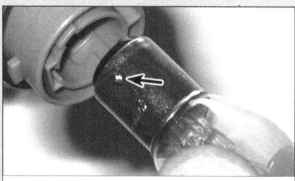

If a bayonet bulb has dual filaments, the posts are staggered, so the bulb can only be installed one way

To remove most overhead interior light bulbs, simply unclip them

20   Remove the two turn signal mounting fasteners and detach the two signal housing clips from the mirror.

21   Rotate the bulb counterclockwise to remove it.

22   Installation is the reverse of removal.

20   Cruise control system - general information

1   The Powertrain Control Module (PCM) controls the cruise control system electronically via the electronic throttle control system. If you have problems with the cruise control system, check for the presence of trouble codes stored in the PCM (see Chapter 6). If that doesn't turn up any problems, have it checked by a dealer service department or other qualified repair shop.

21   Power window system - description and check

Note: *These models are equipped with a Body Control Module (BCM) incorporated with the window system. Several systems are linked to a centralized control module that allows simple and accurate troubleshooting, but only with a professional-grade scan tool. The Body Control Module governs the door locks, the power windows, the ignition lock and security system, the interior lights, the Daytime Running Lights system, the horn, the windshield wipers, the heating/air conditioning system and the power mirrors. In the event of malfunction with this system, have the vehicle diagnosed by a dealership service department or other qualified automotive repair facility.*

1   The power window system operates electric motors, mounted in the doors, which lower and raise the windows. The system consists of the control switches, the motors, regulators, glass mechanisms, the Body Control Module (BCM) and associated wiring.

2   The power windows can be lowered and raised from the master control switch by the driver or by remote switches located at the individual windows. Each window has a separate motor that is reversible. The position of the control switch determines the polarity and therefore the direction of operation.

3   The window motor circuits are protected by a fuse. Each motor is also equipped with an internal circuit breaker; this prevents one stuck window from disabling the whole system.

4   The power window system will only operate when the ignition switch has activated the Retained Accessory Power (RAP) relay. In addition, many models have a window lock-out switch at the master control switch which, when activated, disables the switches at the rear windows and, sometimes, the switch at the passenger's window also. Always check these items before troubleshooting a window problem related to any of the other windows besides the driver's window. Window lock-out does not affect the driver's window.

5   These procedures are general in nature, so if you can't find the problem using them, take the vehicle to a dealer service department or to an independent repair facility that specializes in electrical repairs.

6   If the power windows won't operate, always check the fuse and circuit breaker first.

7   If only the rear windows are inoperative, or if the windows only operate from the master control switch, check the rear window lockout

switch for continuity in the unlocked position. Replace it if it doesn't have continuity.

8   Check the wiring between the switches and fuse panel for continuity. Repair the wiring, if necessary.

9   If only one window is inoperative from the master control switch, try the other control switch at each individual window.

Note: *This doesn't apply to the driver's door window.*

10   If the same window works from one switch, but not the other, check the switch and/or wiring for continuity.

11   If the switch tests OK, check for a short or open in the circuit between the affected switch and the window motor.

12   If one window is inoperative from both switches, remove the switch panel from the affected door. Check for voltage at the motor (refer to Chapter 11 for door panel removal) while the switch is operated.

13   If voltage is reaching the motor, disconnect the glass from the regulator. Move the window up and down by hand while checking for binding and damage. Also check for binding and damage to the regulator. If the regulator is not damaged and the window moves up and down smoothly, replace the motor. If there's binding or damage, lubricate, repair or replace parts, as necessary.

14   If voltage isn't reaching the motor, check the wiring in the circuit for continuity between the switches and the BCM, and between the BCM (if applicable) and the motors. You'll need to consult the wiring diagram. If the circuit is equipped with a relay, check that the relay is grounded properly and receiving voltage.

15   Test the windows to confirm proper repairs.

Note: *To verify if the motor is getting the needed voltage or ground, a simple but effective test is to sit in the car with the ignition on, open a door and look at the dome light. Then operate the switch to the faulty window. If you see the dome light dimming slightly this is a good indication that the motor is getting power and is probably a stuck motor or faulty wiring. Do not hold the switch on for very long when a motor is stuck or wiring is in question or more damage may occur. Finally, in some cases, a good rap on the door panel in the general area of the window motor - while the key is on and the window switch is depressed in the direction the window needs to move - will free up a stuck motor temporarily. You can damage the door panel or more internal components if you hit it too hard or are too aggressive.*

Window express down programming

Note: *Any time the battery or window motor are disconnected, you will need to perform this procedure to reestablish the auto feature.*

16   Lower the window to its lowest position, holding the switch in the Down position for an additional five seconds.

17   Raise the window to its highest position, holding the window switch in the Up position

for an additional five seconds.

18   Verify proper operation.

22   Power door lock and keyless entry system - description and check

Note: *These models are equipped with a Body Control Module (BCM). Several systems are linked to a centralized control module that allows simple and accurate troubleshooting, but only with a professional-grade scan tool. The Body Control Module governs the door locks, the power windows, the ignition lock and security system, the interior lights, the Daytime Running Lights system, the horn, the windshield wipers, the heating/air conditioning system and the power mirrors. In the event of malfunction with this system, have the vehicle diagnosed by a dealership service department or other qualified automotive repair facility.*

1   The power door lock system operates the door lock actuators mounted in each door. The system consists of the switches, actuators, lock and unlock relays, Body Control Module (BCM) and associated wiring. Diagnosis can usually be limited to simple checks of the wiring connections and actuators for minor faults that can be easily repaired.

2   Power door lock systems are operated by bi-directional solenoids located in the doors. The lock actuators are mounted as part of the door latch. Remove the door latch for access to the door lock actuator. The lock switches have two operating positions: Lock and Unlock. These switches send a signal to the BCM, which in turn sends a signal to the door lock relays, the relays then send the needed voltage to each of the door lock solenoids.

3   If you are unable to locate the trouble using the following general steps, consult your dealer service department or qualified independent repair shop.

4   Always check the circuit protection first. Some vehicles use a combination of circuit breakers and fuses.

5   Check for voltage at the switches. If no voltage is present, check the fuse first. If the fuse is good then check the wiring between the fuse panel and the switches for an open lead.

6   If voltage is present, test the switch for continuity. Replace it if there's not continuity in both switch positions. There should be a voltage input and, when switch is depressed, voltage should be going out on the appropriate lead. Follow the wiring diagram for the actual wire and position on the switch. To remove the switch, use a flat-bladed trim tool to pry out the door/window switch assembly.

7   If the switch has continuity, check the wiring between the switch, BCM, door lock relay and the door lock solenoid.

8   If all but one of the lock solenoids operates, remove the trim panel from the affected door and check for voltage at the solenoid while the lock switch is operated. One of the

wires should have positive voltage in the Lock position; the other lead should have positive voltage in the Unlock position.

9   If the inoperative solenoid is receiving positive voltage on one lead and negative on the other, the solenoid is most likely defective. Check the connections for good contact; if the connection is good, replace the solenoid.

10   If the inoperative solenoid isn't receiving voltage or ground, check for an open or short in the wire between the lock solenoid and the relay. A good method of non-destructive testing is to squeeze the rubber corrugated tubing and search with your fingers for an individual wire. Follow the wire as far as possible and feel for any breaks in the leads.

Note: *It's not uncommon for wires to break in the portion of the harness between the body and door (opening and closing the door fatigues and eventually breaks the wires).*

11   On the models covered by this manual, power door lock system communication goes through the Body Control Module. If the above tests do not pinpoint a problem, take the vehicle to a dealer or qualified shop with the proper scan tool to retrieve trouble codes from the BCM. Replacing of some components may result in programming issues. To avoid replacing good components always test thoroughly before any parts are deemed faulty.

Keyless entry system

12   The keyless entry system consists of a remote control transmitter that sends a coded infrared signal to a receiver, which then operates the door lock system.

13   Replace the battery when the transmitter doesn't operate the locks at a distance of 10 feet. Normal range should be about 65 feet.

14   For more information on key fob battery replacement, programming and additional keys see Section 6.

23   Daytime Running Lights (DRL) - general information

1   The Daytime Running Lights (DRL) system used on Canadian models illuminates the headlights whenever the engine is running. The only exception is with the engine running and the parking brake engaged. Once the parking brake is released, the lights will remain on as long as the ignition switch is on, even if the parking brake is later applied.

2   The DRL system supplies reduced power to the headlights so they won't be too bright for daytime use, while prolonging headlight life.

24   Airbag system - general information

1   All models are equipped with a Supplemental Restraint System (SRS), more commonly known as an airbag. This system is

designed to protect the driver, and the front seat passenger, from serious injury in the event of a head-on or frontal collision. It consists of airbag sensors mounted in the front, sides and a sensing/diagnostic module mounted in the center of the vehicle, near the floor console.

Components and operation

Airbag modules

2   The airbag modules consist of a housing incorporating the cushion (airbag) and inflator unit. The inflator assembly is mounted on the back of the housing over a hole through which gas is expelled, inflating the bag almost instantaneously when an electrical signal is sent from the system. The specially wound wire on the driver's side that carries this signal to the driver's module is called a spiral cable. The spiral cable is a flat, ribbon-like electrically conductive tape that is wound many times so that it can transmit an electrical signal regardless of steering wheel position. Airbag modules are located in the steering wheel, on the passenger's side above the glove box, at the upper side of each front seat (side-impact airbags) and head-level airbags located along the roof rails (side curtain airbags). The side-impact airbag modules are mounted inside of the seats, and the seat must be disassembled to gain access to the modules.

Seat belt retractors

3   The seat belt retractors are a pyrotechnic (explosive) unit, which operate both the lap and shoulder belts. During an impact that would trigger the airbag system, the airbag control unit also triggers the seat belt retractors. When the pyrotechnic charges go off, they accelerate the retractors to instantly take up any slack in the seat belt system to more fully prepare the driver and front seat passenger for impact. This pulls the passenger (and driver) back into their seat at the same instant as the airbag module deploys the airbag. This softens the impact into the airbag to further dissipate the collision's energy.

Warning: *Never strike the pillars or floorpan with a hammer or use an impact-driver tool in these areas unless the system is disabled.*

Control unit and sensors

4   The sensing/diagnostic control unit contains an on-board microprocessor which monitors the operation of the system, and the crash sensors. The crash sensors are located in the front and sides of the vehicle. It checks this system every time the vehicle is started, causing the "AIRBAG" light to illuminate for five seconds, then go off, if the system is operating properly. If there is a fault in the system, the light may not come on at all, the light will go on and continue, either illuminated steadily or blinking, and the unit will store fault codes indicating the nature of the fault.

Operation

5   For the airbag(s) to deploy, the impact sensor(s) must be activated. When this condi-

tion occurs, the circuit to the airbag inflator is closed and the airbag inflates.

Self-diagnosis system

6   A self-diagnosis circuit in the SRS unit displays a light on the instrument panel when the ignition switch is turned to the On position. If the system is operating normally, the light should go out after about five seconds. If the light doesn't come on, or doesn't go out after a short time, or if it comes on while you're driving the vehicle, or if it blinks at any time, there's a malfunction in the SRS system. Have it inspected and repaired as soon as possible. Do not attempt to troubleshoot or service the SRS system yourself. Even a small mistake could cause the SRS system to malfunction when you need it.

Servicing components near the SRS system

7   Nevertheless, there are times when you need to remove the steering wheel, radio or service other components on or near the dashboard. At these times, you'll be working around components and wire harnesses for the SRS system.

Warning: *Do not use electrical test equipment on airbag system wires; it could cause the airbag(s) to deploy. ALWAYS DISABLE THE SRS SYSTEM BEFORE WORKING NEAR THE SRS SYSTEM COMPONENTS OR RELATED WIRING.*

Disabling the SRS system

Warning: *Any time you are working in the vicinity of airbag wiring or components, DISABLE THE SRS SYSTEM.*

Warning: *An auxiliary voltage input device (memory saver) must not be used when working near airbag system components.*

8   To disable the airbag system, perform the following steps:

a) *Turn the steering wheel to the straight-ahead position and turn the ignition switch to the Lock position, then remove the key.*

b) *Disconnect the negative battery cable (see Chapter 5).*

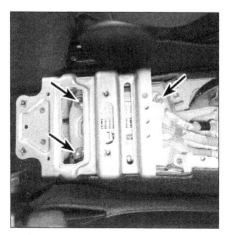

24.11 Airbag control module fasteners

c) *Wait at least two minutes to allow the back-up power supply to become depleted before working around any airbag system component*

d) *Before touching any airbag system component, ground yourself to a metal part of the vehicle to discharge any static electricity built up in your body.*

Enabling the system

9   To enable the airbag system, perform the following steps:

a) *Turn the ignition switch to the On position.*

b) *Make sure nobody is inside the vehicle.*

c) *Connect the negative battery cable.*

d) *Turn the ignition to the Off position, then with your body out of the path of the airbag, turn the ignition switch to the On position. Confirm that the airbag warning light is functioning properly.*

e) *Take the vehicle to a dealer service department or other qualified repair facility and have the airbag system checked and the diagnostic light canceled, if it remains lit.*

Component removal and installation

Warning: *Whenever handling an airbag module, always keep the airbag opening (the trim side) pointed away from your body. Never place the airbag module on a workbench or other surface with the airbag opening (upholstered side) facing the surface. Always place the airbag module in a safe location with the airbag opening facing up.*

Warning: *Never dispose of a live airbag module. Return it to a dealer service department or other qualified repair shop for safe deployment and disposal.*

Note: *Always wait at least 90 seconds after disconnecting the negative battery terminal before working on any airbag component.*

Airbag control module

Note: *The airbag control module can be referred to by several different names, depending on whom you're talking to. It is sometimes referred to as a crash box, smart box, air bag controller, or the restraint control module. All of which refer to the same component. Never use a salvage controller. Always replace with a new (unprogrammed) module.*

10   Remove the center console (see Chapter 11).

11   Locate the airbag control module and remove the fasteners securing it to the vehicle (see illustration).

12   Remove the fasteners, then disconnect the electrical connectors (see illustration).

Caution: *Proper disconnection of the airbag module is critical, otherwise you can damage the connector (see illustration).*

Note: *After replacing the airbag module it will need to be programmed. This will require a factory level scanner. Take your vehicle to the appropriate repair facility to have this procedure performed.*

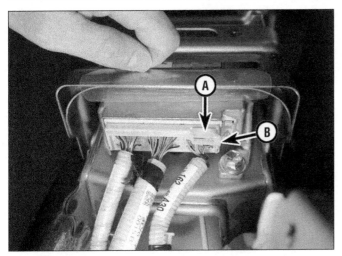

24.12 Airbag controller electrical connection removal procedure

A   *Push down on the white tab slightly*
B   *Now swing the white tab toward the middle of the airbag control module connection to remove the connector*

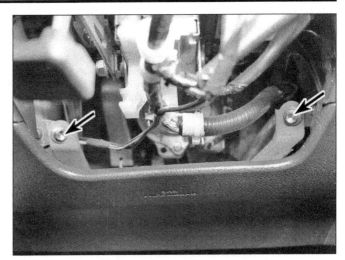

24.17a Remove the upper fasteners . . .

24.17b . . . then the two lower fasteners

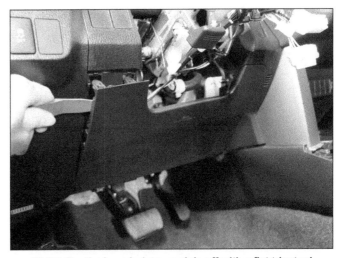

24.17c Pry the knee bolster module off with a flat trim tool

Driver's airbag module and spiral cable removal

13   Refer to Chapter 10, *Steering wheel - removal and installation*, for the driver's side airbag module and spiral cable removal and installation procedures.
Note: *The spiral cable is also referred to as the clockspring.*

Passenger's airbag module

14   Disarm the airbag system as describes previously in this Section.
15   Remove the glove box (see Chapter 11), then unplug the electrical connector and remove the airbag module mounting fasteners. Remove the upper instrument panel (see Chapter 11) and module out as a unit.
Warning: *Slightly bend the mounting hooks back and lift out the front passenger's airbag*

rearwards. Be sure to heed the precautions outlined previously in this Section.
16   Installation is the reverse of removal. Tighten the airbag module mounting fasteners to 15 ft-lbs (20 Nm).

Knee bolster module

17   The knee bolster trim and airbag are basically one component (see illustrations). Follow the illustrations for details on the removal of the knee bolster module.

Other airbag system components (front seat cushion airbag, side curtain airbags, seat belt retractors, occupant detection system)

18   It is not advisable to tamper with any of these components.

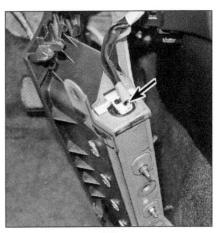

24.17d Disconnect the electrical connection and remove the module

25   Wiring diagrams - general information

1   Since it isn't possible to include all wiring diagrams for every year covered by this manual, the following diagrams are those that are typical and most commonly needed.

2   Prior to troubleshooting any circuits, check the fuse and circuit breakers (if equipped) to make sure they're in good condition. Make sure the battery is properly charged and check the cable connections (see Chapter 1).

3   When checking a circuit, make sure that all connectors are clean, with no broken or loose terminals. When unplugging a connector, do not pull on the wires. Pull only on the connector housings themselves.

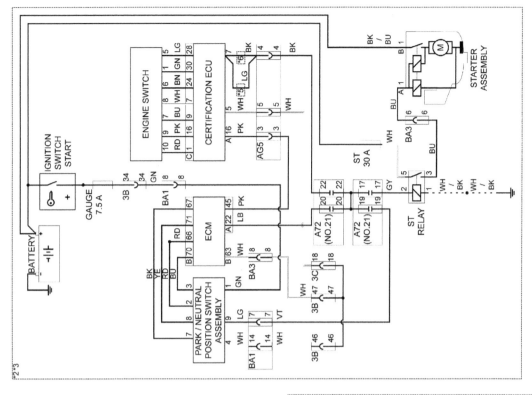

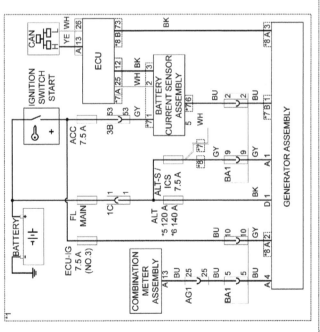

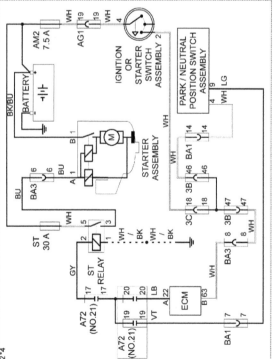

Starting and charging systems

*1 Charging
*2 Starting
*3 With smart key system
*4 Without smart key system
*5 Up to September 2015
*6 From October 2015
*7 For 2WD
*8 For 4WD

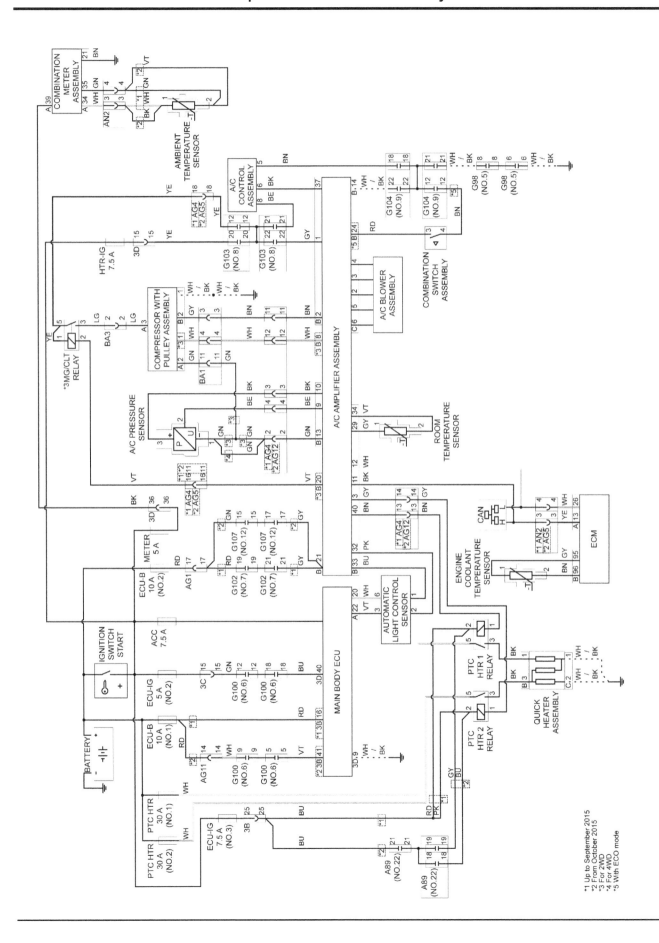

Heating and air conditioning systems (automatic)

*1 Up to September 2015
*2 From October 2015
*3 For 2WD
*4 For 4WD
*5 With ECO mode

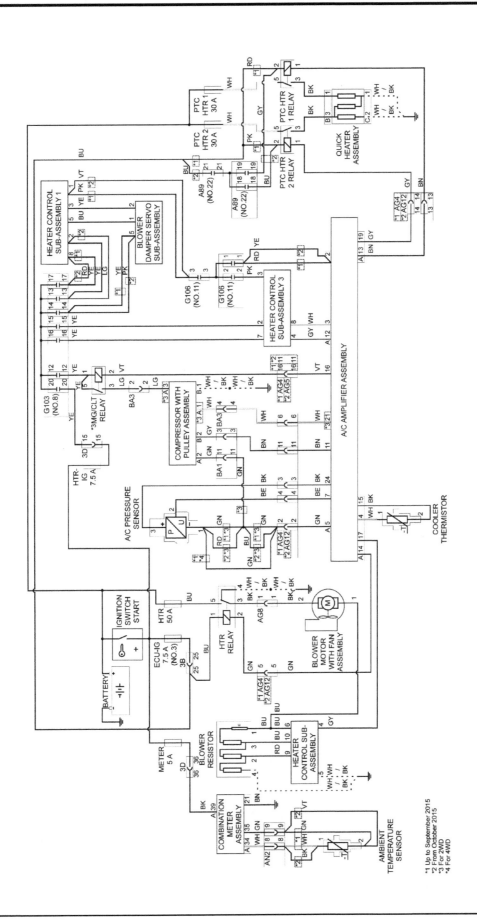

Heating and air conditioning systems (manual)

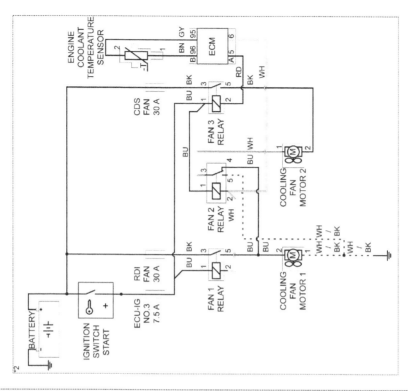

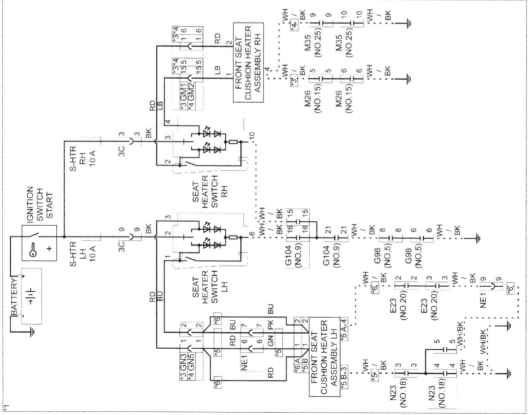

Seat heater and engine cooling fans system

*1 Seat heating
*2 Cooling fans
*3 Up to September 2015
*4 From October 2015
*5 With power seats
*6 Without power seats

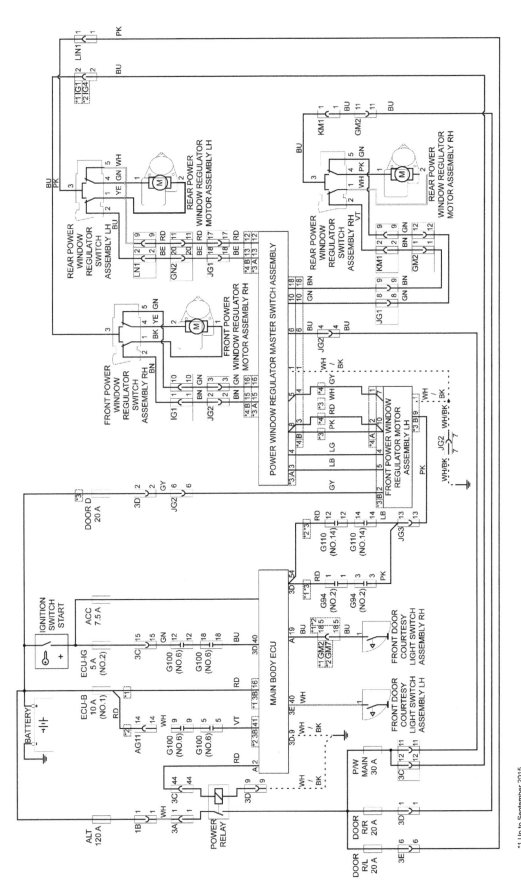

Power window system

*1 Up to September 2015
*2 From October 2015
*3 With Jam protection
*4 Without Jam protection

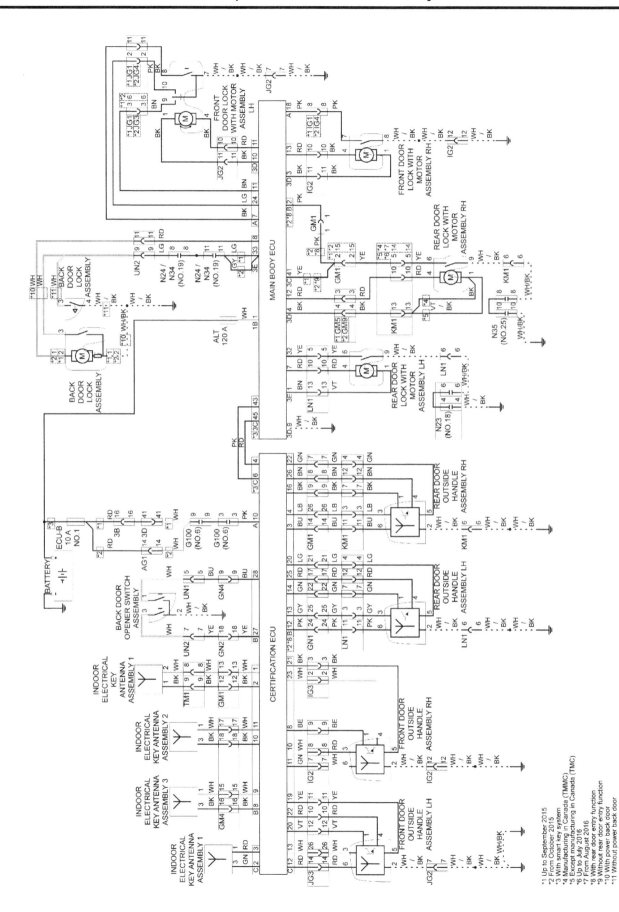

Power door lock system

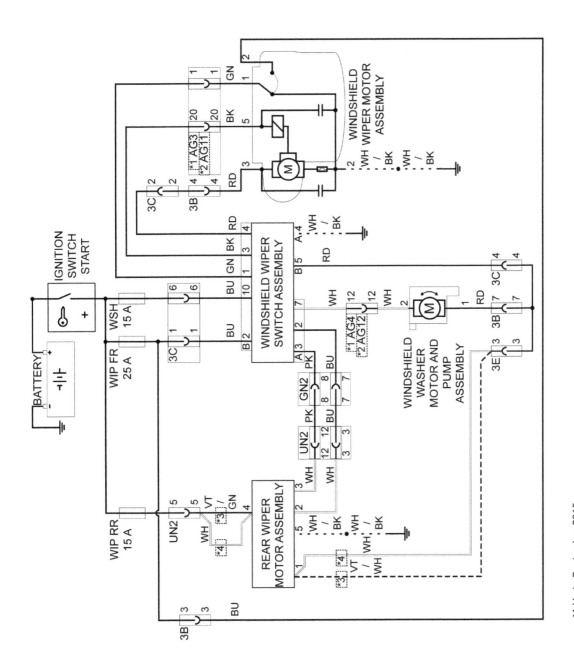

Wiper and washer systems

*1 Up to September 2015
*2 From October 2015
*3 Except manufacturing in Canada (TMC)
*4 Manufacturing in Canada (TMMC)

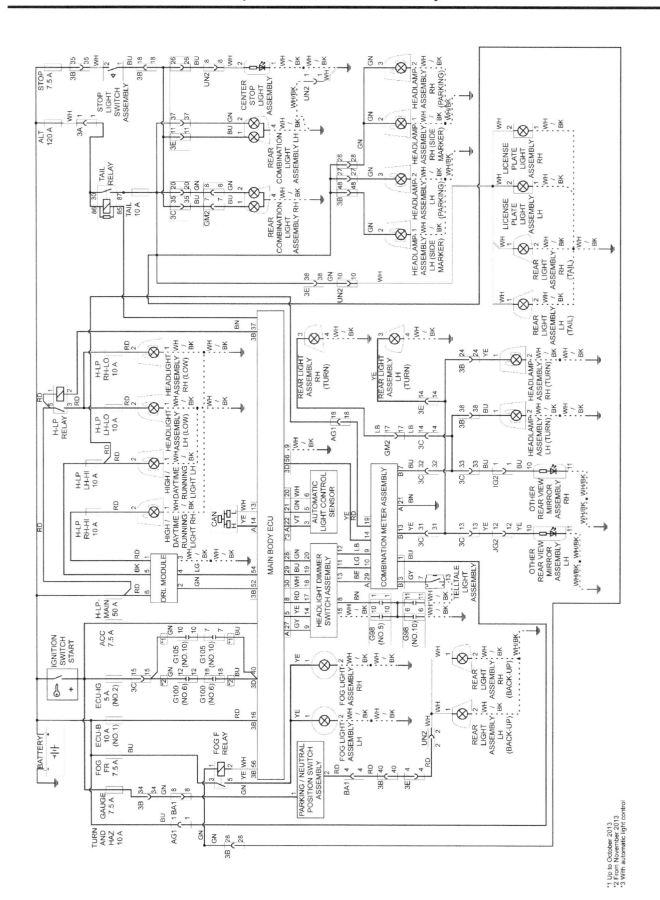

Exterior lighting system (models up to 9/2015)

*1 Up to October 2013
*2 From November 2013
*3 With automatic light control

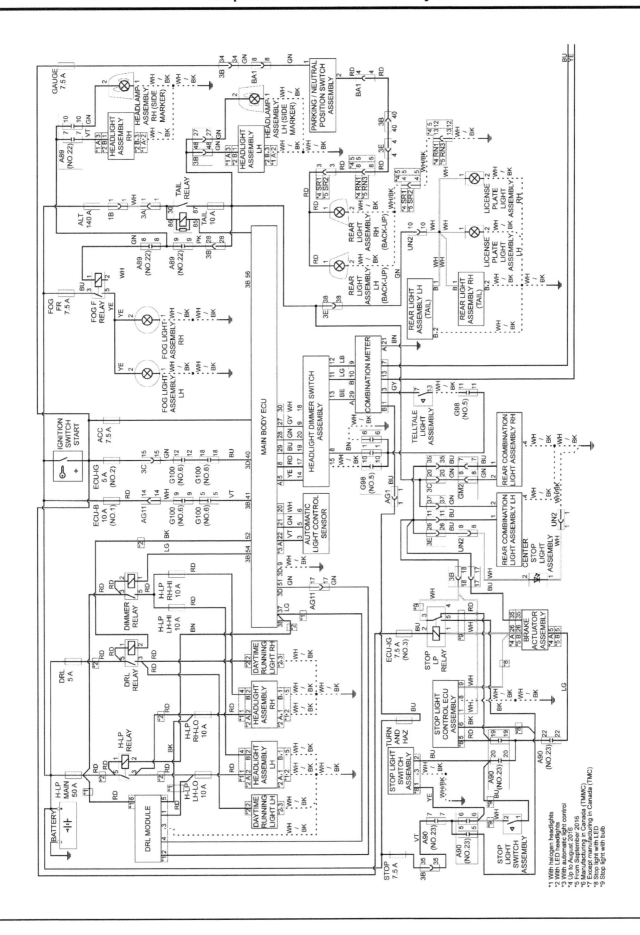

Exterior lighting system (10/2015 and later models) - 1 of 2

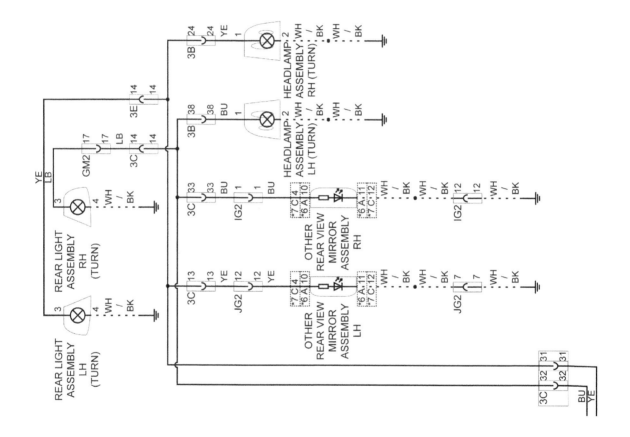

Exterior lighting system (10/2015 and later models) - 2 of 2

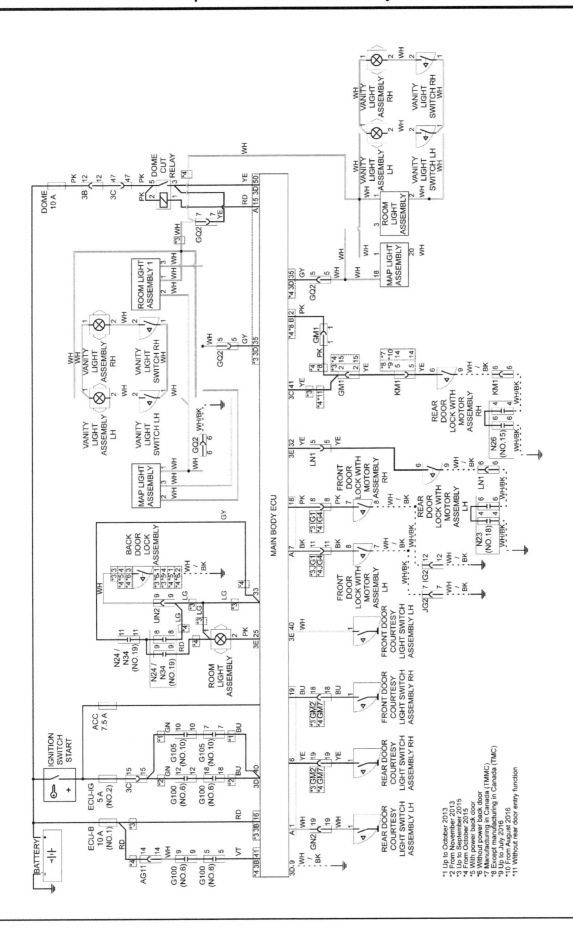

Interior lighting system

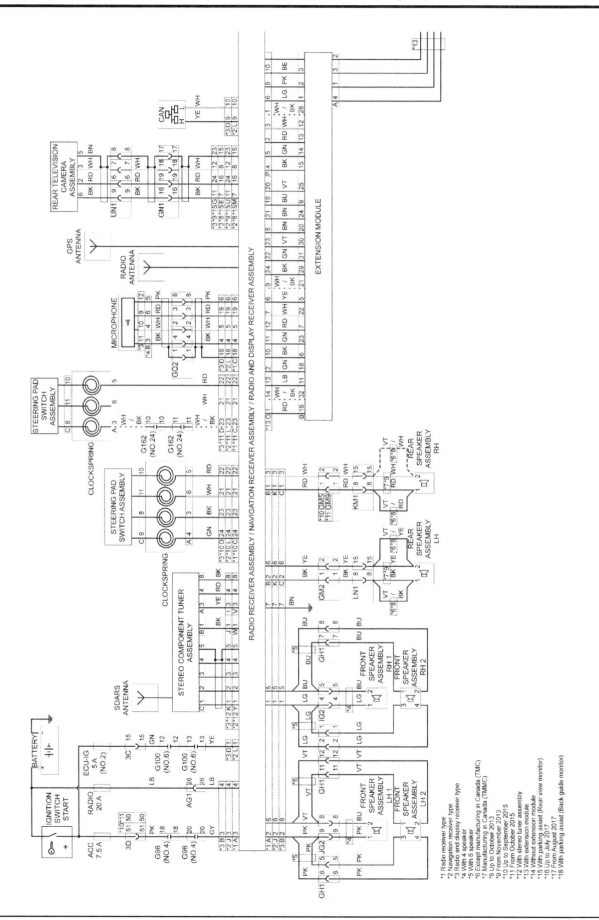

Audio system (with built-in amplifier) - 1 of 2

*1 Radio receiver type
*2 Navigation receiver type
*3 Radio and display receiver type
*4 With 4 speaker
*5 With 6 speaker
*6 Except manufacturing in Canada (TMC)
*7 Manufacturing in Canada (TMMC)
*8 Up to October 2013
*9 From November 2013
*10 Up to September 2015
*11 From October 2015
*12 With stereo tuner assembly
*13 With extension module
*14 Without extension module
*15 With parking assist (Rear view monitor)
*16 Up to July 2017
*17 From August 2017
*18 With parking assist (Back guide monitor)

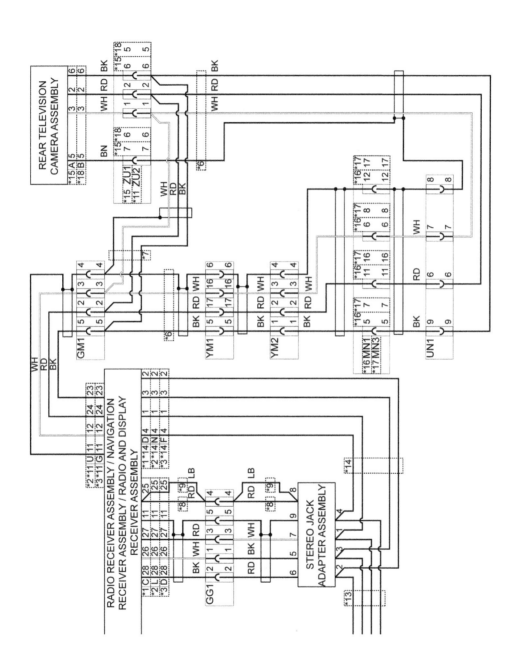

Audio system (with built-in amplifier) - 2 of 2

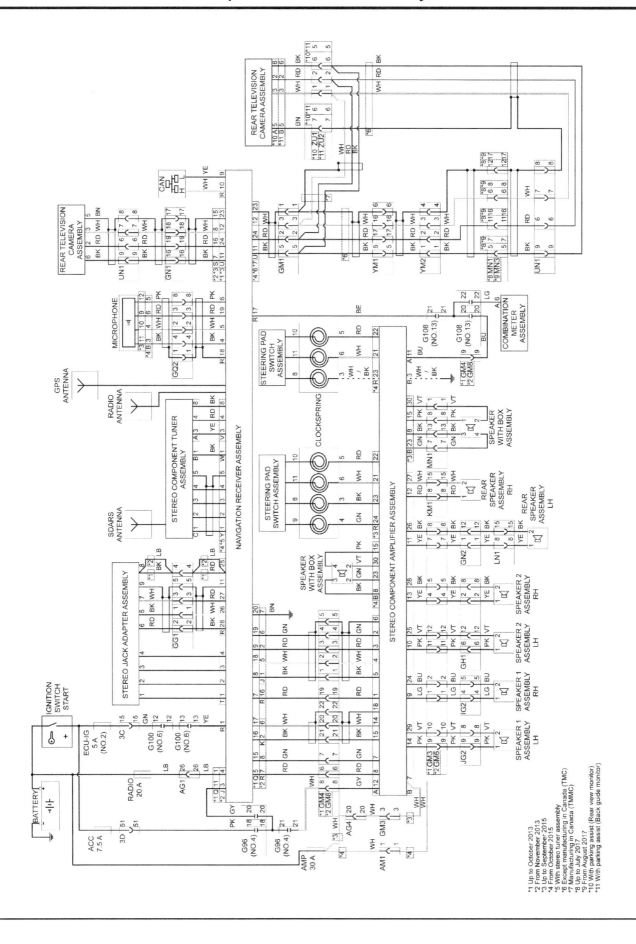

Audio system (with separate amplifier)

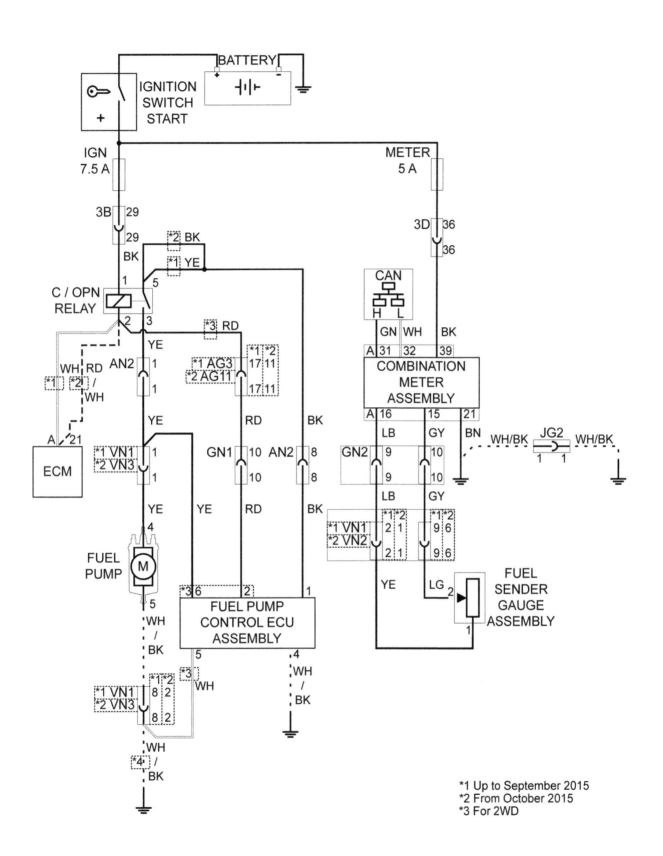

*1 Up to September 2015
*2 From October 2015
*3 For 2WD

FUSE AND RELAY BOX IN ENGINE COMPARTMENT UP TO SEPTEMBER 2015

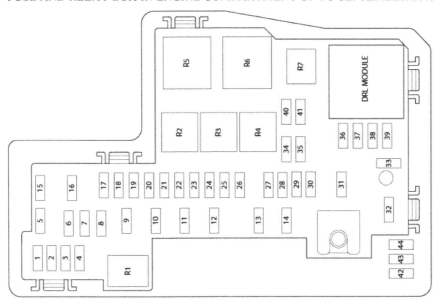

FUSE/RELAY	VALUE	DESCRIPTION	OEM NAME
1	20 A	Audio System or Navigation System, Parking assist	RADIO
2	10 A	4WD, ABS, Auto LSD, Automatic air conditioning, Automatic light control, Back Door Opener, Combination Meter, Cruise Control, Door Lock Control, Electronically Controlled Transmission and A/T Indicator, Engine Control, Front Fog Light, Headlight, Illumination, Immobiliser System (with Smart Key System), Wireless Door Lock Control, Interior Light, Key Reminder (without Smart Key System), Light Auto Turn Off System, Light Reminder, Option Connector (Remote Engine Starter System) (with Smart Key System), Power Back Door, Power Seat, Power Window, Sliding Roof, Smart Key System, SRS, Starting (with Smart Key System), Steering Lock (with Smart Key System), Taillight, Theft Deterrent, Tire Pressure Warning System, TRAC, VSC	ECU-B NO.1
3	10 A	Back Door Opener (with Smart Key System), Interior Light, Option Connector (Remote Engine Starter System) (with Smart Key System), Smart Key System, Starting (with Smart Key System), Steering Lock (with Smart Key System), Theft Deterrent, Wireless Door Lock Control	DOME
4	-	Not used	-
5	30 A	Cooling fan	CDS FAN
6	20 A	Front window de-icer	DEICER
7	-	Not used	-
8	7.5 A	Front fog light	FOG FR
9	30 A	Air conditioning	PTC HTR 2
10	30 A	Air conditioning	PTC HTR 1
11	30 A	Cruise Control, Electronically Controlled Transmission and A/T Indicator, Engine Control, Mirror Heater, Rear Window Defogger	DEF
12	30 A	ABS, Auto LSD, TRAC, VSC	ABS NO.2
13	30 A	Cooling fan	RDI FAN
14	50 A	ABS, Auto LSD, TRAC, VSC	ABS NO.1
15	50 A	Air conditioning	HTR
16	50 A	Automatic Light Control, Back Door Opener (with Smart Key System),	H-LP-MAIN

		Combination Meter, Combination Meter, Headlight, Immobiliser System (with Smart Key System), Light Auto Turn Off System, Light Reminder, Option Connector (Remote Engine Starter System) (with Smart Key System), Smart Key System, Starting (with Smart Key System), Wireless Door Lock Control	
17	30 A	Audio System, Navigation System, Parking Assist on rear View Monitor (Separate Type Amplifier)	AMP
18	30 A	Starting , Smart Key System, Back Door Opener (with Smart Key System), Immobiliser System (with Smart Key System), Option Connector (Remote Engine Starter System) (with Smart Key System), Steering Lock (with Smart Key System), Wireless Door Lock Control (with Smart Key System)	ST
19	20 A	Cruise Control, Engine Control, Electronically Controlled Transmission and A/T Indicator	EFI-MAIN NO.1
20	-	Not used	-
21	15 A	Combination Meter, Cruise Control, Electronically Controlled Transmission and A/T Indicator, Engine Control, Ignition, Immobiliser System (with Smart Key System), Back Door Opener (with Smart Key System), Option Connector (Remote Engine Starter System) (with Smart Key System), Starting (with Smart Key System), Steering Lock (with Smart Key System), Wireless Door Lock Control (with Smart Key System), Smart Key System	IG2
22	10 A	Combination Meter, Turn signal and Hazard warning light	TURN AND HAZ
23	7.5 A	Cruise Control, Engine Control, Back Door Opener (Smart Key System), Electronically Controlled Transmission and A/T Indicator , Immobiliser System (Smart Key System) , Option Connector (Remote Engine Starter System) (Smart Key System) , Smart Key System , Starting, Steering Lock (Smart Key System) , Wireless Door Lock Control (Smart Key System)	AM2
24	10 A	4WD, ABS, Navigation System or Audio System, Parking Assist (Rear View Monitor), Auto LSD, Automatic Air Conditioning, Automatic Light Control, Back Door Opener (with Smart Key System), Blind Spot Monitor System, Charging, Combination Meter, Cruise Control, Electronically Controlled Transmission and A/T Indicator, Engine Control, EPS, Headlight, Illumination, Immobiliser System, Key Reminder (without Smart Key System), Lane Departure Alert, Light Auto Turn Off System, Light Reminder, Manual Air Conditioning, Option Connector (Remote Engine Starter System) (with Smart Key System), Power Back Door, Seat Belt Warning, Sliding Roof, Smart Key System, SRS, Starting (with Smart Key System), Steering Lock (with Smart Key System), Taillight, Theft Deterrent, Tire Pressure Warning System, TRAC, Turn Signal and Hazard Warning Light, VSC, Wireless Door Lock Control	ECU-B NO.2
25	10 A	Back Door Opener (with Smart Key System), Option Connector (Remote Engine Starter System) (with Smart Key System), Immobiliser System (with Smart Key System), Smart Key System, Starting (with Smart Key System), Steering Lock (with Smart Key System), Wireless Door Lock Control (with Smart Key System)	STRG LOCK
26	30 A	Power source	D/C CUT
27	10 A	Horn, Back Door Opener (with Smart Key System), Immobiliser System (with Smart Key System), Option Connector (Remote Engine Starter System) (with Smart Key System), Smart Key System, Starting (with Smart Key System), Steering Lock (with Smart Key System), Theft Deterrent, Wireless Door Lock Control	HORN
28	10 A	Cruise Control, Electronically Controlled Transmission and A/T Indicator, Engine Control	ETCS
29	20 A	Cruise Control, Electronically Controlled Transmission and A/T Indicator,	EFI-MAIN

		Engine Control	NO.2
30	7.5 A	Charging	ALT-S/ICS
31	80 A	EPS	EPS
32	120 A	Automatic Light Control, Back Door Opener (with Smart Key System), Charging, Front Fog Light, Illumination, Immobiliser System (with Smart Key System), Light Auto Turn Off System, Light Reminder, Option Connector (Remote Engine Starter System) (with Smart Key System), Parking Assist (Intuitive Parking Assist), Power Window, Smart Key System, Starting (with Smart Key System), Steering Lock (with Smart Key System), Taillight, Theft Deterrent, Wireless Door Lock Control	ALT
33	-	Not used	-
34	10 A	Cruise Control, Electronically Controlled Transmission and A/T Indicator, Engine Control, Mirror Heater, Rear Window Defogger	MIR HTR
35	-	Not used	-
36	10 A	Cruise Control, Electronically Controlled Transmission and A/T Indicator, Engine Control	EFI NO.1
37	10 A	Cruise Control, Electronically Controlled Transmission and A/T Indicator, Engine Control	EFI NO.2
38	10 A	Combination Meter, Headlight	H-LP LH-HI
39	10 A	Headlight	H-LP RD-HI
40	10 A	Headlight	H-LP LH-LO
41	10 A	Headlight	H-LP RD-LO
42	-	Not used	-
43	-	Not used	-
44	-	Not used	-
R1	-	DEF relay	DEF
R2	-	EFI-MAIN NO.2 relay	EFI-MAIN NO.2
R3	-	IG2 relay	IG2
R4	-	C/OPN relay	C/OPN
R5	-	HTR relay	HTR
R6	-	H-LP relay	H-LP
R7	-	EFI MAIN NO.1 relay	EFI MAIN NO.1

FUSE AND RELAY BOX IN ENGINE COMPARTMENT FROM OCTOBER 2015

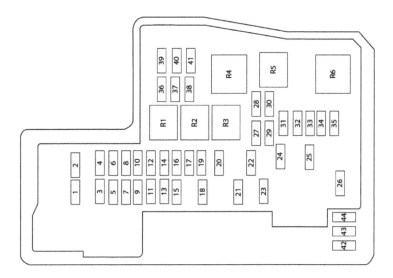

FUSE/RELAY	VALUE	DESCRIPTION	OEM NAME
1	-	Not used	-
2	-	Not used	-
3	20 A	Cruise Control, Engine Control, Electronically Controlled Transmission and A/T Indicator	EFI-MAIN NO.1
4	30 A	Starting, Smart Key System, Back Door Opener (with Smart Key System), Immobiliser System (with Smart Key System), Option Connector (Remote Engine Starter System) (with Smart Key System), Steering Lock (with Smart Key System), Wireless Door Lock Control (with Smart Key System)	ST
5	-	Not used	-
6	30 A	Audio System, Navigation System, Parking Assist (Rear View Monitor or Back Guide Monitor) with separate amplifier	AMP
7	10 A	Back Door Opener (with Smart Key System), Immobiliser System (with Smart Key System), Option Connector (Remote Engine Starter System) (with Smart Key System), Smart Key System, Starting (with Smart Key System), Steering Lock (with Smart Key System), Wireless Door Lock Control (with Smart Key System)	STRG LOCK
8	10 A	Cruise Control, Dynamic Radar Cruise Control, Electronically Controlled Transmission and A/T Indicator, Engine Control	ETCS
9	10 A	4WD, ABS, Wireless Door Lock Control, Audio System, Navigation System, Parking Assist, Auto LSD, Automatic Air Conditioning, Automatic Light Control, Back Door Opener (with Smart Key System), Blind Spot Monitor System, Charging, Combination Meter, Cruise Control, Dynamic Radar Cruise Control, Electronically Controlled Transmission and A/T Indicator, Engine Control, EPS, Front Fog Light, Headlight, Illumination, Immobiliser System, Key Reminder (without Smart Key System), Lane Departure Alert, Light Auto Turn Off System, Light Reminder, Manual Air Conditioning, Multiplex Communication System (CAN), Option Connector (Remote Engine Starter System) (with Smart Key System), Power Back Door, Pre-Collision System, Seat Belt Warning, Sliding Roof, Smart Key System, SRS, Starting (with Smart Key System), Steering Lock (with Smart Key System), Taillight, Theft Deterrent, Tire Pressure Warning System, TRAC, Turn Signal and Hazard Warning Light, VSC	ECU-B NO.2

10	-	Not used	-
11	10 A	Combination Meter, Turn Signal and Hazard Warning Light	TURN AND HZD
12	15 A	Combination Meter, Cruise Control, Electronically Controlled Transmission and A/T Indicator, Engine Control, Ignition, Immobiliser System (with Smart Key System), Option Connector (Remote Engine Starter System) (with Smart Key System), Smart Key System, Starting (with Smart Key System), Steering Lock (with Smart Key System), Wireless Door Lock Control (with Smart Key System)	IG2
13	20 A	Cruise Control, Engine Control, Electronically Controlled Transmission and A/T Indicator	EFI-MAIN NO.2
14	7.5 A	Back Door Opener (with Smart Key System), Cruise Control, Electronically Controlled Transmission and A/T Indicator, Engine Control, Immobiliser System (with Smart Key System), Option Connector (Remote Engine Starter System) (with Smart Key System), Smart Key System, Starting, Steering Lock (with Smart Key System), Wireless Door Lock Control (with Smart Key System)	AM2
15	-	Not used	-
16	7.5 A	Charging	ALT-S/ICS
17	10 A	Back Door Opener (with Smart Key System), Horn, Immobiliser System (with Smart Key System), Option Connector (Remote Engine Starter System) (with Smart Key System), Smart Key System, Starting (with Smart Key System), Steering Lock (with Smart Key System), Theft Deterrent, Wireless Door Lock Control	HORN
18	-	Not used	-
19	-	Not used	-
20	30 A	Power source	D/C CUT
21	30 A	ABS, Auto LSD, TRAC, VSC	ABS NO.2
22	50 A	Automatic Light Control, Back Door Opener (with Smart Key System), Combination Meter, Headlight, Immobiliser System (with Smart Key System), Light Auto Turn Off System, Light Reminder, Option Connector (Remote Engine Starter System) (with Smart Key System), Smart Key System, Starting (with Smart Key System), Steering Lock (with Smart Key System), Theft Deterrent, Wireless Door Lock Control	H-LP-MAIN
23	50 A	ABS, Auto LSD, TRAC, VSC	ABS NO.1
24	-	Not used	-
25	80 A	EPS	EPS
26	140 A	Automatic Light Control, Back Door Opener (with Smart Key System), Charging, Front Fog Light, Illumination, Immobiliser System (with Smart Key System), Light Auto Turn Off System, Light Reminder, Option Connector (Remote Engine Starter System) (with Smart Key System), Parking Assist (Intuitive Parking Assist) (Manual A/C), Power Window, Smart Key System, Starting (with Smart Key System), Steering Lock (with Smart Key System), Taillight, Theft Deterrent, Wireless Door Lock Control	ALT
27	-	Not used	-
28	-	Not used	-
29	-	Not used	-
30	-	Not used	-
31	10 A	Cruise Control, Engine Control, Electronically Controlled Transmission and A/T Indicator	EFI NO.1
32	10 A	Cruise Control, Engine Control, Electronically Controlled Transmission and A/T Indicator	EFI NO.2
33	10 A	Combination Meter, Headlight	H-LP LH-HI
34	10 A	Headlight	H-LP RH-HI

35	-	Not used	-
36	20 A	Audio System or Navigation System, Parking assist	RADIO
37	10 A	4WD, ABS, Auto LSD, Automatic Air Conditioning, Automatic Glare-Resistant EC Mirror, Automatic Light Control, Back Door Opener, Charging, Combination Meter, Cruise Control, Door Lock Control, Electronically Controlled Transmission and A/T Indicator, Engine Control, Front Fog Light, Garage Door Opener, Headlight, Illumination, Immobiliser System (with Smart Key System), Interior Light, Key Reminder (without Smart Key System), Light Auto Turn Off System, Light Reminder, Option Connector (Remote Engine Starter System) (with Smart Key System), Panoramic View Monitor System, Power Back Door, Power Seat (with Seat Position Memory), Power Window, Seat Belt Warning, Sliding Roof, Smart Key System, SRS, Starting (with Smart Key System), Steering Lock (with Smart Key System), Taillight, Theft Deterrent, Tire Pressure Warning System, TRAC, VSC, Wireless Door Lock Control	ECU-B NO.1
38	10 A	Back Door Opener (with Smart Key System), Immobiliser System (with Smart Key System), Interior Light, Option Connector (Remote Engine Starter System) (with Smart Key System), Smart Key System, Starting (with Smart Key System), Steering Lock (with Smart Key System), Theft Deterrent, Wireless Door Lock Control	DOME
39	10 A	Headlight	H-LP LH-LO
40	10 A	Headlight	H-LP RH-LO
41	-	Not used	-
42	-	Not used	-
43	-	Not used	-
44	-	Not used	-
R1	-	EFI-MAIN NO.2 relay	EFI-MAIN NO.2
R2	-	IG2 relay	IG2
R3	-	C/OPN relay	C/OPN
R4	-	DIMMER relay	DIMMER
R5	-	EFI MAIN NO.1 relay	EFI MAIN NO.1
R6	-	H-LP relay (LED headlight) or DRL module (Halogen headlight)	H-LP or DRL

FUSE AND RELAY BOX IN ENGINE COMPARTMENT 2

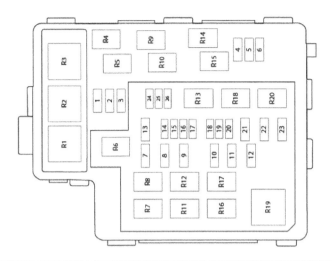

FUSE/RELAY	VALUE	DESCRIPTION	OEM NAME
1	-	Not used	-
2	-	Not used	-
3	-	Not used	-
4	-	Not used	-
5	-	Not used	-
6	-	Not used	-
7	30 A	Air conditioning	PTC HTR NO.1
8	30 A	Air conditioning	PTC HTR NO.2
9	-	Not used	-
10	30 A	Cooling fan	CDS FAN
11	30 A	Cooling fan	RDI FAN
12	50 A	Air conditioning, Charging	HTR
13	30 A	Cruise control, Electronically Controlled Transmission and A/T Indicator, Engine Control, Mirror Heater, Rear Window Defogger	DEF
14	5 A	Headlight, Theft deterrent	DRL
15	-	Not used	-
16	7.5 A	Front Fog Light	FOG FR
17	10 A	Heated Steering Wheel System	NOISE FILTER
18	-	Not used	-
19	15 A	Power Outlet	INV
20	20 A	Front Window De-icer	DEICER
21	-	Not used	-
22	-	Not used	-
23	-	Not used	-
R1	-	Not used	-
R2	-	Not used	-
R3	-	Not used	-
R4	-	STRG HTR relay	STRG HTR
R5	-	INV relay	INV

R6	-	DEF relay	DEF
R7	-	FAN NO.2 relay	FAN NO.2
R8	-	DRL relay	DRL
R9	-	Not used	-
R10	-	MG/CLT relay	MG/CLT
R11	-	FOG FR relay	FOG FR
R12	-	FAN NO.3 relay	FAN NO.3
R13	-	STOP LP relay	STOP LP
R14	-	PTC HTR NO.2 relay	PTC HTR NO.2
R15	-	PTC HTR NO.1 relay	PTC HTR NO.1
R16	-	HORN relay	HORN
R17	-	FAN NO.1 relay	FAN NO.1
R18	-	ST relay	ST
R19	-	HTR relay	HTR
R20	-	DEICER relay	DEICER

RELAY ON ENGINE COMPARTMENT

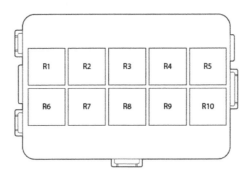

RELAY	VALUE	DESCRIPTION	OEM NAME
R1	-	FOG FR relay	FOG FR
R2	-	MG/CLT relay	MG/CLT
R3	-	PTC HTR NO.2 relay	PTC HTR NO.2
R4	-	DEICER relay	DEICER
R5	-	HORN relay	HORN
R6	-	FAN NO.2 relay	FAN NO.2
R7	-	PTC HTR NO.1 relay	PTC HTR NO.1
R8	-	FAN NO.3 relay	FAN NO.3
R9	-	ST relay	ST
R10	-	FAN NO.1 relay	FAN NO.1

FUSE BOX IN PASSENGER COMPARTMENT

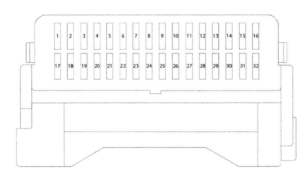

FUSE	VALUE	DESCRIPTION	OEM NAME
1	-	Not used	-
2	7.5 A	4WD, ABS, Auto LSD, Charging, Back Door Opener (with Smart Key System), Cruise Control, Electronically Controlled Transmission and A/T Indicator, Engine Control, Immobiliser System (with Smart Key System), Option Connector (Remote Engine Starter System) (with Smart Key System), Shift Lock, Smart Key System, Starting (with Smart Key System), Steering Lock (with Smart Key System), Stop Light, TRAC, VSC, Wireless Door Lock Control (with Smart Key System)	STOP
3	7.5 A	Sliding roof	S/ROOF
4	5 A	Power source	AM1
5	7.5 A	Data Link Connector 3	OBD
6	20 A	Back Door Opener (with Smart Key System), Door Lock Control, Immobiliser System (with Smart Key System), Option Connector (Remote Engine Starter System) (with Smart Key System), Smart Key System, Starting, Steering Lock (with Smart Key System), Wireless Door Lock Control	D/L NO.2
7	-	Not used	-
8	10 A	Back Door Opener, Immobiliser System (with Smart Key System), Option Connector (Remote Engine Starter System) (with Smart Key System), Smart Key System, Starting, Steering Lock (with Smart Key System), Wireless Door Lock Control	D/L BACK
9	15 A	Power outlet	P/OUTLET NO.1
10	20 A	Power Window	DOOR D
11	20 A	Power Window	DOOR R/R
12	20 A	Power Window	DOOR R/L
13	5 A	Rear Wiper and Washer	WIP RR
14	15 A	Front Wiper and Washer, Rear Wiper and Washer	WSH
15	7.5 A	4WD, ABS, Audio System, Navigation System, Auto LSD, Automatic Glare-Resistant EC Mirror, Back Door Opener, Option Connector (Remote Engine Starter System) (with Smart Key System), Back-Up Light, Blind Spot Monitor System, Combination Meter, Cruise Control, Door Lock Control, Electronically Controlled Transmission and A/T Indicator, Engine Control, Headlight, Immobiliser System (with Smart Key System), Lane Departure Alert, Parking Assist, Power Back Door, Power Seat, Smart Key System, Starting (with Smart Key System), Steering Lock (with Smart Key System), TRAC, VSC, Wireless Door Lock Control, Charging, Charging, Dynamic Radar Cruise Control (if equipped), Garage Door Opener, Pre-Collision System, Panoramic View Monitor System	GAUGE

16	25 A	Front Wiper and Washer	WIP FR
17	5 A	Shift Lock	SFT LOCK-ACC
18	15 A	Power outlet	P/OUTLET NO.2
19	7.5 A	4WD, ABS, Audio System, Navigation System, Auto LSD, Automatic Air Conditioning, Automatic Light Control, Back Door Opener, Charging, Combination Meter, Door Lock Control, Front Fog Light, Headlight, , Illumination, Immobiliser System (with Smart Key System), Interior Light, Key Reminder (without Smart Key System), Light Auto Turn Off System, Light Reminder, Option Connector (Remote Engine Starter System) (with Smart Key System), Parking Assist (Rear View Monitor), Power Back Door, Power Outlet, Power Seat, Power Window, Remote Control Mirror, Sliding Roof, Smart Key System, SRS, Starting, Steering Lock (with Smart Key System), Taillight, Theft Deterrent, Tire Pressure Warning System, TRAC, VSC, Wireless Door Lock Control, Multiplex Communication System (CAN), Seat Belt Warning	ACC
20	7.5 A	Illumination, Parking Assist, Taillight	PANEL
21	10 A	Front Fog Light, Illumination, Taillight	TAIL
22	-	Not used	-
23	5 A	EPS	EPS-IG
24	10 A	4WD, EPS, Auto LSD, Cruise Control, Electronically Controlled Transmission and A/T Indicator, Engine Control, Parking Assist (Intuitive Parking Assist), TRAC, VSC, Power Seat (with Seat Position Memory), Seat Belt Warning	ECU-IG NO.1
25	5 A	4WD, ABS, Audio System, Navigation System, Auto LSD, Automatic Air Conditioning, Automatic Light Control, Back Door Opener, Combination Meter, Combination Meter, , Door Lock Control, Front Fog Light, Headlight, Illumination, Immobiliser System (with Smart Key System), Interior Light, Key Reminder (without Smart Key System), Lane Departure Alert, Light Auto Turn Off System, Light Reminder, Charging, Dynamic Radar Cruise Control, Option Connector (Remote Engine Starter System) (with Smart Key System), Parking Assist (Rear View Monitor), Multiplex Communication System (CAN), Power Back Door, Power Seat, Power Window, Shift Lock, Sliding Roof, Smart Key System, Starting (with Smart Key System), Steering Lock (with Smart Key System), Taillight, Theft Deterrent, Tire Pressure Warning System, TRAC, VSC, Wireless Door Lock Control, Pre-Collision System, Seat Belt Warning	ECU-IG NO.2
26	7.5 A	Air Conditioning, Mirror Heater, Rear Window Defogger, Heated Steering Wheel System (If equipped)	HTR-IG
27	10 A	Seat Heater	S-HTR LH
28	10 A	Seat Heater	S-HTR RH
29	7.5 A	Back Door Opener (with Smart Key System), Cruise Control, Electronically Controlled Transmission and A/T Indicator, Engine Control, Immobiliser System, Option Connector (Remote Engine Starter System) (with Smart Key System), Smart Key System, Starting (with Smart Key System), Steering Lock (with Smart Key System), Wireless Door Lock Control (with Smart Key System), Dynamic Radar Cruise Control, Multiplex Communication System (CAN)	IGN
30	7.5 A	Seat Belt Warning, SRS	A/B
31	5 A	4WD, ABS, Audio System, Navigation System, Auto LSD, Air Conditioning, Back Door Opener (with Smart Key System), Blind Spot Monitor System, Charging, Combination Meter, Cruise Control, Electronically Controlled Transmission and A/T Indicator, Engine Control,	METER

		EPS, Headlight, Illumination, Immobiliser System (with Smart Key System), Key Reminder (without Smart Key System), Lane Departure Alert, Light Auto Turn Off System, Light Reminder, Option Connector (Remote Engine Starter System) (with Smart Key System), Parking Assist (Rear View Monitor), Power Back Door, Seat Belt Warning, Sliding Roof, Smart Key System, SRS, Starting (with Smart Key System), Steering Lock (with Smart Key System), Taillight, Theft Deterrent, Tire Pressure Warning System, TRAC, Turn Signal and Hazard Warning Light, VSC, Wireless Door Lock Control, Dynamic Radar Cruise Control, Front Fog Light, Pre-Collision System	
32	7.5 A	ABS, Auto LSD, Air Conditioning, Charging, Cooling Fan, Cruise Control, Electronically Controlled Transmission and A/T Indicator, Engine Control, Front Window De-icer, Manual Air Conditioning, Mirror Heater, Rear Window Defogger, TRAC, VSC, Dynamic Radar Cruise Control, Engine Control, Lane Departure Alert, Power Outlet (120V), Pre-Collision System, Stop Light	ECU-IG NO.3

FUSE BOX IN PASSENGER COMPARTMENT 2

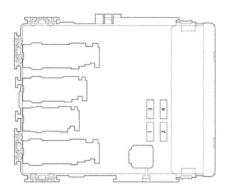

FUSE	VALUE	DESCRIPTION	OEM NAME
1	30 A	Power Seat	P/SEAT F/L
2	30 A	Power Back Door	PBD
3	30 A	Power Seat	P/SEAT F/R
4	30 A	Power Window	P/W-MAIN

RELAY BOX IN PASSENGER COMPARTMENT

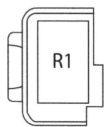

RELAY	VALUE	DESCRIPTION	OEM NAME
R1	-	DOME CUT relay	DOME CUT

Index

V

W

Haynes Automotive Manuals

NOTE: *If you do not see a listing for your vehicle, please visit* **haynes.com** *for the latest product information and check out our* ***Online Manuals!***

ACURA

12020 **Integra** '86 thru '89 **& Legend** '86 thru '90
12021 **Integra** '90 thru '93 **& Legend** '91 thru '95
Integra '94 thru '00 - *see HONDA Civic (42025)*
MDX '01 thru '07 - *see HONDA Pilot (42037)*
12050 **Acura TL** all models '99 thru '08

AMC

14020 **Mid-size models** '70 thru '83
14025 **(Renault) Alliance & Encore** '83 thru '87

AUDI

15020 **4000** all models '80 thru '87
15025 **5000** all models '77 thru '83
15026 **5000** all models '84 thru '88
Audi A4 '96 thru '01 - *see VW Passat (96023)*
15030 **Audi A4** '02 thru '08

AUSTIN-HEALEY

Sprite - *see MG Midget (66015)*

BMW

18020 **3/5 Series** '82 thru '92
18021 **3-Series** incl. Z3 models '92 thru '98
18022 **3-Series** incl. Z4 models '99 thru '05
18023 **3-Series** '06 thru '14
18025 **320i all 4-cylinder models** '75 thru '83
18050 **1500 thru 2002** except Turbo '59 thru '77

BUICK

19010 **Buick Century** '97 thru '05
Century (front-wheel drive) - *see GM (38005)*
19020 **Buick, Oldsmobile & Pontiac Full-size (Front-wheel drive)** '85 thru '05
Buick Electra, LeSabre and Park Avenue; **Oldsmobile** Delta 88 Royale, Ninety Eight and Regency; **Pontiac** Bonneville
19025 **Buick, Oldsmobile & Pontiac Full-size (Rear wheel drive)** '70 thru '90
Buick Estate, Electra, LeSabre, Limited, **Oldsmobile** Custom Cruiser, Delta 88, Ninety-eight, **Pontiac** Bonneville, Catalina, Grandville, Parisienne
19027 **Buick LaCrosse** '05 thru '13
Enclave - *see GENERAL MOTORS (38001)*
Rainier - *see CHEVROLET (24072)*
Regal - *see GENERAL MOTORS (38010)*
Riviera - *see GENERAL MOTORS (38030, 38031)*
Roadmaster - *see CHEVROLET (24046)*
Skyhawk - *see GENERAL MOTORS (38015)*
Skylark - *see GENERAL MOTORS (38020, 38025)*
Somerset - *see GENERAL MOTORS (38025)*

CADILLAC

21015 **CTS & CTS-V** '03 thru '14
21030 **Cadillac Rear Wheel Drive** '70 thru '93
Cimarron - *see GENERAL MOTORS (38015)*
DeVille - *see GENERAL MOTORS (38031 & 38032)*
Eldorado - *see GENERAL MOTORS (38030)*
Fleetwood - *see GENERAL MOTORS (38031)*
Seville - *see GM (38030, 38031 & 38032)*

CHEVROLET

10305 **Chevrolet Engine Overhaul Manual**
24010 **Astro & GMC Safari Mini-vans** '85 thru '05
24013 **Aveo** '04 thru '11
24015 **Camaro V8** all models '70 thru '81
24016 **Camaro** all models '82 thru '92
24017 **Camaro & Firebird** '93 thru '02
Cavalier - *see GENERAL MOTORS (38016)*
Celebrity - *see GENERAL MOTORS (38005)*
24018 **Camaro** '10 thru '15
24020 **Chevelle, Malibu & El Camino** '69 thru '87
Cobalt - *see GENERAL MOTORS (38017)*
24024 **Chevette & Pontiac T1000** '76 thru '87
Citation - *see GENERAL MOTORS (38020)*
24027 **Colorado & GMC Canyon** '04 thru '12
24032 **Corsica & Beretta** all models '87 thru '96
24040 **Corvette** all V8 models '68 thru '82
24041 **Corvette** all models '84 thru '96
24042 **Corvette** all models '97 thru '13
24044 **Cruze** '11 thru '19
24045 **Full-size Sedans** Caprice, Impala, Biscayne, Bel Air & Wagons '69 thru '90
24046 **Impala SS & Caprice and Buick Roadmaster** '91 thru '96
Impala '00 thru '05 - *see LUMINA (24048)*
24047 **Impala & Monte Carlo** all models '06 thru '11
Lumina '90 thru '94 - *see GM (38010)*
24048 **Lumina & Monte Carlo** '95 thru '05
Lumina APV - *see GM (38035)*
24050 **Luv Pick-up** all 2WD & 4WD '72 thru '82
24051 **Malibu** '13 thru '19
24055 **Monte Carlo** all models '70 thru '88
Monte Carlo '95 thru '01 - *see LUMINA (24048)*
24059 **Nova** all V8 models '69 thru '79
24060 **Nova and Geo Prizm** '85 thru '92
24064 **Pick-ups** '67 thru '87 - Chevrolet & GMC
24065 **Pick-ups** '88 thru '98 - Chevrolet & GMC
24066 **Pick-ups** '99 thru '06 - Chevrolet & GMC
24067 **Chevrolet Silverado & GMC Sierra** '07 thru '14
24068 **Chevrolet Silverado & GMC Sierra** '14 thru '19
24070 **S-10 & S-15 Pick-ups** '82 thru '93, **Blazer & Jimmy** '83 thru '94,
24071 **S-10 & Sonoma Pick-ups** '94 thru '04, including **Blazer, Jimmy & Hombre**
24072 **Chevrolet TrailBlazer, GMC Envoy & Oldsmobile Bravada** '02 thru '09
24075 **Sprint** '85 thru '88 **& Geo Metro** '89 thru '01
24080 **Vans - Chevrolet & GMC** '68 thru '96
24081 **Chevrolet Express & GMC Savana** Full-size Vans '96 thru '19

CHRYSLER

10310 **Chrysler Engine Overhaul Manual**
25015 **Chrysler Cirrus, Dodge Stratus, Plymouth Breeze** '95 thru '00
25020 **Full-size Front-Wheel Drive** '88 thru '93
K-Cars - *see DODGE Aries (30008)*
Laser - *see DODGE Daytona (30030)*
25025 **Chrysler LHS, Concorde, New Yorker, Dodge** Intrepid, **Eagle** Vision, '93 thru '97
25026 **Chrysler LHS, Concorde, 300M, Dodge** Intrepid, '98 thru '04
25027 **Chrysler 300** '05 thru '18, **Dodge Charger** '06 thru '18, **Magnum** '05 thru '08 **& Challenger** '08 thru '18
25030 **Chrysler & Plymouth Mid-size** front wheel drive '82 thru '95
Rear-wheel Drive - *see Dodge (30050)*
25035 **PT Cruiser** all models '01 thru '10
25040 **Chrysler Sebring** '95 thru '06, **Dodge Stratus** '01 thru '06 **& Dodge Avenger** '95 thru '00
25041 **Chrysler Sebring** '07 thru '10, **200** '11 thru '17 **Dodge Avenger** '08 thru '14

DATSUN

28005 **200SX** all models '80 thru '83
28012 **240Z, 260Z & 280Z** Coupe '70 thru '78
28014 **280ZX** Coupe & 2+2 '79 thru '83
300ZX - *see NISSAN (72010)*
28018 **510 & PL521 Pick-up** '68 thru '73
28020 **510** all models '78 thru '81
28022 **620 Series Pick-up** all models '73 thru '79
720 Series Pick-up - *see NISSAN (72030)*

DODGE

400 & 600 - *see CHRYSLER (25030)*
30008 **Aries & Plymouth Reliant** '81 thru '89
30010 **Caravan & Plymouth Voyager** '84 thru '95
30011 **Caravan & Plymouth Voyager** '96 thru '02
30012 **Challenger & Plymouth Sapporro** '78 thru '83
30013 **Caravan, Chrysler Voyager & Town & Country** '03 thru '07
30014 **Grand Caravan & Chrysler Town & Country** '08 thru '18
30016 **Colt & Plymouth Champ** '78 thru '87
30020 **Dakota Pick-ups** all models '87 thru '96
30021 **Durango** '98 & '99 **& Dakota** '97 thru '99
30022 **Durango** '00 thru '03 **& Dakota** '00 thru '04
30023 **Durango** '04 thru '09 **& Dakota** '05 thru '11
30025 **Dart, Demon, Plymouth Barracuda, Duster & Valiant** 6-cylinder models '67 thru '76
30030 **Daytona & Chrysler Laser** '84 thru '89
Intrepid - *see CHRYSLER (25025, 25026)*
30034 **Neon** all models '95 thru '99
30035 **Omni & Plymouth Horizon** '78 thru '90
30036 **Dodge & Plymouth Neon** '00 thru '05
30040 **Pick-ups** full-size models '74 thru '93
30042 **Pick-ups** full-size models '94 thru '08
30043 **Pick-ups** full-size models '09 thru '18
30045 **Ram 50/D50 Pick-ups & Raider and Plymouth Arrow Pick-ups** '79 thru '93
30050 **Dodge/Plymouth/Chrysler** RWD '71 thru '89
30055 **Shadow & Plymouth Sundance** '87 thru '94
30060 **Spirit & Plymouth Acclaim** '89 thru '95
30065 **Vans - Dodge & Plymouth** '71 thru '03

EAGLE

Talon - *see MITSUBISHI (68030, 68031)*
Vision - *see CHRYSLER (25025)*

FIAT

34010 **124 Sport Coupe & Spider** '68 thru '78
34025 **X1/9** all models '74 thru '80

FORD

10320 **Ford Engine Overhaul Manual**
10355 **Ford Automatic Transmission Overhaul**
11500 **Mustang '64-1/2 thru '70 Restoration Guide**
36004 **Aerostar Mini-vans** all models '86 thru '97
36006 **Contour & Mercury Mystique** '95 thru '00
36008 **Courier Pick-up** all models '72 thru '82
36012 **Crown Victoria & Mercury Grand Marquis** '88 thru '11
36014 **Edge** '07 thru '19 **& Lincoln MKX** '07 thru '18
36016 **Escort & Mercury Lynx** all models '81 thru '90
36020 **Escort & Mercury Tracer** '91 thru '02
36022 **Escape** '01 thru '17, **Mazda Tribute** '01 thru '11, **& Mercury Mariner** '05 thru '11
36024 **Explorer & Mazda Navajo** '91 thru '01
36025 **Explorer & Mercury Mountaineer** '02 thru '10
36026 **Explorer** '11 thru '17
36028 **Fairmont & Mercury Zephyr** '78 thru '83
36030 **Festiva & Aspire** '88 thru '97
36032 **Fiesta** all models '77 thru '80
36034 **Focus** all models '00 thru '11
36035 **Focus** '12 thru '14
36045 **Fusion** '06 thru '14 **& Mercury Milan** '06 thru '11
36048 **Mustang V8** all models '64-1/2 thru '73
36049 **Mustang II** 4-cylinder, V6 & V8 models '74 thru '78
36050 **Mustang & Mercury Capri** '79 thru '93
36051 **Mustang** all models '94 thru '04
36052 **Mustang** '05 thru '14
36054 **Pick-ups & Bronco** '73 thru '79
36058 **Pick-ups & Bronco** '80 thru '96
36059 **F-150** '97 thru '03, **Expedition** '97 thru '17, **F-250** '97 thru '99, **F-150 Heritage** '04 **& Lincoln Navigator** '98 thru '17
36060 **Super Duty Pick-ups & Excursion** '99 thru '10
36061 **F-150** full-size '04 thru '14
36062 **Pinto & Mercury Bobcat** '75 thru '80
36063 **F-150** full-size '15 thru '17
36064 **Super Duty Pick-ups** '11 thru '16
36066 **Probe** all models '89 thru '92
Probe '93 thru '97 - *see MAZDA 626 (61042)*
36070 **Ranger & Bronco II** gas models '83 thru '92
36071 **Ranger** '93 thru '11 **& Mazda Pick-ups** '94 thru '09
36074 **Taurus & Mercury Sable** '86 thru '95
36075 **Taurus & Mercury Sable** '96 thru '07
36076 **Taurus** '08 thru '14, **Five Hundred** '05 thru '07, **Mercury Montego** '05 thru '07 **& Sable** '08 thru '09
36078 **Tempo & Mercury Topaz** '84 thru '94
36082 **Thunderbird & Mercury Cougar** '83 thru '88
36086 **Thunderbird & Mercury Cougar** '89 thru '97
36090 **Vans** all V8 Econoline models '69 thru '91
36094 **Vans** full size '92 thru '14
36097 **Windstar** '95 thru '03, **Freestar & Mercury Monterey Mini-van** '04 thru '07

GENERAL MOTORS

10360 **GM Automatic Transmission Overhaul**
38001 **GMC Acadia** '07 thru '16, **Buick Enclave** '08 thru '17, **Saturn Outlook** '07 thru '10 **& Chevrolet Traverse** '09 thru '17
38005 **Buick Century, Chevrolet Celebrity, Oldsmobile Cutlass Ciera & Pontiac 6000** all models '82 thru '96
38010 **Buick Regal** '88 thru '04, **Chevrolet Lumina** '88 thru '04, **Oldsmobile Cutlass Supreme** '88 thru '97 **& Pontiac Grand Prix** '88 thru '07
38015 **Buick Skyhawk, Cadillac Cimarron, Chevrolet Cavalier, Oldsmobile Firenza, Pontiac J-2000 & Sunbird** '82 thru '94
38016 **Chevrolet Cavalier & Pontiac Sunfire** '95 thru '05
38017 **Chevrolet Cobalt** '05 thru '10, **HHR** '06 thru '11, **Pontiac G5** '07 thru '09, **Pursuit** '05 thru '06 **& Saturn ION** '03 thru '07
38020 **Buick Skylark, Chevrolet Citation, Oldsmobile Omega, Pontiac Phoenix** '80 thru '85
38025 **Buick Skylark** '86 thru '98, **Somerset** '85 thru '87, **Oldsmobile Achieva** '92 thru '98, **Calais** '85 thru '91, **& Pontiac Grand Am** all models '85 thru '98
38026 **Chevrolet Malibu** '97 thru '03, **Classic** '04 thru '05, **Oldsmobile Alero** '99 thru '03, **Cutlass** '97 thru '00, **& Pontiac Grand Am** '99 thru '03
38027 **Chevrolet Malibu** '04 thru '12, **Pontiac G6** '05 thru '10 **& Saturn Aura** '07 thru '10
38030 **Cadillac Eldorado, Seville, Oldsmobile Toronado & Buick Riviera** '71 thru '85
38031 **Cadillac Eldorado, Seville, DeVille, Fleetwood, Oldsmobile Toronado & Buick Riviera** '86 thru '93
38032 **Cadillac DeVille** '94 thru '05, **Seville** '92 thru '04 **& Cadillac DTS** '06 thru '10
38035 **Chevrolet Lumina APV, Oldsmobile Silhouette & Pontiac Trans Sport** all models '90 thru '96
38036 **Chevrolet Venture** '97 thru '05, **Oldsmobile Silhouette** '97 thru '04, **Pontiac Trans Sport** '97 thru '98 **& Montana** '99 thru '05
38040 **Chevrolet Equinox** '05 thru '17, **GMC Terrain** '10 thru '17 **& Pontiac Torrent** '06 thru '09

GEO

Metro - *see CHEVROLET Sprint (24075)*
Prizm - *'85 thru '92 see CHEVY (24060), '93 thru '02 see TOYOTA Corolla (92036)*
40030 **Storm** all models '90 thru '93
Tracker - *see SUZUKI Samurai (90010)*

(Continued on other side)

Haynes North America, Inc. • (805) 498-6703 • www.haynes.com

Haynes Automotive Manuals (continued)

NOTE: If you do not see a listing for your vehicle, please visit **haynes.com** *for the latest product information and check out our* **Online Manuals!**

GMC
Acadia - *see GENERAL MOTORS (38001)*
Pick-ups - *see CHEVROLET (24027, 24068)*
Vans - *see CHEVROLET (24081)*

HONDA
42010 **Accord CVCC** all models '76 thru '83
42011 **Accord** all models '84 thru '89
42012 **Accord** all models '90 thru '93
42013 **Accord** all models '94 thru '97
42014 **Accord** all models '98 thru '02
42015 **Accord** '03 thru '12 **& Crosstour** '10 thru '14
42016 **Accord** '13 thru '17
42020 **Civic 1200** all models '73 thru '79
42021 **Civic 1300 & 1500 CVCC** '80 thru '83
42022 **Civic 1500 CVCC** all models '75 thru '79
42023 **Civic** all models '84 thru '91
42024 **Civic & del Sol** '92 thru '95
42025 **Civic** '96 thru '00, **CR-V** '97 thru '01 **& Acura Integra** '94 thru '00
42026 **Civic** '01 thru '11 **& CR-V** '02 thru '11
42027 **Civic** '12 thru '15 **& CR-V** '12 thru '16
42030 **Fit** '07 thru '13
42035 **Odyssey** all models '99 thru '10
Passport - *see ISUZU Rodeo (47017)*
42037 **Honda Pilot** '03 thru '08, **Ridgeline** '06 thru '14 **& Acura MDX** '01 thru '07
42040 **Prelude CVCC** all models '79 thru '89

HYUNDAI
43010 **Elantra** all models '96 thru '19
43015 **Excel & Accent** all models '86 thru '13
43050 **Santa Fe** all models '01 thru '12
43055 **Sonata** all models '99 thru '14

INFINITI
G35 '03 thru '08 - *see NISSAN 350Z (72011)*

ISUZU
Hombre - *see CHEVROLET S-10 (24071)*
47017 **Rodeo** '91 thru '02, **Amigo** '89 thru '94 & '98 thru '02 **& Honda Passport** '95 thru '02
47020 **Trooper** '84 thru '91 **& Pick-up** '81 thru '93

JAGUAR
49010 **XJ6** all 6-cylinder models '68 thru '86
49011 **XJ6** all models '88 thru '94
49015 **XJ12 & XJS** all 12-cylinder models '72 thru '85

JEEP
50010 **Cherokee, Comanche & Wagoneer Limited** all models '84 thru '01
50011 **Cherokee** '14 thru '19
50020 **CJ** all models '49 thru '86
50025 **Grand Cherokee** all models '93 thru '04
50026 **Grand Cherokee** '05 thru '19 **& Dodge Durango** '11 thru '19
50029 **Grand Wagoneer & Pick-up** '72 thru '91 Grand Wagoneer '84 thru '91, Cherokee & Wagoneer '72 thru '83, Pick-up '72 thru '88
50030 **Wrangler** all models '87 thru '17
50035 **Liberty** '02 thru '12 **& Dodge Nitro** '07 thru '11
50050 **Patriot & Compass** '07 thru '17

KIA
54050 **Optima** '01 thru '10
54060 **Sedona** '02 thru '14
54070 **Sephia** '94 thru '01, **Spectra** '00 thru '09, **Sportage** '05 thru '20
54077 **Sorento** '03 thru '13

LEXUS
ES 300/330 - *see TOYOTA Camry (92007, 92008)*
ES 350 - *see TOYOTA Camry (92009)*
RX 300/330/350 - *see TOYOTA Highlander (92095)*

LINCOLN
MKX - *see FORD (36014)*
Navigator - *see FORD Pick-up (36059)*
59010 **Rear-Wheel Drive Continental** '70 thru '87, **Mark Series** '70 thru '92 **& Town Car** '81 thru '10

MAZDA
61010 **GLC (rear-wheel drive)** '77 thru '83
61011 **GLC (front-wheel drive)** '81 thru '85
61012 **Mazda3** '04 thru '11
61015 **323 & Protegé** '90 thru '03
61016 **MX-5 Miata** '90 thru '14
61020 **MPV** all models '89 thru '98
Navajo - *see Ford Explorer (36024)*
61030 **Pick-ups** '72 thru '93
Pick-ups '94 thru '09 - *see Ford Ranger (36071)*
61035 **RX-7** all models '79 thru '85
61036 **RX-7** all models '86 thru '91
61040 **626 (rear-wheel drive)** all models '79 thru '82
61041 **626 & MX-6 (front-wheel drive)** '83 thru '92
61042 **626** '93 thru '01 **& MX-6/Ford Probe** '93 thru '02
61043 **Mazda6** '03 thru '13

MERCEDES-BENZ
63012 **123 Series Diesel** '76 thru '85
63015 **190 Series** 4-cylinder gas models '84 thru '88
63020 **230/250/280** 6-cylinder SOHC models '68 thru '72
63025 **280 123 Series** gas models '77 thru '81
63030 **350 & 450** all models '71 thru '80
63040 **C-Class:** C230/C240/C280/C320/C350 '01 thru '07

MERCURY
64200 **Villager & Nissan Quest** '93 thru '01
All other titles, see FORD Listing.

MG
66010 **MGB** Roadster & GT Coupe '62 thru '80
66015 **MG Midget, Austin Healey Sprite** '58 thru '80

MINI
67020 **Mini** '02 thru '13

MITSUBISHI
68020 **Cordia, Tredia, Galant, Precis & Mirage** '83 thru '93
68030 **Eclipse, Eagle Talon & Plymouth Laser** '90 thru '94
68031 **Eclipse** '95 thru '05 **& Eagle Talon** '95 thru '98
68035 **Galant** '94 thru '12
68040 **Pick-up** '83 thru '96 **& Montero** '83 thru '93

NISSAN
72010 **300ZX** all models including Turbo '84 thru '89
72011 **350Z & Infiniti G35** all models '03 thru '08
72015 **Altima** all models '93 thru '06
72016 **Altima** '07 thru '12
72020 **Maxima** all models '85 thru '92
72021 **Maxima** all models '93 thru '08
72025 **Murano** '03 thru '14
72030 **Pick-ups** '80 thru '97 **& Pathfinder** '87 thru '95
72031 **Frontier** '98 thru '04, **Xterra** '00 thru '04, **& Pathfinder** '96 thru '04
72032 **Frontier & Xterra** '05 thru '14
72037 **Pathfinder** '05 thru '14
72040 **Pulsar** all models '83 thru '86
72042 **Roque** '08 thru '20
72050 **Sentra** all models '82 thru '94
72051 **Sentra & 200SX** all models '95 thru '06
72060 **Stanza** all models '82 thru '90
72070 **Titan pick-ups** '04 thru '10, **Armada** '05 thru '10 **& Pathfinder Armada** '04
72080 **Versa** all models '07 thru '19

OLDSMOBILE
73015 **Cutlass** V6 & V8 gas models '74 thru '88
For other OLDSMOBILE titles, see BUICK, CHEVROLET or GENERAL MOTORS listings.

PLYMOUTH
For PLYMOUTH titles, see DODGE listing.

PONTIAC
79008 **Fiero** all models '84 thru '88
79018 **Firebird** V8 models except Turbo '70 thru '81
79019 **Firebird** all models '82 thru '92
79025 **G6** all models '05 thru '09
79040 **Mid-size Rear-wheel Drive** '70 thru '87
Vibe '03 thru '10 - *see TOYOTA Corolla (92037)*
For other PONTIAC titles, see BUICK, CHEVROLET or GENERAL MOTORS listings.

PORSCHE
80020 **911** Coupe & Targa models '65 thru '89
80025 **914** all 4-cylinder models '69 thru '76
80030 **924** all models including Turbo '76 thru '82
80035 **944** all models including Turbo '83 thru '89

RENAULT
Alliance & Encore - *see AMC (14025)*

SAAB
84010 **900** all models including Turbo '79 thru '88

SATURN
87010 **Saturn** all S-series models '91 thru '02
Saturn Ion '03 thru '07 - *see GM (38017)*
Saturn Outlook - *see GM (38001)*
87020 **Saturn L-series** all models '00 thru '04
87040 **Saturn VUE** '02 thru '09

SUBARU
89002 **1100, 1300, 1400 & 1600** '71 thru '79
89003 **1600 & 1800** 2WD & 4WD '80 thru '94
89080 **Impreza** '02 thru '11, **WRX** '02 thru '14, **& WRX STI** '04 thru '14
89100 **Legacy** all models '90 thru '99
89101 **Legacy & Forester** '00 thru '09
89102 **Legacy** '10 thru '16 **& Forester** '12 thru '16

SUZUKI
90010 **Samurai/Sidekick & Geo Tracker** '86 thru '01

TOYOTA
92005 **Camry** all models '83 thru '91
92006 **Camry** '92 thru '96 **& Avalon** '95 thru '96
92007 **Camry, Avalon, Solara, Lexus ES 300** '97 thru '01

92008 **Camry, Avalon, Lexus ES 300/330** '02 thru '06 **& Solara** '02 thru '08
92009 **Camry, Avalon & Lexus ES 350** '07 thru '17
92015 **Celica Rear-wheel Drive** '71 thru '85
92020 **Celica Front-wheel Drive** '86 thru '99
92025 **Celica Supra** all models '79 thru '92
92030 **Corolla** all models '75 thru '79
92032 **Corolla** all rear-wheel drive models '80 thru '87
92035 **Corolla** all front-wheel drive models '84 thru '92
92036 **Corolla & Geo/Chevrolet Prizm** '93 thru '02
92037 **Corolla** '03 thru '19, **Matrix** '03 thru '14, **& Pontiac Vibe** '03 thru '10
92040 **Corolla Tercel** all models '80 thru '82
92045 **Corona** all models '74 thru '82
92050 **Cressida** all models '78 thru '82
92055 **Land Cruiser FJ40, 43, 45, 55** '68 thru '82
92056 **Land Cruiser FJ60, 62, 80, FZJ80** '80 thru '96
92060 **Matrix** '03 thru '11 **& Pontiac Vibe** '03 thru '10
92065 **MR2** all models '85 thru '87
92070 **Pick-up** all models '69 thru '78
92075 **Pick-up** all models '79 thru '95
92076 **Tacoma** '95 thru '04, **4Runner** '96 thru '02 **& T100** '93 thru '08
92077 **Tacoma** all models '05 thru '18
92078 **Tundra** '00 thru '06 **& Sequoia** '01 thru '07
92079 **4Runner** all models '03 thru '09
92080 **Previa** all models '91 thru '95
92081 **Prius** all models '01 thru '12
92082 **RAV4** all models '96 thru '12
92085 **Tercel** all models '87 thru '94
92090 **Sienna** all models '98 thru '10
92095 **Highlander** '01 thru '19 **& Lexus RX330/330/350** '99 thru '19
92179 **Tundra** '07 thru '19 **& Sequoia** '08 thru '19

TRIUMPH
94007 **Spitfire** all models '62 thru '81
94010 **TR7** all models '75 thru '81

VW
96008 **Beetle & Karmann Ghia** '54 thru '79
96009 **New Beetle** '98 thru '10
96016 **Rabbit, Jetta, Scirocco & Pick-up** gas models '75 thru '92 & Convertible '80 thru '92
96017 **Golf, GTI & Jetta** '93 thru '98, **Cabrio** '95 thru '02
96018 **Golf, GTI, Jetta** '99 thru '05
96019 **Jetta, Rabbit, GLI, GTI & Golf** '05 thru '11
96020 **Rabbit, Jetta & Pick-up** diesel '77 thru '84
96021 **Jetta** '11 thru '18 **& Golf** '15 thru '19
96023 **Passat** '98 thru '05 **& Audi A4** '96 thru '01
96030 **Transporter 1600** all models '68 thru '79
96035 **Transporter 1700, 1800 & 2000** '72 thru '79
96040 **Type 3 1500 & 1600** all models '63 thru '73
96045 **Vanagon Air-Cooled** all models '80 thru '83

VOLVO
97010 **120, 130 Series & 1800 Sports** '61 thru '73
97015 **140 Series** all models '66 thru '74
97020 **240 Series** all models '76 thru '93
97040 **740 & 760 Series** all models '82 thru '88
97050 **850 Series** all models '93 thru '97

TECHBOOK MANUALS
10205 **Automotive Computer Codes**
10206 **OBD-II & Electronic Engine Management**
10210 **Automotive Emissions Control Manual**
10215 **Fuel Injection Manual** '78 thru '85
10225 **Holley Carburetor Manual**
10230 **Rochester Carburetor Manual**
10305 **Chevrolet Engine Overhaul Manual**
10320 **Ford Engine Overhaul Manual**
10330 **GM and Ford Diesel Engine Repair Manual**
10331 **Duramax Diesel Engines** '01 thru '19
10332 **Cummins Diesel Engine Performance Manual**
10333 **GM, Ford & Chrysler Engine Performance Manual**
10334 **GM Engine Performance Manual**
10340 **Small Engine Repair Manual,** 5 HP & Less
10341 **Small Engine Repair Manual,** 5.5 thru 20 HP
10345 **Suspension, Steering & Driveline Manual**
10355 **Ford Automatic Transmission Overhaul**
10360 **GM Automatic Transmission Overhaul**
10405 **Automotive Body Repair & Painting**
10410 **Automotive Brake Manual**
10411 **Automotive Anti-lock Brake (ABS) Systems**
10420 **Automotive Electrical Manual**
10425 **Automotive Heating & Air Conditioning**
10435 **Automotive Tools Manual**
10445 **Welding Manual**
10450 **ATV Basics**

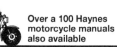

Over a 100 Haynes motorcycle manuals also available

10/22